T0122487

Lecture Notes in Networks and Systems

Volume 387

The series "Lecture Notes in Networks and Systems" publishes the latest developments in Networks and Systems—quickly, informally and with high quality. Original research reported in proceedings and post-proceedings represents the core of LNNS.

Volumes published in LNNS embrace all aspects and subfields of, as well as new challenges in, Networks and Systems.

The series contains proceedings and edited volumes in systems and networks, spanning the areas of Cyber-Physical Systems, Autonomous Systems, Sensor Networks, Control Systems, Energy Systems, Automotive Systems, Biological Systems, Vehicular Networking and Connected Vehicles, Aerospace Systems, Automation, Manufacturing, Smart Grids, Nonlinear Systems, Power Systems, Robotics, Social Systems, Economic Systems and other. Of particular value to both the contributors and the readership are the short publication timeframe and the world-wide distribution and exposure which enable both a wide and rapid dissemination of research output.

The series covers the theory, applications, and perspectives on the state of the art and future developments relevant to systems and networks, decision making, control, complex processes and related areas, as embedded in the fields of interdisciplinary and applied sciences, engineering, computer science, physics, economics, social, and life sciences, as well as the paradigms and methodologies behind them.

Indexed by SCOPUS, INSPEC, WTI Frankfurt eG, zbMATH, SCImago.

All books published in the series are submitted for consideration in Web of Science.

For proposals from Asia please contact Aninda Bose (aninda.bose@springer.com).

More information about this series at https://link.springer.com/bookseries/15179

Carlos Jahn · László Ungvári · Igor Ilin
Editors

Algorithms and Solutions Based on Computer Technology

5th Scientific International Online Conference Algorithms and Solutions based on Computer Technology (ASBC 2021)

Editors
Carlos Jahn
Institute of Maritime Logistics
Hamburg University of Technology
Hamburg, Germany

László Ungvári
Rektor Wildau
Deutsch-Kasachische Universität
Wildau, Brandenburg, Germany

Igor Ilin
Graduate School of Business
and Management
Peter the Great Saint Petersburg
Polytechnic University
St. Petersburg, Russia

ISSN 2367-3370 ISSN 2367-3389 (electronic)
Lecture Notes in Networks and Systems
ISBN 978-3-030-93871-0 ISBN 978-3-030-93872-7 (eBook)
https://doi.org/10.1007/978-3-030-93872-7

This Springer imprint is published by the registered company Springer Nature Switzerland AG
The registered company address is: Gewerbestrasse 11, 6330 Cham, Switzerland

Preface

5th Scientific International Online Conference "Algorithms and solutions based on computer technology" was held on June 8–9th, 2021 in the Peter the Great St. Petersburg Polytechnic University (SPbPU), Saint Petersburg, Russia.

The conference was devoted to discussing a wide range of issues of algorithms and IT- and digital solutions development and implementation for various fields of economy and science. The following issues were discussed at the conference: supercomputers and exointelligent platforms—current state and future development; cybersecurity technologies of the digital industry; applied computer technologies in area of the digital manufacturing, healthcare and bio-medical systems, logistics and management; digital technologies for visualization and prototyping of physical objects.

The conference was held in online format and welcomed the researchers from Russia, Germany, Greece, Hungary, Kazakhstan the Netherlands, Portugal, Poland.

The keynote speaker of the conference was Prof. Askar Akaev, Foreign Member of the Russian Academy of Science, Lomonosov Moscow State University (Russia) with the presentation "Algorithm for modeling technological progress in the digital economy era". The program of the plenary session included presentations of Prof. Andrey Kozlov, Vavilov Institute of General Genetics (Russia), Vladimir Zaborovsky, Prof., Head of the Research Laboratory, «Supercomputer Technologies and Machine Learning», Peter the Great St. Petersburg Polytechnic University (Russia), Prof. Tessaleno Devezas, Prof. University of Beira Interior (Portugal), Prof. Igor Ilin, Prof. Dr. Sc., Head of the Graduate School of Business and Management, Peter the Great Saint Petersburg Polytechnic University (Russia), Prof. László Ungvári, Prof., Technical University of Applied Science Wildau (Germany), Prof. Dmitry Zegzhda, Prof., Peter the Great St. Petersburg Polytechnic University (Russia).

Hamburg, Germany	Carlos Jahn
Wildau, Germany	László Ungvári
St. Petersburg, Russia	Igor Ilin

Contents

Mismatch-Resistant Intrusion Detection with Bioinspired Suffix Tree Algorithm

Haejin Cho, Alexey Andreev, Maxim Kalinin, Dmitry Moskvin, and Dmitry Zegzhda

Abstract Signature-based intrusion detection is the fastest kind of the protection facility. It's work principle is the matching of command sequences executed by the monitored system to the database of attack patterns (signatures). Different mismatches in the attack sequences, including replacement of the equivalent commands, rearranging the command chain, inserting the blank or garbage operations in the chain, reduce the efficiency of signature-based intrusion detection. The paper reviews the methods of computational bioinformatics that solve the similar task of genetic sequences matching. A new bioinspired approach based on the suffix tree algorithm is applied to build the mismatch-resistant intrusion detection approach. The use of suffix tree allows us to achieve 95% of accuracy for intrusion detection with 10% of mutations in the command sequences. The bioinspired suffix tree algorithm is effective in memory usage and sequence processing speed, as well as sustainable to mismatches between the monitored command sequences and the signatures. This makes the developed method applicable for the modern reconfigurable networks that has limited computing and storage power (Internet of Things, connected cyber-physical devices, wireless sensor networks).

Keywords Bioinformatics · Intrusion detection · Mismatch · Mutation · Polymorphism · Signature · Suffix tree

1 Introduction

Intrusion detection system (IDS) performs the function of passive protection of individual network hosts (host IDS) or network connections (network IDS) by monitoring and analyzing security events occurring within the system at the level of system calls

H. Cho
LG Electronics, Inc, Seoul, Korea

A. Andreev · M. Kalinin (✉) · D. Moskvin · D. Zegzhda
Peter the Great St.Petersburg Polytechnic University, St.Petersburg, Russia
e-mail: max@ibks.spbstu.ru

C. Jahn et al. (eds.), *Algorithms and Solutions Based on Computer Technology*, Lecture Notes in Networks and Systems 387,
https://doi.org/10.1007/978-3-030-93872-7_1

or network flows [1]. The IDS detects signs of various types of malicious activity, such as exploitation of software vulnerabilities, DDoS attacks, port scanning, and network penetration attempts. Regardless of the IDS type, high requirements are imposed on the quality of detection and the speed of recognizing attacks, for the damage inflicted by an attack depends on that system's work characteristics [2].

The IDS permanently gathers various security-relevant data and builds a set of identifying features, analyzing which makes a conclusion about the security state of the monitored system. For intrusion detection, the signature-based IDS compares the current set of the system commands or other security-relevant attributes with defined patterns of unsafe states—the signatures. If the current sequence matches one of the given signatures, the IDS alarms the detected intrusion. Comparing with anomaly-based IDS, in which deviations from the average statistical profile of the controlled system's/user's behavior are determined, the signature-based IDS has the advantages—the higher quality and detection performance, the lower level of type I errors (false positive, FP) and type II errors (false negative, FN), and no requirement for a dedicated phase of building a behavioral profile [3–5].

As for the signature-based IDS the only way to detect an attack is to find an identical sequence in the signature database, even minor mismatch in command sequence prevent the IDS from recognizing polymorphic attacks using the signatures. Let's define the protected system as a set of entities that interact with each other. Informational interaction between entities in the system is realized by executing commands. As a result of such communication, the state of the system changes. The knowledge that the commands were executed in a certain sequence during the time interval determines all the states of the system obtained from the initial one. A behavior that is presented as a sequence of commands that brings the protected system into an insecure state is called an attack. The intrusion detection task for a signature-based IDS is equivalent to matching the command sequences with the known signatures of attacks, represented by command sequences leading to an insecure state. When matching the sequences, the signature-based method meets the problem of mutations, or polymorphism, at the attack sequences [6–8]. Attack mutations include such changes in the command sequence as: replacing commands with equivalent ones, rearranging commands and their blocks in the sequence, adding garbage and empty commands. For example, mismatches in the command sequences for a chain $S = abcdefgh$ can be: $S_1 = aBcdefgh$ (this is a replacement with equivalent command), $S_2 = abcedfgh$ (this is a rearrangement of commands), $S_3 = abc\_edfgh$ (this is an addition of garbage commands, empty commands or time gaps). Such sequence mismatches complicate intrusion detection, as they require large signature databases for all actual variations of all possible attacks, excessive resource consumption for searching and matching signatures with the monitored sequences, and constant replenishment of attack signatures.

Therefore, the signature-based IDS are vulnerable to mismatches and depend on the filling of the signature database. The solution to this issue can be a high-speed mechanism that can process and compare incoming command sequences with a large set of available signatures with high accuracy, despite mismatches in the current sequence. To solve this task, the paper reviews the existing algorithms of computational bioinformatics, which allow solving a similar biological task—to localize the mismatches in the sequences of biological codes of the genome (Sect. 2); and proposes our new signature-based method for detecting the attacks based on the mechanism of suffix trees used in assembling and verifying the similarity of genomic sequences (Sect. 3). Section 4 discusses the achieved results; and, finally, Sect. 5 concludes our work and sets the further plan.

2 Materials and Methods

2.1 Bioinformatics Algorithms for Genomic Sequences Matching

Requirements close to the signature-based IDSs regarding the detection of polymorphic attacks are imposed on computational bioinformatics algorithms. The purpose of bioalgorithms is the fast assembly and comparison of big genomic sequences. The task of processing genomic sequences in bioinformatics is the restoration and ordering of large nucleotide-coded chains up to billions of symbols based on information obtained as a result of sequencing [9–11]. Sequencing is the general name for bioinformation methods that allow you to establish the sequence of nucleotides in a genomic sequence. Currently, there is not a single sequencing method that would work on genomic sequences as a whole—they are all arranged in such a way that first a large number of nucleotide blocks are prepared (the genome is cloned and cut into blocks, the reads), which are then processed. The methods of genomic reads processing differ in the types of parallelism and the organization of computations on the data structures of the encoded representation of the reads.

Reads are the sequences obtained during sequencing and containing information about fragments of the genome. Reed is represented by a string of the 4-symbol alphabet of nucleotides corresponding to a fragment of the genome. Sequencers, depending on the mechanism of their work, make mistakes, the most common of which is the replacement of one nucleotide with another (for example, there is nucleotide T in the read, and A in the corresponding position of the genome), as well as errors of skipping a number of nucleotides and insertion foreign nucleotides in read. When assembling a genome from reads, it is necessary to eliminate sequencing mismatches. For this, various algorithms for sequence alignment and search for homologically similar genomic codes are used. Thus, the assembly and sequence alignment of genomes is similar to the task of searching for similar sequences among IDS signatures. The purpose of genome assembly algorithms, as well as signature IDS, is to find matches or mismatches of a region in a sequence of system commands with one of the regions from a large signature database. Bioinformatics algorithms fully meet the requirements for an IDS resistant to mismatches in sequences, since changes introduced to attacks appear in the command sequence as the insertion of

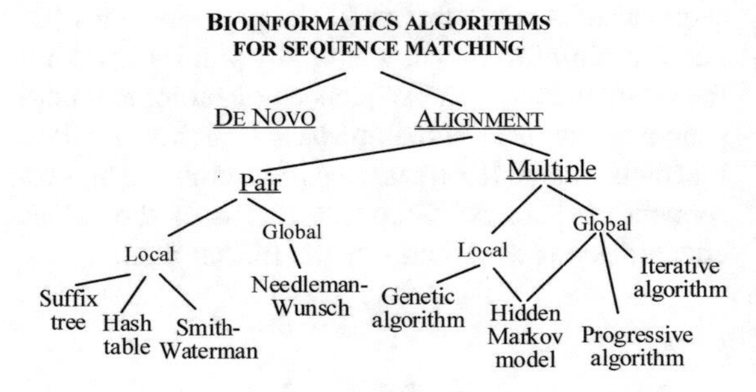

Fig. 1 Taxonomy of bioinformatics algorithms for genomic sequence matching

additional elements into the sequence and are similar to errors in the insertion of extra nucleotide blocks by a sequencer. The quality and performance of bioinformation algorithms are close to the high accuracy and fast sequence processing required for IDS [12].

A taxonomy of bioinformatics algorithms applicable for matching the genomic sequences has been built (Fig. 1).

The analysis of De Novo *algorithms* [13], e.g. *Overlap Layout Consensus* algorithms [14] and algorithms on *de Bruijn graphs* [15, 16], has shown that they have quadratic complexity and therefore are not suitable for fast processing of large sequences.

Alignment algorithms, in addition to a set of reads, receive a previously assembled genome [17]. Assessing the similarity of sequences, performed in this case, simplifies the process of finding the mismatches. *Global alignment* [18] assumes that sequences are inherently homologous and takes this similarity into account throughout the alignment process. For example, the *Needleman-Wunsch algorithm* [19, 20] is based on an estimate of the collinearity of two sequences and works best when comparing very similar sequences. The correspondence of the aligned symbols is specified by the similarity matrix, the values of the cells of which increase when they match and decrease when the elements do not match. Correct operation of the algorithm is due to the additivity property, according to which the optimal choice is made at each step.

Local alignment searches for similar blocks in sequences and aligns the sequences with respect to the blocks. Accordingly, there may be several local alignments for a pair of sequences [21]. The *Smith-Waterman algorithm* [22, 23] takes into account the local alignments of similar regions that fall into different sets of sequences. Unlike the Needleman-Wunsch algorithm, the values in the similarity matrix cannot fall below a certain minimum, so the final alignment is split into several optimal regions.

Multiple alignment of three or more sequences suggests that the input set oa sequenceы has an evolutionary relationship. Due to the greater computational complexity, many implementations of multiple alignment are based on heuristic algorithms. There are the following multiple alignment methods:

- the progressive alignment [24, 25] implements two stages: (a) construction of a binary guiding tree, in which the leaves are sequences; and (b) construction of multiple alignment by adding sequences to the growing alignment according to the guiding tree. Alignment may fail in case of a set of sequences that are very distant from each other. Errors received at any stage of growing multiple alignment reach the resulting alignment. The algorithm is demanding on the similarity of the initial sequences, which is not suitable for obtaining our target;

- the iterative alignment [24, 26] works like the progressive one, but it constantly rearranges the original alignments when adding new sequences. The algorithm can revert to the originally calculated pairwise alignments and sub-alignments containing subsets of the sequences from the query, and thus optimize the target function and improve the quality of result. Progressive and iterative alignments are efficient enough for simultaneous processing of a large number (100...1,000) of sequences, but as the method is heuristic, it does not guarantee finding the globally optimal alignment;
- hidden Markov model [11, 27] is a probabilistic model that can estimate the likelihood for all possible combinations of omissions, matches or mismatches in order to determine the most probable multiple alignment. This model can compute a single highly weighted alignment, but can also generate a set of possible alignments, which can then be estimated by their weight. This method can be used to obtain both global and local alignments. Despite the fact that this method has appeared not so far, it is a method with significant improvements in computational complexity, especially for sequences containing overlapping regions. Like the progressive algorithms, this method is demanding on the classification of signatures, but represents an alignment closer to a specific signature;
- optimization methods of artificial intelligence are also applied to construct multiple alignments. The genetic algorithm implements a hypothetical evolutionary division of a series of possible chains into fragments and their restructuring with the introduction of mismatches in different locations [26]. The genetic algorithm solves the optimization problem of step-by-step approximation to the pattern, which satisfies the purpose of genome assembly, but does not allow quick assessment of the similarity of sequence pairs, which makes it unsuitable for intrusion detection.

Separately, there are fast local pairwise alignment methods—the *short read aligners*:

- *hash-table algorithm* that uses a hash function that transforms a string into a fast search key [29]. The simplest way would be to split a genome sequence into words that match the length of the read, but such approach does not work right, because long words are usually unique and their storage requires too much memory. Instead, hashing of shorter blocks is used. Once the appropriate positions have been obtained using the hash function, the remainder of the read can be mapped onto the genome. The approach of dividing the read into several blocks allows the substitutions in the algorithm, i.e. a reed can be broken into multiple blocks with a shift of several nucleotides. Thus, it is possible to resist the mismatch errors;
- hash-table algorithm does not do well with repetitions, as the number of reads that need to be checked increases dramatically. To solve this issue, a *suffix tree*

Table 1 Comparison of sequence matching algorithms (m—length of analyzed block, n—length of a genomic sequence, k—number of signatures, c—size of alphabet)

Algorithm	Alignment of mutated sequences	Requirement for big signature database	Data structures building	Search	Memory usage	Mismatch resilience (exact alignment)
Needleman-Wunsch	–	+	$O(nmk)$	$O(nmk)$	$O(nmk)$	–
Smith-Waterman	+	–	$O(nmk)$	$O(nmk)$	$O(nmk)$	–
Progressive	–	+	$O(n^2k^2)$	$O(m)$	$O(nk)$	–
Iterative	+	+	$O(n^2k^2)$	$O(m)$	$O(nk)$	+
Hidden Markov model	+	+	$O(nc^2)$	$O(m)$	$O(nk)$	+
Hash table	+	–	$O(nmk)$	$O(k)$	$O(nk)$	+
Suffix tree	+	–	$O(nk)$	$O(mk)$	$O(nk)$	+

has been designed [30, 31]. The advantage of suffix trees is that repetitions do not increase the running time of the algorithm, because repeating sections collapse in the suffix tree. This mechanism works extremely fast.

The above reviewed algorithms are summarized in Table 1.

For our further research, a suffix tree mechanism was selected, which was supplemented by the prior division of sequences into smaller blocks to better work with mutations in attack sequences. This solution eliminates the influence of mismatches, omissions and command reordering, because, in case of mutations, it allows you to completely restore one of the closest signatures from the signature database.

2.2 Suffix Tree Method for Mismatch-Resistant Intrusion Detection

Trie is a data structure for storing a set of encoded sequences with symbols on the edges. Sequences, as the strings of symbols, are obtained by sequential writing of all characters stored on the edges between the root and the terminal vertex. The size of that trie is linearly dependent on the sum of the lengths of all strings. Searching in the trie takes time proportionally to the length of the sequence. Consider a trie containing a certain set of strings $s_1, ..., s_k$. The number of trie's vertices can reach the total string length $|s_1| + ... + |s_k|$.

To reduce the number of vertices, consider a chain of vertices of the trie such that a single edge emanates from each vertex into the next one, and squeeze such a chain into one edge, and instead of a symbol, write on it the entire sequence of symbols from the edges that we replaced. This sequence of symbols is a substring of some string s_i from the set, so we write on the edge only the line number, as well as the beginning and end of the corresponding substring. Compressing all strings in a trie allows us to construct a compressed trie—a rooted tree, on each edge of which a non-empty string is written, which has the following properties: no vertex leaves two edges, the lines for which begin with the same character; and, if a vertex is not a root or a leaf of a tree, at least two edges go out of it.

The number of vertices in a compressed tree is $O(k)$, where k is the number of strings in the set. The total number of vertices does not exceed $2k$. Compressed tree takes up $O(k)$ of memory, however, for operations with it, it is necessary to explicitly store all strings s_i, therefore, a similar gain in memory is not achieved.

The suffix tree of a string s is a compressed tree built on all suffixes of s. Such a tree takes up $O(|s|)$ of memory. However, we do not need to explicitly store all suffixes separately: they are all present in the string s. It means that the suffix tree allows us to answer the questions: "Is a string t a suffix of s?" and "Is a string t a suffix prefix, i.e. a substring of s?" in time of $O(|t|)$. It also allows us to get the following information about a string s and its substrings: number of different substrings of a string s; length of the longest common prefix for two substrings of a string s; lexicographic order of suffixes of string s.

An implicit suffix tree of a string s is a tree obtained from a suffix tree $s\$$ by removing all occurrences of the terminal symbol $ from the labels of the edges of the tree, then removing the unlabeled edges and then removing the vertices that have less than two children. The implicit suffix tree for prefix $s[1...i]$ of string s is derived similarly from the suffix tree for $s[1\ldots i]\$$ by removing $, arcs, and vertices.

An implicit suffix tree for any string s will have less leaves than a suffix tree for a string $s\$$ if and only if at least one of the suffixes of s is a prefix of another suffix. A terminal symbol $ is added to s just to avoid this case. If s ends with a character that does not appear anywhere else in s, then the implicit suffix tree for s will have a leaf for each suffix and therefore will be a real suffix tree.

Although an implicit suffix tree for any string may not have leaves for all suffixes, all suffixes for s are encoded in it—each suffix is named by symbols of some path from the root of this implicit suffix tree. However, if this path does not end with a leaf, then there will be no marker indicating the end of the path. Thus, implicit suffix trees are themselves uninformative.

The generalized suffix tree for a set of string $s_1 \ldots s_n$ is a suffix tree containing all the suffixes of each of n strings. When building such a suffix tree, each string must be padded with a unique non-alphabetical symbol (or string) to ensure that the suffix is not a substring of another suffix and is represented by a unique terminal vertex.

To solve the given target of intrusion detection, a generalized suffix tree for attack signatures is created based on the signature database. For our need, the Ukkonen algorithm [31] is applied for building a generalized suffix tree. The basic algorithm sequentially builds a suffix tree for all prefixes in the original text $S = s_1 s_2 \ldots s_n$. At the step i, the implicit suffix tree τ_{i-1} for the prefix $s[1...i-1]$ is completed up to τ_i for the prefix $s[1...i]$. To do this, for each suffix of the substring $s[1...i-1]$, descend from the root of the tree to the end of the suffix and add the symbol s_i. The algorithm consists of n stages, on each of which all the suffixes of the current prefix of the string are extended that takes $O(n^2)$ time. The general asymptotics of the algorithm is $O(n^3)$.

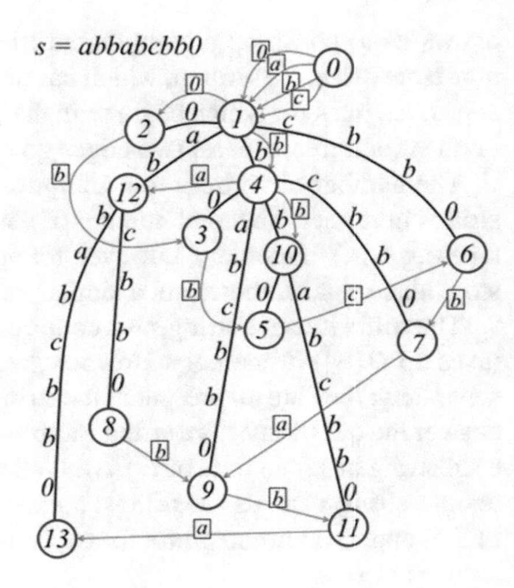

Fig. 2 Sample of suffix links in the suffix tree (1 is a suffix tree root, 0 is a dummy vertex)

Let $x\alpha$ denotes an arbitrary string, where x is its first character, and α is the remaining (possibly empty) substring. If for an inner vertex v with a path $x\alpha$ there is another vertex $s(v)$ with a path α, then the link from v to $s(v)$ is called a suffix link. For any internal vertex v of the suffix tree, there is a suffix link leading to some internal vertex u (sample of suffix links topology is presented in Fig. 2).

Let's look at the use of suffix links. Let the suffix $s[j...i-1]$ has just been extended to the suffix $s[j...i]$. Now, using the constructed suffix links, you can find the end of the suffix $s[j+1...i-1]$ in the suffix tree to extend it to the suffix $s[j+1...i]$. To do this, you need to go up through the tree to the nearest inner vertex v, to which the path $s[j...r]$ leads.

A vertex v always has a suffix link leading to the vertex u, to which the path $s[j+1...r]$ corresponds. Further, from the vertex u one should go down the tree to the end of the suffix $s[j+1...i-1]$ and extend it to the suffix $s[j+1...i]$. In this case, the substring $s[j+1...i-1]$ is a suffix of the substring $s[j...i-1]$. Therefore, after going through the suffix link to the vertex marked with a path $s[j+1...r]$, one can reach the location to which the mark $s[r+1...i-1]$ corresponds by comparing not the symbols on the edges, but only the length of the edge by the first symbol of the part of the substring and the length of this substring.

In the process of building a suffix tree, the suffix links that have been already built are not changed. Therefore, we will consider the construction of suffix links for the created vertices. Let's take a new internal vertex v, that was created as a result of the extension of the suffix $s[j...i-1]$. Instead of looking for where the suffix link of the vertex v should point to when walking the path to the root of the tree, we extend the next suffix $s[j+1...i-1]$. At this time, we can put a suffix reference for the vertex v. It will indicate either to the existing vertex, if the next suffix ends in it; or to a just created one. Thus, at the next step of the algorithm, for a vertex, there will definitely be an internal vertex to which the suffix link has to point to.

The depth $d(v)$ of a vertex v is the number of edges on the path from the root to the vertex. When we follow the suffix link, the depth decreases by no more than one. Number of edge jumps within a step i of algorithm is $O(i)$. At the beginning of each phase, only one descent from the root is done, and then jumps are performed by the suffix links. Amount of jumps within the algorithm phase are estimated as $O(i)$. The algorithm's phase consists of i iterations, and cumulatively we get that $O(1)$ actions will be performed at one iteration. Therefore, the resulting aymptotics of the improved algorithm is $O(n^2)$.

For further optimization of this algorithm to $O(n)$, it has been proposed to use a linear amount of memory. We save the label of each edge as two numbers: the position of its left-most character in the source text and the position of right-most character in the source text. If at some time in the algorithm's operation a leaf with a label i is created (for a suffix that starts at the position i of a string s), it will remain a leaf in all sequential trees created by the algorithm. If the extension rule is applied in the continuation of the suffix starting in the position j, it will be applied in all further extensions (from $j + 1$ to i) until the end of the algorithm step. Therefore, in each phase i, the algorithm works with suffixes from the range $j \ldots k$, $k \leq i$ instead of the range $1 \ldots i$.

Using a suffix tree, the basic algorithm allows detecting only those sequences of attacks, exact patterns of which are included in the signature database. Mismatches of the analyzed sequence of commands constituting the attack will lead to high level of false negative (FN) errors. This issue has been solved by dividing the sequences into smaller blocks, which eliminates the mismatches (influence of mutations, omissions, and command reordering), as in case of mismatches it allows restoring from the sequence one of the closest signatures in the signature database.

Let us denote the analyzed sequence of commands—$a = C_{a1}C_{a2}C_{a3} \ldots C_{an}$, and the known signatures: $c_1 = C_{c_1,1}C_{c_1,2}C_{c_1,3} \ldots C_{c_1,n_1}, c_2 = C_{c_2,1}C_{c_2,2}C_{c_2,3} \ldots C_{c_2,n_2}, c_m = C_{c_m,1}C_{c_m,2}C_{c_m,3} \ldots C_{c_m,n_m}$. Each attack vector from a known set of attacks is divided into intersecting fragments of length k. A suffix tree is constructed on the obtained sections. To do this, divide the next signature $c_i = C_{c_i,1}C_{c_i,2}C_{c_i,3} \ldots C_{c_i,n_i}$ into blocks of length k :$C_{c_i,1}C_{c_i,2} \ldots C_{c_i,k}$; $C_{c_i,2}C_{c_i,3} \ldots C_{c_i,k+1}$; and add the resulting subsequences to the existing suffix tree. When comparing the received sequence a with signatures, which have already been used to build a suffix tree, we will also divide it into blocks of length k, after which we will search for each separate block in the suffix tree and calculate the percentage of matches. If this share exceeds the specified threshold value, then the sequence a will be referred to as an intrusion. The training of the final algorithm contains the selection of the optimal value of the length k and the threshold for alarming the attack.

Let us consider an example of the algorithm discussed above. Let the following attack signatures be specified in the signature database:

```
(1) open, read, open, write, execute, connect, write, write,
connect, execute;
(2) execute, execute, connect, open, write, execute, connect, open,
read, write.
```

Let us build a suffix tree on the blocks of length $k = 3$ selected from the listed above signatures (Fig. 3, the commands are marked by the first character).

```
open, read, open; read, open, write; open, write, execute; write,
execute, connect; execute, connect, write; connect, write, write;
write, write, connect; write, connect, execute.
```

We split the second signature into blocks and add them to our suffix tree (Fig. 4):

```
execute, execute, connect; execute, connect, open; connect, open,
write; open, write, execute; write, execute, connect; execute,
connect, open; connect, open, read; open, read, write.
```

Let us analyze the following received sequence of commands for similarity to the given signature database:

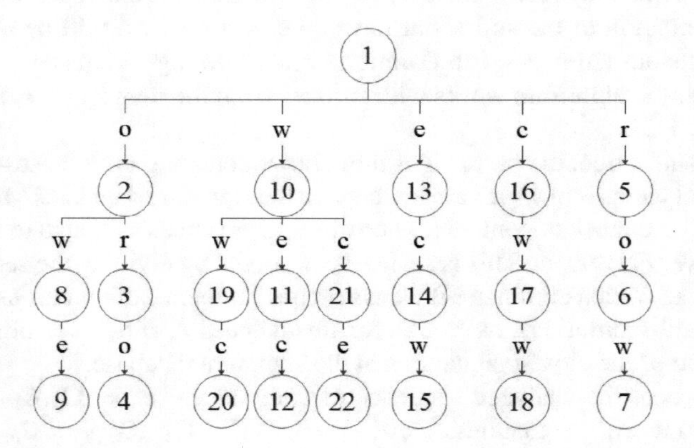

Fig. 3 Suffix tree for the first attack sequence

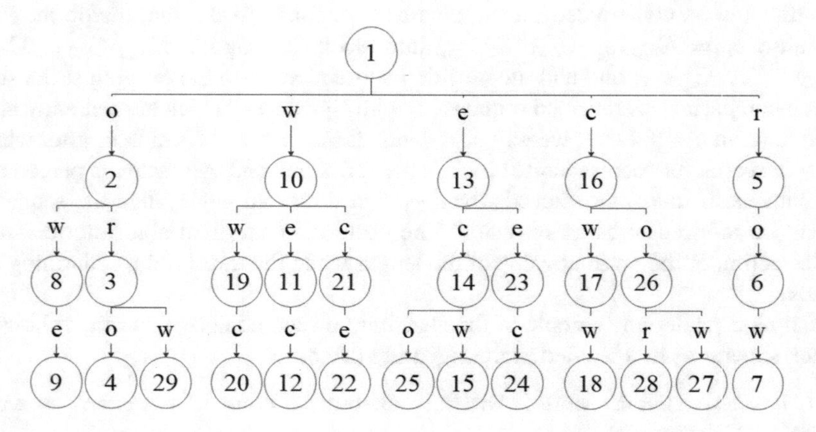

Fig. 4 Suffix tree extended with the second attack sequence

`connect, open, write, execute, connect, write, open, write, execute, connect.`

We split this input string in to blocks of the same length $k = 3$:

`connect, open, write; open, write, execute; write, execute, connect; execute, connect, write; connect, write, open; write, open, write; open, write, execute; write, execute, connect.`

A search in the suffix tree reveals that 6 out of 8 blocks (75%) are contained in the suffix tree. The blocks `connect, write, open` and `write, open, write` are not presented in the suffix tree. If a threshold is 70%, the input sequence can be determined as an intrusion.

3 Results

To approve the suffix tree method for mismatch-resistant intrusion detection, we have experimentally estimated the level of errors and resource costs of the algorithm when working on testbench.

For experiments, the test dataset KDD Cup 1999 [33] was utilized, which is a marked dataset of network packet sequences for normal and abnormal traffic. The following network attacks from KDD Cup 1999 dataset were checked: back dos; multihop r2l; satan probe; buffer_overflow u2r; neptune dos; smurf dos; ftp_write r2l; nmap probe; spy r2l; guess_passwd r2l; perl u2r; teardrop dos; imap r2l; phf r2l; warezclient r2l; ipsweep proe; pod dos; warezmaster r2l; land dos; portsweep probe; loadmodule u2r; rootkit u2r. The testbench was deployed on Intel Core i7 hardware, 32 GB of memory, 1 TB SSD, MS Windows 10 platform. The intrusion detection algorithm based on suffix tree is implemented in C programming language.

The level of errors in detecting KDD Cup 1999 attacks depending on the size of the signature database is presented in Fig. 5 (an example is given for the test parameters: block length $k = 4$, intrusion detection threshold—80%). With increase of size of the signature database, the level of false positive (FP) errors increases within tenths of a percent, not exceeding 1%. In this case, the probability of false negative (FN) errors is reduced to the level of 0.08.

The level of resource consumption of the suffix algorithm depending on the size of the signature database is shown in Figs. 6 and 7. The time spent on checking the sequences and the amount of memory occupied by the suffix tree change little with the filling of the signature base, which is determined by the limited depth of the suffix tree in which the search is performed and the determinism of transitions within the tree. The construction time of a suffix tree has an almost linear dependence on the size of the data on which the tree is built, which confirms the theoretical estimations of the algorithm. Repetition of the same string elements and their blocks within the signature database allows the suffix tree to compactly store data and efficiently use the memory resource.

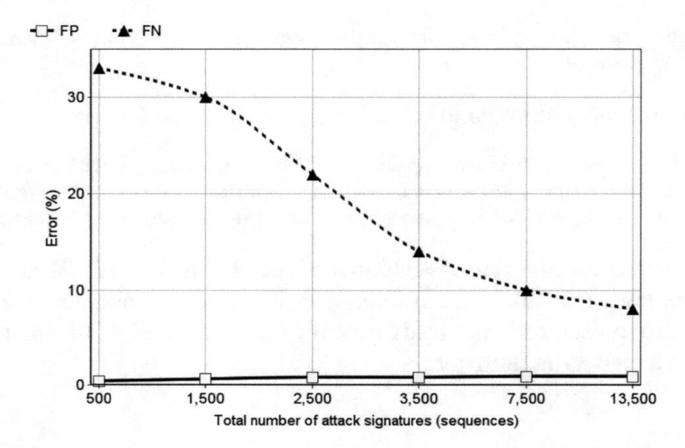

Fig. 5 Levels of false positive (FP) and false negative (FN) errors for the suffix tree algorithm (for case of block length $k = 4$, intrusion detection threshold—80%)

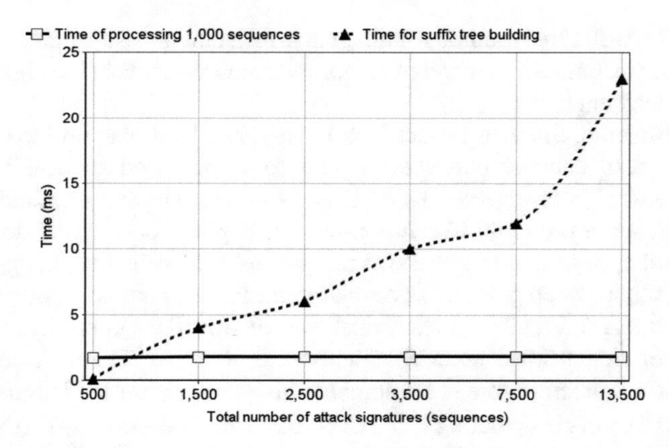

Fig. 6 Estimation of time spent on building of a suffix tree structure and sequence matching with a built suffix tree

Fig. 7 Memory consumption for storing the suffix tree structure

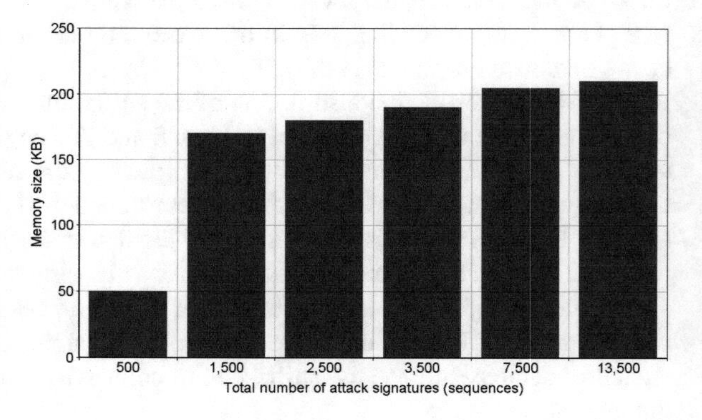

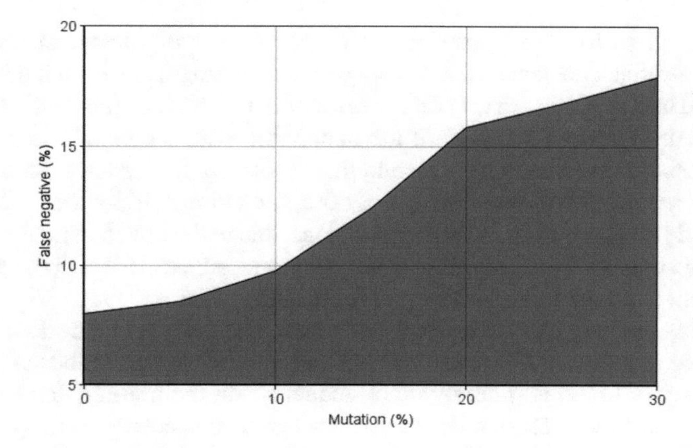

Fig. 8 Impact of mutations on the algorithm's false negative level

The level of FN errors, depending on the proportion of the mismatched portion of the analyzed sequence (level of mutations), is shown in Fig. 8 for the signature database of 13,500 records for the KDD Cup 1999 testset with artificial introduction of mutations (the range of the proportion of mutations is 0...30%).

4 Discussion

The suffix tree method is characterized with a set of advantages. It is that repetitions in the signature database do not influent the running time and memory volume used for the algorithm, because repeating sections are 'collapsed' in the suffix tree. The complexity of algorithms is $O(n)$. The experimental study has approved this.

This mechanism works fast with big sequences and large databases of signatures. At the same time suffix tree-based solution eliminates the influence of mismatches (mutations, omissions and command reordering in the chains), as, in case of mutations, it allows us to completely restore one of the closest signatures from the signature database. Therefore, recognizing the polymorphic attacks, this method is resistant to the mismatches in sequences.

The algorithm allows skipping mismatched elements, and, consequently, improving the signature-based IDS, providing us a new opportunity to recognize new intrusions, the exact signatures of which are not clearly contained in the signature database.

To improve the algorithm accuracy when working with mutating sequences, a next modification has been done, as a result of which the accuracy of intrusion detection was increased to 95% at a mutation level of 10%. The modified algorithm differs from the original one by combining two suffix trees based on the signatures of abnormal and normal behavior at once.

The second suffix tree is applied, built on the known patterns of normal (not containing an attack) sequences. At the sequence matching stage, even if the sequence being tested has a high match rate for the first suffix tree built by the attack signatures, it is not identified as an attack until the match rate with the second tree exceeds a certain threshold specified for the second tree. This way, for instance, before modification with a threshold for the first tree of 60%, for a threshold for the second tree of 90%, a single first suffix tree is ineffective, since, due to the low threshold, the system classifies the overwhelming number of sequences as attacks (FP is about 90%), and when adding a second tree, the FP level is reduced to 2%.

Obviously, the tandem of suffix trees slows down the original algorithm. The performance reduces by *N* times, where *N* is the ratio of the number of vectors of normal behavior to the number of attack sequences in the training dataset. But due to the initial linear complexity, this slowdown can be considered as insignificant.

Except the effective use of memory for storing signature databases, the technology of intrusion detection using the suffix tree presents us a unique property—the ability to dynamically expand the signature database during the IDS's operation. The complexity of adding a new branch of the suffix tree linearly depends on the length of the vector being added, so this action does not affect performance and does not require rewriting the entire suffix tree. This makes it possible to immediately adapt the IDS to new attacks added to the database of already known signatures.

5 Conclusions

The presented research has analyzed the existing solutions in the field of computational bioinformatics—the fast methods of assembly and matching of genomic sequences. Their applicability for solving the urgent problem of polymorphic attacks detection by the signature-based IDS is considered. As the result, a bioinformatics algorithm for processing sequences based on a suffix tree has been proposed which allows to achieve high accuracy of intrusion detection—more than 90% with best opportunities in terms of effective usage of memory storage, high speed and genuine resistance to possible mutations in attack vectors.

To improve the accuracy characteristics, a number of modifications of the developed algorithm were carried out, as a result of which the accuracy of detecting attacks was increased to 95% with the level of mutations in the sequence up to 10%.

The developed method for detecting mutating attacks using a suffix tree and mismatch-resistant IDSs that implement this method can be applied to detect intrusions both in classical computer networks and in modern reconfigurable network infrastructures with limited resources, e.g. the Internet of Things, cyber-physical networks, wireless sensor networks (WSN).

Our further work is targeted to optimize the proposed algorithm for the mobile network environment.

Acknowledgements The reported study was supported by LG Electronics, Inc., in Sect. 2.

The reported study was funded as the part of the State Task for Basic Research (code of theme 0784-2020-0026); suppl. agreement to the Agreement for the financial support №075-03-2020-158/2, 17.03.2020 (int. №075-GZ/SCH4575/784/2), in Sects. 3–5.

References

1. Khraisat, A., Gondal, I., Vamplew, P., Kamruzzaman, J.: Survey of intrusion detection systems: techniques, datasets and challenges. Cybersecurity **2**(1) (2019)
2. Jatti, S.A.V., Kishor Sontif, V.J.K.: Intrusion detection systems. Int. J. Recent Technol. Eng. **8**(2), special issue 11, 3976–3983 (2019)
3. Lakshminarayana, D.H., Philips, J., Tabrizi, N.: A survey of intrusion detection techniques. In: 18th IEEE International Conference on Machine Learning and Applications, ICMLA, pp. 1122–1129 (2019)
4. Platonov, V.V., Semenov, P.O.: An adaptive model of a distributed intrusion detection system. Autom. Control. Comput. Sci. **51**(8), 894–898 (2017)
5. Platonov, V.V., Semenov, P.O.: Detection of abnormal traffic in dynamic computer networks with mobile consumer devices. Autom. Control. Comput. Sci. **52**(8), 959–964 (2018)
6. Aljawarneh, S.A., Moftah, R.A., Maatuk, A.M.: Investigations of automatic methods for detecting the polymorphic worms signatures. Futur. Gener. Comput. Syst. **60**, 67–77 (2016)
7. Khonde, S.R., Venugopal, U.: Hybrid architecture for distributed intrusion detection system. Ingenierie des Systemes d'Information **24**(1), 19–28 (2019)
8. Zhang, W.A., Hong, Z., Zhu, J.W., Chen, B.: A survey of network intrusion detection methods for industrial control systems. Kongzhi yu Juece/Control and Decision **34**(11), 2277–2288 (2019)
9. Seoane Fernández, J.A., Miguélez Rico, M.: Bio-Inspired Algorithms in Bioinformatics I. Encycl Artif. Intell. (2011)
10. Coull, S., Branch, J., Szymanski, B., Breimer, E.: Intrusion detection: a bioinformatics approach. In: Annual Computer Security Applications Conference, ACSAC, pp. 24–33 (2003)
11. Lavrova, D., Zaitceva, E., Zegzhda, P.: Bio-inspired approach to self-regulation for industrial dynamic network infrastructure. In: CEUR Workshop Proceedings, pp. 34–39 (2019)
12. Miller, W.: An introduction to bioinformatics algorithms. J. Am. Stat. Assoc. **101**(474), 855–855 (2006)
13. Sohn, J., Nam, J.W.: The present and future of de novo whole-genome assembly. Brief. Bioinform. **19**(1), 23–40 (2018)
14. Recanati, A., Brüls, T., D'Aspremont, A.: A spectral algorithm for fast de novo layout of uncorrected long nanopore reads. Bioinformatics **33**(20), 3188–3194 (2017)
15. Rizzi, R., et al.: Overlap graphs and de Bruijn graphs: data structures for de novo genome assembly in the big data era. Quant. Biol. **7**(4), 278–292 (2019)
16. Wittler, R.: Alignment- and reference-free phylogenomics with colored de Bruijn graphs. Algorithms Mol. Biol. **15**(1) (2020)
17. Tan, T.W., Lee, E.: Sequence alignment. In: Beginners Guide to Bioinformatics for High Throughput Sequencing, pp. 81–115 (2018)
18. Muhamad, F.N., Ahmad, R.B., Asi, S.M., Murad, M.N.: Performance analysis of Needleman-Wunsch algorithm (Global) and Smith-Waterman algorithm (Local) in reducing search space and time for DNA sequence alignment. J. Phys. Conf. Ser. **1019**(1) (2018)
19. Lee, Y.S., Kim, Y.S., Uy, R.L.: Serial and parallel implementation of Needleman-Wunsch algorithm. Int. J. Adv. Intell. Inform. **6**(1), 97–108 (2020)
20. Čavojský, M., Drozda, M., Balogh, Z.: Analysis and experimental evaluation of the Needleman-Wunsch algorithm for trajectory comparison. Expert Syst. Appl. **165** (2021)

21. Alesinskaya, T.V., Arutyunova, D.V., Orlova, V.G., Ilin, I.V., Shirokova, S.V.: Conception BSC for investment support of port and industrial complexes. Acad. Strateg. Manag. J. **16**(Specialissue1), 10–20
22. Sun, J., Chen, K., Hao, Z.: Pairwise alignment for very long nucleic acid sequences. Biochem. Biophys. Res. Commun. **502**(3), 313–317 (2018)
23. Zou, H., Tang, S., Yu, C., Fu, H., Li, Y., Tang, W.: ASW: accelerating smith-waterman algorithm on coupled CPU-GPU architecture. Int. J. Parallel Prog. **47**(3), 388–402 (2019)
24. Chowdhury, B., Garai, G.: A review on multiple sequence alignment from the perspective of genetic algorithm. Genomics **109**(5–6), 419–431 (2017)
25. Dijkstra, M.J.J., Van Der Ploeg, A.J., Feenstra, K.A., Fokkink, W.J., Abeln, S., Heringa, J.: Tailor-made multiple sequence alignments using the PRALINE 2 alignment toolkit. Bioinformatics **35**(24), 5315–5317 (2019)
26. Chen, S., Yang, S., Zhou, M., Burd, R., Marsic, I.: Process-oriented iterative multiple alignment for medical process mining. In: IEEE International Conference on Data Mining Workshops, ICDMW, pp. 438–445 (2017)
27. Ye, N.: Markov chain models and hidden Markov models. Data Min., 287–305 (2021)
28. Behera, N., Jeevitesh, M.S., Jose, J., Kant, K., Dey, A., Mazher, J.: Higher accuracy protein multiple sequence alignments by genetic algorithm. Proced. Comput. Sci. **108**, 1135–1144 (2017)
29. Cui, X., Shi, H., Zhao, J., Ge, Y., Yin, Y., Zhao, K.: High accuracy short reads alignment using multiple hash index tables on FPGA platform. In: Information Technology and Mechatronics Engineering Conference, pp. 567–573 (2020)
30. Marçais, G., Delcher, A.L., Phillippy, A.M., Coston, R., Salzberg, S.L., Zimin, A.: MUMmer4: a fast and versatile genome alignment system. PLoS Comput. Biol. **14**(1) (2018)
31. Kay, M.: Substring alignment using suffix trees. Lecture Notes in Computer Science, vol. 2945, pp. 275–282 (2004)
32. Ukkonen, E.: On-line construction of suffix trees. Algorithmica **14**(3), 249–260 (1995)
33. KDD Cup 1999 Data homepage, kdd.ics.uci.edu/databases/kddcup99/kddcup99.html. Accessed 2021/04/01

Automatic Control Approach to the Cyber-Physical Systems Security Monitoring

Maria Poltavtseva and Andrea Tick

Abstract Monitoring the security of cyber-physical systems (CPS), including IoTh components, is an important task for modern information security. Modern approaches to the protection of cyber-physical systems are based on the theory of control and sustainability, but the CPS is not considered from this point of view as an object of evaluation and analysis (monitoring). The novelty of the work is that the cyber-physical system is considered as an object of management (control) of information security based on the approaches of the theory of automatic control. The article presents the concept of a cyber-physical system as an object of protection, formalizes the characteristics of controllability, observability and identifiability of the system in relation to security management. An approach to the evaluation of these characteristics is given. A practical example is the characteristics of a monitoring system based on the work of Peter the Great St. Petersburg Polytechnic University. The proposed approach develops the theory of protection of cyber-physical systems on the basis of stability.

Keywords Information security · Cyber-physical systems · CPS · Security monitoring · Security management · Automatic control

1 Introduction

Today, the security of computer systems that manage physical processes (cyber-physical systems [1]—CPS) is an actual task in the field of information security. A new look at the safety characteristics of CPS is required, due to the peculiarities of occurring in them irreversible processes [2]. The effectiveness of CPS security management systems depends directly on monitoring systems that provide data for

M. Poltavtseva (✉)
Peter the Great St. Petersburg Polytechnic University, Saint Petersburg, Russia
e-mail: poltavtseva@ibks.spbstu.ru

A. Tick
Óbuda Unuversity, Budapest, Hungary

C. Jahn et al. (eds.), *Algorithms and Solutions Based on Computer Technology*, Lecture Notes in Networks and Systems 387,
https://doi.org/10.1007/978-3-030-93872-7_2

detecting attacks and preventing their consequences. Completeness, reliability, and timeliness of information are key to effective security management. At the same time, the definition of these concepts, the characteristics of security monitoring, and the assessment of the quality of data collection and processing remain insufficiently developed.

The specifics and diversity of CPS [3] require fundamentally new approaches and solutions from researchers in the field of information security [4, 5], including the approach to security management based on sustainability proposed by a number of researchers [6, 7]. It consists in redefining the concept of stability of dynamic systems in relation to their functioning in the conditions of unknown cyber threats. In this article, the cyber-physical system is considered from the perspective of the theory of automatic control. The approach to monitoring the security of cyber-physical systems based on control theory is used to determine the properties of the monitoring system and the completeness of data in terms of detecting attacks.

Modern monitoring in the field of information security is based on the assessment of security events [8]. On this approach, there is a separate class of systems—SIEM. SIEM systems and first-generation monitoring systems collected and registered events by aggregating data from various sources [9, 10]. Klasa and Fray [11], Gertner et al. [12], Farrand [13], as well as Saenko et al. [14] discussed the problem of decentralized monitoring data processing.

The direction of security monitoring related to the monitoring of digital industrial systems (cyber-physical systems) has appeared quite recently. Works in this area are presented by staged publications that define new issues (analyzed in [15]), works on detecting attacks based on signatures [16] and big data processing [14, 17, 18]. The authors search for new threat models [19], and consider monitoring individual types of systems [20–22] or classes of objects [23–25]. These solutions, as well as in domestic works, used traditional approaches to security monitoring, without taking into account the specifics of cyber-physical systems as a special class of security objects [26]. In most works, monitoring systems for managing the security of cyber-physical systems are adaptations of traditional network monitoring systems. The works that emphasize the specifics of the CPS [19, 23, 27] use a knowledge-based approach and assess the parameters of the object of protection. However, they do not determine the reliability and completeness of data, and there are no objective criteria for evaluating the monitoring characteristics and quality.

CPS information security researchers establish a relationship between security aspects (including confidentiality, availability, and data integrity) and the properties of a technological system [28]. Works [4–7], consider a security breach of a cyber-physical system associated with a violation of its target function, without dividing it into a physical process and a control process. Zegzhda D. P. introduced the concept of stability in relation to the CPS information security in the work [4]. In the theory of automatic control, CPS information security is closely related to other properties of a dynamic system: observability, controllability, and identifiability [29]. The approach to ensuring the safety of CPS based on sustainability and homeostasis [4] defines sustainability as the ability of a system to continue performing its target function under destructive influences. Interpretation of properties other than sustainability is not given in modern research. The CPS representation as a set of separate control systems [26] allows describing the CPS properties and its security management system based on the properties and control systems of dynamic objects. From this point of view, security management is a specific area of dynamic system management. There is a need to set the properties of a dynamic system (observability, manageability, sustainability, identifiability) in accordance with the protection. Defining concepts and formalizing computable criteria for observability, controllability, and identifiability in relation to security objects—cyber-physical systems—will make it possible to take a step towards building a consistent theoretical and practical base for managing the security of CPS, defining the concepts of completeness, reliability, and timeliness of monitoring.

The application of the proposed methods and assessments to monitoring the CPS security approved on the works of Peter the Great St. Petersburg Polytechnic University. Evaluation of characteristics based on the theory of automatic control is carried out for a monitoring system by the work of the Department of cybersecurity and information protection of Peter the Great St. Petersburg Polytechnic University [30–43].

2 Materials and Methods

Cyber-physical systems by themselves manage some of physical objects and processes. For this purpose, they include control and communication components and can be represented as a superposition of simpler automated control systems [26] connected by network. For cyber-physical systems, the object of control is a physical process, and the components that perform the control function are controllers and digital modules. To correctly perform control functions in automated control systems, the concept of monitoring the control object and such indicators as observability, manageability, stability and identifiability are introduced [44]. In turn, from the point of view of information security, the object of management is the entire cyber-physical system, including its control and communication components. In this case, the concept of the monitoring object is shifted from the physical process to the cyber-physical system as a whole. Since cyber-physical systems are an evolutionary development of process control systems [26], the main properties of automated process control systems defined in the theory of automatic control are applicable to them [29]. When assessing stability as a criterion for safe operation, it also becomes important to determine whether the system is observable and manageable from the point of view of information security, and not from the point of view of the physical process. The stability-based approach to ensuring the security of cyber-physical systems allows one to compare the properties of a physical process from the point of view of control theory and the properties of a cyber-physical system from the point of view of its protection.

The cyber-physical system representation as a superposition of traditional technical automatic control systems and the description of the CPS from the point of view of control theory in terms of observability, controllability and sustainability imply the formalization of destructive impacts on it in the same terms, and, consequently, the appropriate threat model development. A cyber-physical system is a component of physical process controlling using a digital control system. Approaches and methods of control theory are applicable for industrial CPS [43]. The purpose of their functioning is to perform a certain target function [34] under conditions of destructive external influences $Y = F(X, P, t)$. Here: X—a set of input parameters of the system; *P-a* set of external disturbances (impacts); $P = P_0 \cup P_a$, where P_a—destructive effects on the system by an attacker, and P_0—all the others; t—the operating time $\in [0, +\infty)$; Y—a set of output parameters (the result of functioning) of the system. At any given time, a cyber-physical system is described by a set of parameters that determine its state $\Sigma_t = \mathrm{C}_t\{c_1, \ldots, c_N\}$ where c_i for $\in [1, N]$ is a parameter, that reflects the state of the cyber-physical system at a given time t. In order for the state of the system Σ to be determined at any given time, all the its parameters $c_i \in C$ must also be defined at each point in time. An alternative approach is to represent the set C as a set of variable length parameter vectors $C = \{C_t||C_t| \neq \mathrm{const}\}$. However, this method is more time-consuming than representing the discrete parameters of the system as discrete functions on the interval $t \in [0, +\infty)$. In general, the state of the system is determined by a combination of three types of parameters $\Sigma = \{X, Y, S\}$ where $X = \mathrm{C}_X\{c_1, \ldots, c_M\}$—system input parameters, M—total number of input parameters; $Y = \mathrm{C}_Y\{c_M, \ldots, c_K\}$—system output parameters, K—total number of input and output parameters; $S = \mathrm{C}_S\{c_K, \ldots, c_N\}$ S—internal (structural) system parameters, N—total number of system parameters. Where ci for $i \in [1, N]$—discrete function describing the state parameter of a cyber-physical system. For an ordinary automatic control object, an example of internal parameters can be the type of transfer function of the control node. For a cyber-physical system, internal parameters include the number of nodes, the communication graph in the system, and so on. In fact, the set of parameters s characterizes the model of converting the input data of the system into output data or the target function $Y = F(X, P, t)$.

All parameters of the system with this approach are observable (fixed during monitoring and management of the CPS) and unobservable. To include unobservable parameters that are not fixation subject or fixation which is impossible. Let's define the state of the system from the observer's point of view:

$$\Sigma' = \{X', Y', S'\} \text{ где } \begin{cases} X' = C'_X\{c_1, \ldots, c_m\}, & m \leq M \\ Y' = C'_Y\{c_m, \ldots, c_k\}, & k \leq K \\ S' = C'_S\{c_k, \ldots, c_n\}, & n \leq N \end{cases} \tag{1}$$

Since the state of the system is determined at a specific time (2), the value can be determined at each time $\Sigma'_t = \frac{d\Sigma'}{dt}$ and parameter values $\mathrm{c} \in X', \mathrm{c} \in Y', \mathrm{c} \in S'$. In this case, Σ'_t is the observed state of the system at time t.

3 Results

3.1 *Observability, Controllability and Identifiability of CPS in the Management of Information Security*

The main properties to be evaluated in the design of control systems and monitoring of automated control systems, in addition to sustainsability, are observability and controllability. These properties also allow one to evaluate the quality of the monitoring and control system for an automated control object. Let's highlight the main differences between the security management of the CPS and the management of the physical process. First, consideration as an object of management of the CPS as a whole, including management and communication components. Second, the complexity of the system in terms of internal structure, the lack of a complex mathematical model of the object. And the third—the focus of attention is possible destructive impacts on the system, their identification, prevention and neutralization of consequences.

In terms of information security, the properties of observability, controllability, and identifiability are determined based on the concepts of control theory [29] and the specifics of security tasks. The stability property and its application in security management systems are discussed in detail in [4–7, 34]. Observability reflects the extent to which the operation of a management object is subject to external evaluation based on the data collected about it. Observability from the point of view of information security is a characteristic that reflects the ability of a security management system to detect malicious impact. In this case, observability (interpretation of the input parameters of the system from its output) is interpreted as the ability of the control system to establish the fact of a destructive effect on the measured data. Then, to ensure the observability of the system based on its observed state (4), a binary function must exist:

$$F_{\mathrm{H}}\left(\Sigma'_t\right) = \begin{cases} 0, \forall \Sigma'_t | \nexists p_i(t) \in P_a \\ 1, \forall \Sigma'_t | \exists p_i(t) \in P_a \end{cases} \tag{2}$$

where $p_i(t)$—external destructive impact on the system at time t. Value $F_{\mathrm{H}}\left(\Sigma'_t\right) = 0$ when there is no destructive effect, 1– if there is one, and the FH function itself is defined on the interval $t \in [0, +\infty)$. Controllability shows how much the object is subject to correction by the control system. The manageability of a security object is the ability to bring the system to a safe state (if it is violated) in a finite time. This property is defined by the existence for each state Σ_{ti} of the function $\Sigma_t = F_Y(\Sigma_{t0})$. This function brings the system out of an unsafe state Σ_{t0} in safe state Σ_t and the time $T = t - t_0$. Studies of one of the approaches to controllability based on stability preservation are given in more detail in the works [4, 31, 32].

In terms of security monitoring, an important property of the system (the security management system of the CFS) is identifiability. From the point of view of security management, identifiability is the ability to determine the type of destructive impact and the objects affected by it from the results of measuring output values over a finite time interval. It is divided into:

1. ***Component identification***—identification of components affected by the destructive impact of the protection object;
2. ***Impact identification***—determining the type of destructive impact.

The properties of identifiability (component identifiability and impact identifiability) are determined based on the observed state of the CPS and the set of types of possible destructive effects (attacks) $A = \{a_1, \ldots, a_L\}$. The system's identifiability is determined by the existence of two functions.

Function for identifying the type of impact:

$$F_T(\Sigma'_t, A) = \begin{cases} 0, \text{ при } F_{\text{Н}}(\Sigma'_t) = 0 \\ a_i \in A, i \in [1, |A|], \text{при } F_{\text{Н}}(\Sigma'_t) = 1 \end{cases} \quad (3)$$

Function for identifying affected objects:

$$F_T(\Sigma'_t) = \begin{cases} 0, \text{ при } F_{\text{Н}}(\Sigma'_t) = 0 \\ O_{a_i} \in O_\Sigma, \text{ при } F_{\text{Н}}(\Sigma'_t) = 1 \end{cases} \quad (4)$$

where O_Σ—the set of all hierarchically organized identifiable objects in the system, a O_{a_i}—a subset of objects affected by the i-th type of attack at time t in the system state Σ'_t.

3.2 Characteristics of the Security Monitoring System

The monitoring system should provide the information security management subsystem with sufficient information to make a decision on detecting relevant impacts and predicting system behavior. That is, to ensure the observability and identifiability of the CPS from in terms of information security management. This information must be sufficiently complete, reliable, and timely.

The completeness of security monitoring is determined on the results of monitoring and its mapping to the overlying components that perform security functions. If the monitoring data is sufficient to perform all the assigned security functions, then the monitoring is complete. Thus, the completeness property is determined by the amount of data delivered and the mathematical methods used to ensure the observability and identifiability of the CPS. According to (5), observability is estimated based on the vector of destructive effects P_a. The observability of a system over the entire set of its states according to (5) can be defined as a binary vector $V = \{v_1, \ldots, v_{|P_a|}\}$ где $v_i = \begin{cases} 0, \nexists Fp_i | p_i \in P_a \\ 1, \exists Fp_i | p_i \in P_a \end{cases}$, where $p_i(t)$—i destructive impact and Fp_i—its detection function.

The ***identifiability*** of the CPS based on (6) and (7) is determined based on the attack vector as a binary vector of identifiability of components $I_k = \{i_{k,1}, \ldots, i_{k,|A|}\}$ and a binary vector of identifiability of impact $I_A = \{i_{A,1}, \ldots, i_{A,|A|}\}$. The vector elements take the values {0,1} depending on whether the object can be identified. Depending on the approach, the attack vector A and the vector of destructive effects P_a can match or have a mapping $A \rightarrow P_a$. This mapping is now used in information security to identify the type of attack based on the observed impacts (security events).

The ***reliability*** of security monitoring is defined as the compliance of the data provided by security monitoring with the actual state in the security object. To ensure the reliability of monitoring, two conditions must be met:

- Incoming data should be correspond to the data flows (real situation) in the CPS;
- Incoming data should not be distorted after entering the security monitoring system.

Distortion of data before entering the security monitoring system, for example, because of an attack, is not a violation of the reliability of security monitoring. Distorted data entering the monitoring system can detect a security breach. When using various methods for detecting anomalies in the monitoring system, the reliability of monitoring corresponds to the accuracy (reliability) of these methods. The reliability of security monitoring is a cumulative assessment of the reliability of observability and identifiability, taking into account possible data distortions in the monitoring system:

$$T = \begin{cases} Tv = \{tv_1, \ldots, tv_{|P_a|}\} \\ Ti_k = \{ti_{k,1}, \ldots, ti_{k,|A|}\} \\ Ti_A = \{ti_{A,1}, \ldots, ti_{A,|A|}\} \\ Tt = p_{cor} \end{cases} \quad (5)$$

where Tv—probability of detection of the corresponding destructive effects, Ti_k—probabilities of identifying components affected by the corresponding attack and Ti_A—probabilities of determining the type of attack based on the set of detection methods used. Tt—the probability of data corruption within the security monitoring system, determined based on the threat model and the methods used to ensure data availability and integrity. The reliability of safety monitoring is necessary to ensure the observability and identifiability of CPS. In other words, one can talk about observability and identifiability to the extent that confidence is provided.

The timeliness of security monitoring is determined by the ability of the security management system to generate a response to a security breach. To ensure timeliness, the ratio must be met $T_A > T_M + T_R$ where TA—time of the destructive impact (attack), TM—the time of the corresponding monitoring procedures calculated from the beginning of exposure, and TR is the response time of the safety management system. Then the time margin will be $T_\Delta = T_A - (T_M + T_R)$ where T_Δ is a margin of timeliness.

Based on the position of completeness, reliability and timeliness, it is possible to form a comprehensive assessment of safety monitoring, as a set of characteristic indicators. For practical use, the indicators of observability and identifiability in terms of completeness and reliability of monitoring should be combined. Then the value of the observability and identifiability vectors corresponds to 0 for non-observable (unidentifiable) attacks and impacts. $p > 0$ determines the probability of a reliable observation (identification). As a result, the comprehensive assessment will take the form:

$$
\mathrm{Q} = \begin{cases} V = \{v_1, \ldots, v_{|P_a|}\} \\ I_k = \{i_{k,1}, \ldots, i_{k,|A|}\} \\ I_A = \{i_{A,1}, \ldots, i_{A,|A|}\} \\ Tt = p_{cor} \\ T_\Delta = T_A - (T_M + T_R) \end{cases} \tag{6}
$$

This estimate can be used independently or as a basis for calculating a weighted value.

3.3 Estimation of the CPS Security Monitoring Assessment

The graph model [34–36] is used as a model of the security object. An example object graph is shown in Fig. 1. It shows communication and worker nodes, functions of the

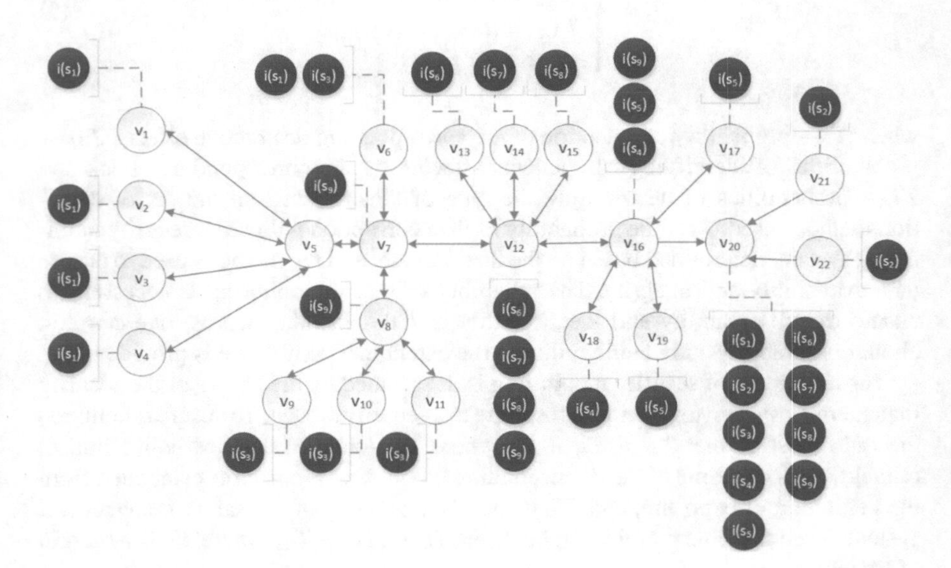

Fig. 1 Example of a security object graph

operating units. The vector of destructive actions P_a is defined as a set of operations on the graph $SP_a = \{p_{E+}, p_{E-}, p_E, p_{V+}, p_{V-}, p_V\}$ where p_{E+}, p_{E-}, p_E—adding, deleting, and modifying a graph edge, and p_{V+}, p_{V-}, p_V—adding, deleting, and changing a graph vertex. Examples of attacks [40] with mapping $A \rightarrow P_a$ presented in Table 1.

Inverse mapping $P_a \rightarrow A$ not unambiguously. Since all changes are observable on the graph and the model of impacts has theoretical completeness [37], provided that all types of impacts are detected [38]. The probability of reliable observation can be estimated based on the probabilities of reliable detection of anomalies in the work [34, 37–40] and will be $V = \{0.95, 0.97, 0.94, 0.95, 0.97, 0.94\}$ with precision of the order of $\varepsilon = 0.01$. Since anomalies on each component are evaluated separately, the identifiability of components in this case will correspond to the observability (however, without taking into account the link to a specific attack).

Table 1 Mapping $A \rightarrow P_a$

A	P_a
Denial of service attack (DoS, DDoS), disabling a system component	$p_{V-}, \{p_{E-}\}$
A denial-of-service attack aimed at disabling a device or disabling it, resulting in a change in the operating rules or system settings that prohibits a group of system components from communicating with each other	
An attacker adds a device that pretends to be legitimate	p_{V+}
Changes to operating rules or system settings that prohibit data transfer between system components	p_{E-}
Changes to the operating rules or system settings that prohibit data exchange in a group of devices	
Changes to the operating rules or system settings that allow previously prohibited data exchange between devices	p_{E+}
A denial-of-service attack, disabling a device or disabling it	p_{V-}
Changing the operating rules or system settings that prohibit data exchange in a group of devices, and then disabling this group of devices	$p_{E-}, \{p_{V-}\}$
The attack "sinkhole", which is characteristic for wireless sensor networks. A compromised network node "listens" to requests for routes and responds to nodes that it "knows" the shortest route to the base station	$p_{V+}, \{p_{E+}\} \{p_{V-}\}$
Changes to operating rules or system settings that initiate direct data exchange for devices that have not previously exchanged data directly, but are only indirectly connected to each other	$\{p_{E+}\}$
A complex attack that combines a "sinkhole" attack and a ban on connecting between nodes	$p_{E-}, p_{V+}, \{p_{E+}\} \{p_{V-}\}$
Man-in-the-Middle (MITM) attack, intruder interference in the data transfer process, data interception	$p_{V+}, p_{E-}, \{p_{E+}\}$

To detect the actual attacks, it is necessary to use intelligent methods that allow determining the type of complex impact based on the knowledge base of precedents [41]. The accuracy of such detection in the tested knowledge base will be for the component $I_k = \{i_{k,j} | i_{k,j} = 0.78 \pm 0.1 \forall j = [1, |A|]\}$ and for the attack $I_A = \{i_{A,j} | i_{A,j} = 0.92 \pm 0.1 \forall j = [1, |A|]\}$ [41–43]. For the test system, the probability of data distortion $p_{\mathrm{cor}} = 0$. In real conditions, this is the probability of a successful attack by an attacker on the monitoring system itself, taking into account the security features.

The time of self-regulation of the FSC, or reaction time T_R, is 12–38% of the time of the attack T_A. In this case, the exposure detection time T_M should be up to 62% of the attack time. T_M—up to 16% of the attack time [36]. So the minimum value $T_\Delta = 46\%$ from the time of the attack.

4 Discussion

The theory of automatic control allow to use a new approach to the security problems of cyber-physical systems, taking into account their features. Formalization of properties (observability, controllability, sustainability, identifiability) makes it possible to build a new model of CPS threats. By means of systematization of threats, objects and types of impact are identified and classified, protection against which ensures the preservation of the system properties in terms of managing the physical process, and therefore performing its target function. An important feature is the coordination of parameters used to control the cybernetic system as an object of the automated process control system and information security parameters. This is how traditional process management can be aligned with the new requirements for cyber attack sustainability.

The paper mathematically defines the state of the CPS as an object of security management and the properties of observability and identifiability (exposure identification and component identification). A mapping of the properties of the CPS to the characteristics of the monitoring system, such as completeness, reliability and timeliness, is constructed. This formulation of the problem allows one to monitor security using the concepts and approaches of the automated process control. This possibility is due to the general task of the industrial cyber-physical systems security monitoring and process control. This is the task of ensuring sustainable functioning, in one case—under the influence of external disturbances, in the second—in the conditions of targeted cyber-attacks. A comprehensive assessment of the monitoring system was formed based on the accepted model of the object of protection and destructive impacts, the list (model) of attacks and their mutual display. The assessment includes indicators of observability of impacts, identifiability of attacks and affected components, as well as an indicator of reliability (absence of data distortions within the monitoring system itself) and timeliness. An example of calculating a comprehensive assessment is based on the work of the Great St. Petersburg Polytechnic University school of cybersecurity.

The proposed performance evaluation monitoring security of cyber-physical systems based on the characteristics of the APCS is the result of the security assessment on the basis of terms and models that apply to specific system monitoring. Flexibility and comparability are important properties of the proposed approach. Basic provisions for evaluating security monitoring cyber-physical systems allow one to interpret them based on various threat models and descriptions of the object of protection. The resulting integral estimates: the completeness of security monitoring; the reliability of security monitoring; the timeliness of security monitoring. They are common to any type of system and are comparable to each other. In this way, security monitoring systems based on different attack detection models and methods can be compared.

The complexity is the complexity of constructing an estimate for each hotel system and the dependence of the final score on the considered attack pool. If the source system does not have data on the statistics of detecting attacks of a certain type, then a comparison can not be made for these attacks. Today, the system must be fully open to the researcher in order to conduct such an assessment.

5 Conclusions

The work performed allows for a comprehensive assessment and comparison of CPS monitoring systems based on one object model (impacts) or on different ones, with an emphasis on the list of detected attacks. The approach based on automated management allows one to take into account the features of the CPS as a security object and apply proven assessments of dynamic systems in the field of information security. Important further areas of work are the development of more universal approaches to attack vectors and the development of approaches and methods for comparing industrial information security monitoring systems in conditions of incomplete data on them.

Acknowledgements The reported study was funded by Russian Ministry of Science (information security), project number 2/2020.

References

1. Sanfelice, R.G.: Analysis and design of cyber-physical systems. A hybrid control systems approach. In: Cyber-Physical Systems: From Theory to Practice, pp. 3–31. CRC Press (2016)
2. Zegzhda, D.P.: Problems of cyber stability of digital production. In: Proceedings 26th Scientific and Technical Conference "Methods and Technical Means of Ensuring Information Security", St. Petersburg, Russia, pp. 85–86. Polytechnic publishing House, St. Petersburg (2017)
3. Zegzhda, D.P., Poltavtseva, M.A., Lavrova, D.S.: Systematization and security assessment of cyber-physical systems. Aut. Control Comp. Sci. **51**, 835–843 (2017). https://doi.org/10.3103/S0146411617080272

4. Aleksandrova, E.B., Shtyrkina, A.A., Iarmak, A.V.: Post-quantum primitives in information security. Nonlinear Phenom. Complex Syst. **22**(3), 269–276 (2019)
5. Aleksandrova, E.B., Shtyrkina, A.A., Yarmak, A.V.: Post-quantum group-oriented authentication in IoT. Nonlinear Phenom. Complex Syst. **23**(4), 405–413 (2020). https://doi.org/10.33581/1561-4085-2020-23-4-405-413
6. Zegzhda, D.P.: Sustainability as a criterion for information security in cyber-physical systems. Aut. Control Comp. Sci. **50**, 813–819 (2016). https://doi.org/10.3103/S0146411616080253
7. Petrenko, S.A.: Management of cyber stability: problem statement. Inf. Prot. Inside **3**(87), 16–24 (2019)
8. Luckham, D.: The power of events: an introduction to complex event processing in distributed enterprise systems. In: Rule Representation, Interchange and Reasoning on the Web. RuleML. LNCS, vol. 5321, pp. 3–3. Springer (2008)
9. Barker, G.T., Alexander, B., Talley, P.: US6542075B2. System and method for providing configurable security monitoring utilizing an integrated information portal (2000). https://patents.google.com/patent/US6542075B2/en?oq=US6542075B2
10. Andersen, C., Alexander, B., Bahneman, L.: US7627665B2. System and method for providing configurable security monitoring utilizing an integrated information system (2015). https://patents.google.com/patent/US7627665B2/en
11. Klasa, T., Fray, I.: El Load-balanced integrated information security monitoring system. In: Communication Papers of the 2017 Federated Conference on Computer Science and Information Systems, Prague, Czech Republic, ACSIS, 2017, vol. 13, pp. 213–221 (2017)
12. Gertner, Y., Herz, F.S.M., Labys, W.P.: US9503470B2 Distributed agent based model for security monitoring and response (2002). https://patents.google.com/patent/US9503470B2/en
13. Farrand, T.E.: US9633547B2. Security monitoring and control (2014). https://patents.google.com/patent/US9633547B2/en
14. Saenko, I.B., Kushnerevich, A.G., Kotenko, I.V.: Implementation of a distributed parallel computing platform for collecting and preprocessing big monitoring data in cyber-physical systems. In: Materials of the International Scientific Congress. International Congress on Informatics: Information Systems and Technologies (CSIST-2016). Republic of Belarus, Minsk, 24–27 October 2016–2016, pp. 641–645 (2016)
15. Sajid, A., Abbas, H., Saleem, K.: Cloud-assisted IoT-based SCADA systems security: a review of the state of the art and future challenges. IEEE Access **4**, 1375–1384 (2016). https://doi.org/10.1109/ACCESS.2016.2549047
16. Knapp, E., Langill, J.: Security Monitoring of Industrial Control Systems. Industrial Network Security. 2nd ed. Syngress Publishing (2014)
17. Marchal, S., Jiang, X., State, R., Engel, T.: A big data architecture for large scale security monitoring. In: 2014 IEEE International Congress on Big Data, Anchorage, USA, 2014, pp. 56–63. IEEE (2014)
18. Manogaran, G., et al.: A new architecture of Internet of Things and big data ecosystem for secured smart healthcare monitoring and alerting system. Future Gener. Comput. Syst. **82**, 375–387 (2018). https://doi.org/10.1016/j.future.2017.10.045
19. Coletta, A., Armando, A.: Security monitoring for industrial control systems. In: Security of Industrial Control Systems and Cyber Physical Systems. CyberICS 2015, WOS-CPS 2015. LNCS, vol. 9588, pp. 48–62. Springer (2015)
20. Trihinas, D., Pallis, G., Dikaiakos, M.: Low-cost adaptive monitoring techniques for the internet of things. IEEE Trans. Serv. Comput. (2018)
21. Lv, F., Wen, C., Liu, M.: Representation learning based adaptive multimode process monitoring. Chemom. Intell. Lab. Syst. **181**, 95–104 (2018). https://doi.org/10.1109/TSC.2018.2808956
22. Shang, C., Yang, F., Huang, B., Huang, D.: Recursive slow feature analysis for adaptive monitoring of industrial processes. IEEE Trans. Industr. Electr. **65**, 8895–8905 (2018). https://doi.org/10.1109/TIE.2018.2811358
23. Hansch, G., Schneider, P., Brost, G.S.: Deriving impact-driven security requirements and monitoring measures for industrial IoT. In: Proceedings of the 5th on Cyber-Physical System Security Workshop (CPSS '19). Association for Computing Machinery, New York, NY, USA, pp. 37–45 (2019). https://doi.org/10.1145/3327961.3329528

24. Wolf, J., et al.: Adaptive modelling for security analysis of networked control systems. In: 4th International Symposium for ICS & SCADA Cyber Security Research 2016, Electronic Workshops in Computing, pp. 64–73 (2016). https://doi.org/10.14236/ewic/ICS2016.8
25. Brost, G.S., et al.: An ecosystem and IoT device architecture for building trust in the industrial data space. In: Proceedings of the 4th ACM Workshop on Cyber-Physical System Security (CPSS '18). Association for Computing Machinery, New York, NY, USA, pp. 39–50 (2018). https://doi.org/10.1145/3198458.3198459
26. Vasil'ev, Y.S., Zegzhda, D.P., Poltavtseva, M.A.: Problems of security in digital production and its resistance to cyber threats. Aut. Control Comp. Sci. **52**, 1090–1100 (2018). https://doi.org/10.3103/S0146411618080254
27. Zegzhda, D.P., Pavlenko, E.Y.: Digital manufacturing security indicators. Aut. Control Comp. Sci. **52**, 1150–1159 (2018). https://doi.org/10.3103/S0146411618080333
28. Gorbachev, I.E., Glukhov, A.P.: Modeling of information security violations of critical infrastructure. In: Proceedings of SPIIRAN, vol. 38, pp. 112–135 (2015)
29. Emelyanov, S.V.: Mathematical Methods of Control Theory. Problems of Sustainability, Controllability and observability, 200 p. FIZMATLIT, Moscow (2014)
30. Pavlenko, E.Y., Yarmak, A.V., Moskvin, D.A.: Hierarchical approach to analyzing security breaches in information systems. Aut. Control Comp. Sci. **51**, 829–834 (2017). https://doi.org/10.3103/S0146411617080144
31. Pavlenko, E., Zegzhda, D., Shtyrkina, A.: Criterion of cyber-physical systems sustainability In: 10th Anniversary International Scientific and Technical Conference on Secure Information Technologies, BIT 2019; Moscow, vol. 2603, pp. 60–64 (2019)
32. Zegzhda, D.P., Pavlenko, E.Y.: Cyber-physical system homeostatic security management. Aut. Control Comp. Sci. **51**, 805–816 (2017). https://doi.org/10.3103/S0146411617080260
33. Zegzhda, D.P., Usov, E.S., Nikol'skii, A.V., et al.: Use of Intel SGX to ensure the confidentiality of data of cloud users. Aut. Control Comp. Sci. **51**, 848–854 (2017). https://doi.org/10.3103/S0146411617080284
34. Lavrova, D.S.: An approach to developing the SIEM system for the internet of things. Aut. Control Comp. Sci. **50**, 673–681 (2016). https://doi.org/10.3103/S0146411616080125
35. Zegzhda, D., Zegzhda, P., Pechenkin, A., et al.: Modeling of information systems to their security evaluation. In: Proceedings of the 10th International Conference on Security of Information and Networks (SIN'17). Association for Computing Machinery, New York, NY, USA, pp. 295–298 (2017). https://doi.org/10.1145/3136825.3136857
36. Lavrova, D.S., Zaitseva, E.A., Zegzhda, D.P.: Approach to presenting network infrastructure of cyberphysical systems to minimize the cyberattack neutralization time. Aut. Control Comp. Sci. **53**, 387–392 (2019). https://doi.org/10.3103/S0146411619050067
37. Lavrova, D., Zegzhda, D., Yarmak, A.: Using GRU neural network for cyber-attack detection in automated process control systems. In: 2019 IEEE International Black Sea Conference on Communications and Networking (BlackSeaCom), Sochi, Russia, pp. 1–3 (2019). https://doi.org/10.1109/BlackSeaCom.2019.8812818
38. Poltavtseva, M.A., Zegzhda, D.P., Pavlenko, E.Y.: High-performance NIDS architecture for enterprise networking. In: 2019 IEEE International Black Sea Conference on Communications and Networking (BlackSeaCom), Sochi, Russia, pp. 1–3 (2019). https://doi.org/10.1109/BlackSeaCom.2019.8812808
39. Lavrova, D., Zegzhda, D., Yarmak, A.: Predicting cyber attacks on industrial systems using the Kalman filter. In: 2019 Third World Conference on Smart Trends in Systems Security and Sustainablity (WorldS4), London, United Kingdom, pp. 317–321 (2019). https://doi.org/10.1109/WorldS4.2019.8904038
40. Zegzhda, D., Lavrova, D., Poltavtseva, M.: Multifractal security analysis of cyberphysical systems. Nonlinear Phenom. Complex Syst. **22**, 196–204 (2019)
41. Stepanova, T., Pechenkin, A., Lavrova, D.: Ontology-based big data approach to automated penetration testing of large-scale heterogeneous systems. In: Proceedings of the 8th International Conference on Security of Information and Networks (SIN '15). Association for Computing Machinery, New York, NY, USA, pp. 142–149 (2015). https://doi.org/10.1145/2799979.2799995

42. Zegzhda, P.D., Poltavtseva, M.A., Pechenkin, A.I., et al.: A use case analysis of heterogeneous semistructured objects in information security problems. Aut. Control Comp. Sci. **52**, 918–930 (2018). https://doi.org/10.3103/S0146411618080278
43. Zaitseva, E.A., Zegzhda, D.P., Poltavtseva, M.A.: Use of graph representation and case analysis to assess the security of computer systems. Aut. Control Comp. Sci. **53**, 937–947 (2019). https://doi.org/10.3103/S0146411619080327
44. Neusypin, K.A., Proletarsky, A.V., Kuznetsov, I.A.: Investigation of the degree of identifiability of parameters of dynamic systems. Bulletin of the Moscow state technical University named after G. I. Nosov., vol. 2, no. 50 (2015)
45. Humayed, A., et al.: Cyber-physical systems security—a survey. IEEE Internet Things J. **4**(6), 1802–1831 (2017). https://doi.org/10.1109/JIOT.2017.2703172
46. Giraldo, J., et al.: A survey of physics-based attack detection in cyber-physical systems. ACM Comput. Surv. **51**, 1–36 (2018). https://doi.org/10.1145/3203245

Detection of Malicious Program for the Android Platform by Deep Training

Aleksandr Suprun, Tatiana Tatarnikova, Igor Sikarev, and Anastasia Shmeleva

Abstract The article deals with threats to the Google Android mobile operating system, methods of attacks. The advantages and disadvantages of malware detection mechanisms are analyzed. To solve the problem of classification of threats, it is proposed to use the apparatus of neural networks. The complex application of static and dynamic analysis of Android operating system programs allows you to accurately classify the programs under study with a high degree of probability. The classification problem is solved using the neural network apparatus. The results showed that all malicious programs that were boycotted by obfuscation were correctly identified as malicious.

Keywords Neural networks · Threats · Malware · Viruses · Android · Dynamic analysis

1 Introduction

The Google Android operating system (OS) currently occupies a leading position in the mobile operating system market [1]. At the same time, the number of malicious programs on the Android OS is also growing. For example, the SMSVova spyware disguised as a system update was downloaded over three million times and remained undetected for more than two years. Mobile viruses can spy on users, be used to generate illegal income, for example, send paid SMS messages or encrypt user data for the purpose of extortion [2].

A. Suprun (✉)
Peter the Great St. Petersburg Polytechnic University (SPbPU), Polytechnicheskaya, 29, St. Petersburg 195251, Russia
e-mail: Afs54@inbox.ru

T. Tatarnikova · I. Sikarev
Russian State Hydrometeorological University (RSHU), Voroneghskaya str., St. Petersburg 192007, Russia

A. Shmeleva
Marriott International Yerevan, Yerevan, Armenia

C. Jahn et al. (eds.), *Algorithms and Solutions Based on Computer Technology*,
Lecture Notes in Networks and Systems 387,
https://doi.org/10.1007/978-3-030-93872-7_3

In addition to these threats, there are other problems associated with insufficient verification of applications implemented by Google Play and other stores, when downloading which, users install malicious software disguised as games and programs. Moreover, in order to save money, users often resort to third-party sources of programs for the Android OS, for example, torrents, where applications may not be checked at all. Users are also exposed to personal data leaks, hidden downloads on websites, and other dangers [3].

Measures to counteract the "hidden" installation of malicious software can be different. With increasing diversity. methods of machine learning and artificial neural networks themselves have become popular, which in particular allow solving problems of pattern recognition, classification and analysis of hidden patterns in data [4, 5].

The problem of detecting malicious software will be considered as a classification problem, the solution of which is aimed at increasing the probability of detecting malicious software on the Android OS.

2 Materials and Methods

In the Android OS, there are 2 main mechanisms for detecting malware: the Google Bouncer testing system and antivirus programs.

As a rule, by default, Android devices do not support the installation of applications from unknown sources, which include applications that are not signed with a Google certificate. If the app developer has received a certificate, it can be published in the Google Play Store. Before publishing, the application is checked by the Google Bouncer system, which is the anti-virus functionality of Google. Verification consists in testing the application for malicious code using a cloud service, that is, the application is installed on a virtual machine and dynamic analysis is performed. If suspicious activity is detected, the application is not verified and is not published, and the developer's account may be blocked [6].

Another approach to detecting malware is to use antivirus software. The antivirus programs that exist for the Android OS are applications, not OS components, which actually determines their main drawback—the isolation of antivirus programs from other applications, which allows only static analysis [7].

In addition to the existing approaches, it is worth noting the presence of a permission mechanism, the essence of which is that, firstly, no Android OS application by default has permissions for operations that can affect the receipt of personal data, the operating system or other applications, and secondly, applications are assigned privileges and it depends only on the user whether to give access to a particular application to the data.

In Android OS up to version 6, all permissions are granted during installation. The user is shown a list of permissions available to the application, and to continue the installation, the user must confirm that he allows the application to use the listed system components. In the new versions of Android, a group of special permissions, called dangerous, is allocated, which will be requested immediately before use, thereby preventing the program from covertly performing an undesirable action on the device. All permissions used by the application must be specified in the manifest [8].

The main disadvantage of dynamic analysis, which is implemented on the Google Bouncer system, is that the application can launch a malicious program after some time after the analysis. There are many methods for determining the completion of the analysis on a virtual device—all sorts of sensors, camera, microphone, Wi-Fi module, which are equipped with modern smartphones, allow you to bypass the dynamic analysis procedure. There are also known hypervisor detection techniques that can be applied by malware without accessing any sensors or mobile device data. In addition to the general disadvantages associated directly with dynamic analysis, the disadvantages of the approach based on the Google Bouncer testing system include the fact that only applications distributed through Google Play are subject to mandatory verification. To test applications from other sources, you need to enable the third-party application validation option and connect to the Internet.

To evaluate the effectiveness of mobile antiviruses, an experiment was conducted, the essence of which is as follows:

- for malware from an open source, the indicator of its detection on the Virustotal service was determined, which was 41 out of 56;
- the program was decompiled;
- various changes were made to the program;
- the program was recompiled;
- new detection indicators were evaluated on the Virustotal service.

The results obtained are shown in Table 1. The experiment showed that antiviruses cannot successfully detect programs whose code has been obfuscated.

The disadvantage of antiviruses lies in the limitations of static application analysis, the use of which requires the presence of signatures, that is, new malware for which signatures have not yet been released cannot be detected. In addition, obfuscation can be applied to the malware code to complicate the analysis [9].

Table 1 Indicators for detecting modified malware

Change made	Detection rate
The program is unchanged	41/56
Re-sign the app	29/56
Renaming an application package	23/56
Manifest changes	22/56
Resource changes	25/56
Encrypting strings in the program's bytecode	13/56
Renaming classes in bytecode	21/56
Renaming classes, methods, and encrypting strings in program bytecode	8/56

The main drawback of the permission mechanism is the large number of permissions that the user must read when installing the application, and the threats of a particular permission are not always obvious. Only the standard permissions in the Android OS are about 300. The feature of dynamically requested permissions, which appeared in Android OS version 6, does not solve these problems either. To support applications written earlier, the developers of the Android OS left the ability to get all the necessary permissions during installation. To do this, you need to specify in the manifest that the program was developed for Android version less than 6.

3 Results

The task of detecting Android malware can be considered as a classification task, which involves the use of neural network tools.

All applications of the Android OS within the framework of this task will be divided into two classes: safe and malicious. For classification, it is necessary to identify the features by which you can determine whether a program belongs to one of the classes. The actions that the program performs on the device can be used as classification features. You can use the manifest, application bytecode, and dynamic analysis to get the features of an Android OS application.

You can get the following attributes from the application manifest:

- list of app permissions;
- list of broadcast intentions being processed.

From the list of permissions, you can determine the functions that the program is allowed to perform on the device. However, as mentioned earlier, the list of permissions can be very large, and it is not necessarily that the application uses all the received permissions. From the entire list of possible permissions, it will be necessary to select the most important and more characteristic of malicious programs. From the list of broadcast intents being processed, you can determine which events on the device the program is tracking. These events can be: connecting to the Internet, receiving an SMS message, an incoming or outgoing call, turning on the device, and others. Recipients of broadcast intentions are used by many malicious applications, for example, to intercept SMS messages confirming a purchase from a bank card, or to wait for an Internet connection event to transmit stolen personal data.

By using the bytecode of the application, you can determine what actions the program performs on the device. Calls to the Android API and kernel libraries can be distinguished as features obtained from the bytecode. The lists of calls to the API or kernel libraries obtained as a result of bytecode analysis are more reliable than the list of permissions, but obtaining features from the bytecode can be difficult using dynamic code loading, reflection mechanisms, and obfuscation. Dynamic code loading allows you to hide almost all the functionality of the program from bytecode analysis, but the fact that the application code is downloaded from the Internet or an encrypted file is suspicious and can be used as a sign for classification.

Signs can also be obtained by dynamic analysis when running the program under study on a real or virtual device, all the actions of the program that it performs without the user's participation will be tracked.

Using the data obtained by static and dynamic analysis at the same time can eliminate their shortcomings, thereby increasing the likelihood of malware detection. Thus, in comparison with the considered malware detection mechanisms of the Android OS, the proposed approach has a number of advantages:

- the ability to detect previously unknown malware;
- no need for a signature database;
- resistance to obfuscation.

The ability to detect previously unknown malware is achieved by using neural networks, which allow you to identify complex dependencies between input and output data. When training on existing programs, the neural network identifies patterns that are characteristic of applications of each class, thereby being able to correctly determine which of the classes the program that is not in the training sample belongs to.

The proposed approach does not require signatures for analysis since all the features by which applications are classified can be obtained from the application as a result of the analysis. The combined use of static and dynamic analysis allows you to get rid of some of their shortcomings. For example, if a malicious program detects that it is being analyzed, it may suspend its work, preventing dynamic analysis. However, this ability of the program will not affect the operation of the static analyzer. Static analysis can be significantly hindered by obfuscation, but dynamic analysis works a level lower in the Android OS software stack and detects real program actions that cannot be hidden by obfuscation. Resistance to obfuscation is also provided by the ability of the neural network to obtain correct results on distorted or incomplete input data.

To evaluate the effectiveness of the proposed approach, a software implementation was created, which is a static and dynamic analyzer, as well as a classifier based on a neural network.

The static analyzer unpacks the application package, analyzes the bytecode, and the application manifest. Decompilation is applied to the application bytecode, and deserialization is applied to the manifest. The smali program was used for decompilation.

The result of the static analyzer is:

- A list of Android API functions and kernel libraries that can be called from the program code during operation.
- A list of app permissions that are requested during installation.
- A list of broadcast intents that can be processed by the program.

The dynamic analyzer is a modified version of the Android OS emulator droidbox, which allows you to track calls to the Android API [8]. The app can only interact with the device via the Android API, so all program actions will be detected. The program to be tested runs on a virtual device and runs for 5 min. While the program was running, the following actions were monitored:

- sending data to the network;
- read and write operations of files on the device;
- Using DexClassLoader;
- sending and reading SMS messages;
- read contacts saved on the device;
- making calls;
- using the built-in cryptographic provider.

A neural network with a multilayer perceptron architecture with two hidden layers is used as a classifier. The training was carried out by the method of back propagation of the error [10, 11]. As examples of applications belonging to the safe class, 100 of the 1000 most popular applications from the Google Play store were taken. Programs belonging to the malware class were obtained from the Android Malware Genome Project.

4 Discussion

To evaluate the work of the proposed solution, existing programs for the Android OS were analyzed. The analysis was carried out on 25 malicious and 50 secure programs. The results obtained can be seen in Table 2.

As a result of the analysis, 84% of malware was correctly classified. It should be noted that the analyzed programs were not used in training the network. Thus, the analysis confirms the ability of the proposed solution to correctly classify previously unknown programs.

To assess the ability of the proposed solution to detect malware that has been obfuscated, we analyzed the modified versions of the malware used in the experiment with antiviruses. All instances were correctly identified as malicious. The results of the analysis are presented in Table 3. The results show the ability of the proposed solution to detect obfuscated versions of malware.

Table 2 Program analysis results

Result characteristics	Quantity
Programs analyzed	75
Malicious software	25
Secure Programs	50
Correct classification result	64
False negative results	4
False positive results	3

Table 3 Results of the analysis of the modified malware

Change made	Result
The program is unchanged	+
Re-sign the app	+
Renaming an application package	+
Manifest changes	+
Resource changes	+
Encrypting strings in bytecode	+
Renaming classes in bytecode	+
Renaming classes, methods, and encrypting strings in bytecode	+

5 Conclusions

The existing mechanisms for detecting malicious programs of the Android operating system are not enough, this is shown by the analysis of information sources and the experiment conducted on programs whose code has been obfuscated or their signatures are missing.

The complex application of static and dynamic analysis of programs of the Android operating system allows us to obtain a sufficient set of features that allow us to accurately classify the programs under study with a high degree of probability.

The classification problem is solved using the neural network apparatus. The results showed that all malicious programs that were boycotted by obfuscation were correctly identified as malicious. For new malware, 84% of them were correctly classified by the neural network.

References

1. Worldwide Smartphone Forecast Update, 2016–2020: December 2016. International Data Corporation (IDC). http://www.idc.com/getdoc.jsp?containerId=US42060116
2. Bezopasnost' Android: vzglyad vnutr'/securitylab.ru. http://www.securitylab.ru/blog/personal/Informacionnaya_bezopasnost_v_detalyah/325122.php
3. Soglashenie Google Play o rasprostranenii programmnyh produktov./ Google. https://play.google.com/intl/ALL_ru/about/developer-distribution-agreement.html
4. Rutkovskaya, D., Pilin'skij, M., Rutkovskij, L.: Nejronnye seti, geneticheskie algoritmy i nechetkie sistemy. M.: Goryachaya liniya–Telekom, 384 p. (2013)
5. Using DroidBox for dynamic malware analysis. https://hackmag.com/uncategorized/droidbox-for-dynamic-malware-analysis/
6. Ilin, I., Levina, A., Lepekhin, A., Kalyazina, S.: Business requirements to the IT Architecture: a case of a healthcare organization. Adv. Intell. Syst. Comput. **983**, 287–294 (2019)
7. Android Platform Architecture./developer.android.com: информационный ресурс. https://developer.android.com/guide/platform/index.html
8. Android App permissions explained. http://www.androidauthority.com/android-app-permissions-explained-642452/

9. Tatarnikova, T.M.: Zashchishchennye korporativnye seti. razdel: Zadachi po zashchite informacii. SPb: RGGMU, 113 p. (2012)
10. Tatarnikova, T.M.: Zadacha sinteza kompleksnoj sistemy zashchity informacii v GIS/Uchenye zapiski Rossijskogo gosudarstvennogo gidrometeorologicheskogo universiteta, no. 30, pp. 204–211 (2013)
11. Tarhov, D.A.: Nejrosetevye modeli i algoritmy. Spravochnik. M.: Radiotekhnika, 349 p. (2014)
12. Worldwide Smartphone Forecast Update, 2016–2020: December 2016. [Electronic resource]/International Data Corporation (IDC). http://www.idc.com/getdoc.jsp?containerId=US42060116

Algorithm for Predicting the Dynamics of Physical and Human Capital

Askar Akaev, Tessaleno Devezas, Askar Sarygulov, and Aleksander Petryakov

Abstract The task of predicting the future worries not only futurologists, but also many researchers engaged in solving practical problems. The tools for such foresight are quite diverse, as are the tasks themselves for the solution of which they are used. One of the branches of knowledge where predictive tools have become widespread is economics. In this article, we propose an algorithm for predicting the dynamics of physical and human capital based on the ideas of N. Kaldor and T. Piketty about the nature of capital accumulation. In particular, our algorithm assumes the approximate constancy of the ratio of capital to output. The algorithm implies a sequential assessment of the investment rate, capital return and capital stock depreciation rate for further predicting capital dynamics using a logistic trajectory. As an example of using the algorithm, verification was carried out using statistical data of the US economy. The algorithm presented in the article can be used both to determine the average annual growth rate of potential GDP, and in related studies of the future dynamics of the economy capital resources.

Keywords Physical capital · Human capital · Mathematical modeling · Capital stock depreciation rate

A. Akaev
Institute for Mathematical Research of Complex Systems, Lomonosov Moscow State University, Lenin Hills 1, Moscow, Russia

T. Devezas
Atlântica School of Management Sciences, Health, IT and Engineering, Barcarena, Lisbon, Portugal

A. Sarygulov · A. Petryakov (✉)
St. Petersburg State University of Economics, 21, Sadovaya street, St. Petersburg, Russia

C. Jahn et al. (eds.), *Algorithms and Solutions Based on Computer Technology*, Lecture Notes in Networks and Systems 387,
https://doi.org/10.1007/978-3-030-93872-7_4

1 Introduction

In modern research, the use of predictive algorithms is very wide: from machine learning algorithms for predicting the incidence of pests and diseases of Coffea Arabica to genetic algorithms for predicting solar cycles and weather forecasting based on machine learning algorithms [1–3]. The same applies to economic research, which, as the tasks become more complex, more and more widely use complex calculation schemes implemented based on computer technology [4]. Indeed, the problems solved on the basis of predictive algorithms is becoming wider: price movements forecasting based on machine learning algorithms for such a complex technological innovation in the modern financial system as bitcoin [4–6]; predictive algorithms and optimization strategies for building energy management and demand response management [7]; supply chain forecasting based on big data and machine learning algorithms [8, 9]; advanced analytics platforms, in-memory computing and artificial intelligence (AI) tools, including machine learning to predict the future of business as an economic activity [10]. In a broader context, predictive algorithms are effectively used to study nonstationary time series [11, 12]. Machine learning (ML) algorithms have been proposed in the academic literature as an alternative to statistical methods for forecasting time series [13]. With the increasingly widespread use of artificial intelligence elements in the manufacturing and service sectors, algorithmic forecasting is becoming one of the tools for studying dynamic systems: as more and more data enter the algorithmic model, the model automatically "learns" more about the scenario, and its predictions become more and more accurate over time [14].

In this article, algorithms, as a research tool, are used to analyze the Kaldor pattern, which suggests that the ratio of physical capital to output is approximately constant in the end, when the consequences of various economic and financial shocks and crises are smoothed out [15]. New trends in capital accumulation, economic growth and income inequality emerging at the beginning of the twenty-first century were comprehensively studied by the French economist T. Piketty. He obtained interesting results detailed in his work [13, 16]. First, Piketty convincingly showed that in the most developed countries (USA, Great Britain, Germany, France, etc.) capital intensity made a huge UU-curve and at the beginning of the twenty-first century returned to its maximum values, close to those observed at the end of the 19th. At the same time, changes in capital intensity in the United States in the twentieth century were of a very limited scale than in Western Europe, and, therefore, created the impression that the first empirical pattern was still taking place, which N. Kaldor recorded. The return of capital intensity in developed countries in the twenty-first century again to its maximum value means that it will now stabilize again, at least until the middle of the century. Therefore, it follows that in the first half of the twenty-first century the first empirical rule of Kaldor remains valid, and therefore the proposed algorithm is based on it.

2 Materials and Methods

Capital accumulation and innovative technologies of the 4th industrial revolution will be the driving force influencing economic development in the first half of the twenty-first century [17]. The accumulation of capital was the most important feature of capitalism in the XIX and XX centuries. It will accelerate in the twenty-first century, as Piketty argues.

To describe the algorithm for predicting the dynamics of physical and human capital, let us assume that the above empirical regularity of Kaldor is formalized as follows:

$$(a)\ K = \sigma_K \cdot Y, \sigma_K = const;\ \ (b)\ Y = æ_K \cdot K, æ_K = const \quad (1)$$

where σ_K—physical capital intensity ratio; $æ_K$—capital-return ratio of physical capital.

For human capital, similar equations can be obtained:

$$(a)\ H = \sigma_H \cdot Y, \sigma_H = const;\ \ (b)\ Y = æ_H \cdot H, æ_H = const \quad (2)$$

where σ_H—human capital intensity ratio; $æ_H$—human capital productivity ratio.

The movement of physical $K(t)$ and human $H(t)$ capital is described by the classical equation of capital accumulation [18, 19]:

$$(a)\ \dot{K}(t) = I_K(t) - \mu_K \cdot K(t);\ \ (b)\ \dot{H}(t) = I_H(t) - \mu_H \cdot H(t) \quad (3)$$

where $I_K(t)$ and $I_H(t)$—gross investment in physical and human capital; μ_K and μ_H—rates of deprecitation of physical and human capital.

Since $I_K(t) = s_K \cdot Y(t)$, $I_H(t) = s_H \cdot Y(t)$, where s_K and s_H—the rate of investment in physical and human capital, respectively, Eqs. (3) are transformed in the form:

$$(a)\ \dot{K}(t) = s_K \cdot Y(t) - \mu_K \cdot K(t);\ \ (b)\ \dot{H}(t) = s_H \cdot Y(t) - \mu_H \cdot H(t) \quad (4)$$

Taking into account (1), Eqs. (4) will be simplified further:

$$(a)\ \dot{K}(t) = (s_K \cdot æ_K - \mu_K) \cdot K(t);\ \ (b)\ \dot{H}(t) = (s_H \cdot æ_H - \mu_H) \cdot H(t) \quad (5)$$

The solution of the simplest differentiated Eqs. (5) has the form:

$$(a)\ K(t) = K_0 \cdot \exp[(s_K \cdot æ_K - \mu_K) \cdot (t - T_0)]$$

$$(b)\ H(t) = H_0 \cdot exp[(s_H \cdot æ_H - \mu_H) \cdot (t - T_0)] \quad (6)$$

where K_0 and H_0—the values of the volumes of physical and human capital at the initial moment of time T_0.

Therefore, under the conditions of the first empirical pattern of Kaldor, in the first half of the twenty-first century there will be an exponential growth of accumulated capital. However, given that, in accordance with economic theory, at the downward stage of the business cycle, the effect of capital saturation should occur, then capital accumulation will occur along a logistic trajectory:

$$K(t) = \frac{K_1}{1 + u_K \cdot exp[-\vartheta_K \cdot (t - T_0)]} \tag{7}$$

where K_1, u_K and ϑ_K—constant parameters.

Similarly, to describe the process of accumulating human capital, we obtain the following equation:

$$H(t) = \frac{H_1}{1 + u_H \cdot exp[-\vartheta_H \cdot (t - T_0)]} \tag{8}$$

where H_1, u_H and ϑ_H—constant parameters.

The set of Eqs. (6–8) is used by us to predict the dynamics of physical and human capital in the first half of the XXI century.

Thus, summarizing the above, we highlight the main stages of the algorithm for predicting physical and human capital:

1. Determination of the investment rate based on the ratio of gross savings to GDP;
2. Calculation of the capital return ratio through the ratio of GDP to capital;
3. Searching the rate of capital stock depreciation by numerical methods based on retrospective data;
4. Construction of predicted capital values, taking into account the equality of the exponential and logistic predictive trajectory at the upward stage of the business cycle.

One of the key points of the algorithm is an accurate assessment of the parameters of models describing the accumulation of physical and human capital for making forecasts.

3 Results

To verify the algorithm, the numerical values of physical capital (K) for the US economy were taken from the World Bank database. These data were verified and refined with data from the University of Groningen [20] and the US Bureau of Economic Analysis [21]. Christian [22] borrowed data on human capital (H) from the article.

Based on the statistics of the World Bank [23] for $K(t)$ of the US economy in the framework of 1982–2018 using the least squares method, estimates of the coefficients for models of physical capital accumulation in the twenty-first century were obtained. The parameters of the logistic function (7) were found from the condition that the obtained trajectory of growth of capital accumulation coincides with the trajectory (6a) at the initial section of the upward stage (2018–2034). The results are summarized in Table 1.

To demonstrate the quality of parameter estimation, the actual and modeled values of physical capital are shown in Fig. 1.

The result of the last step of the algorithm, a long-term forecast of the dynamics of physical capital, is shown in Fig. 2.

Christian's article presents a study on assessing the human capital of the US economy, the result of which is a retrospective series from 1975 to 2013 for human capital, in connection with which the calculations using H are limited to the specified period [22].

By analogy with the calculation of values for physical capital based on data from the source [22] to describe the dynamics $H(t)$ for 1975–2013 estimates of the coefficients for models of human capital accumulation in the twenty-first century were obtained (see Table 2).

Table 1 Calculation results for models of physical capital accumulation

Parameters estimation	Accuracy assessment
σ_K= 3.31; $æ_K$= 0.302; μ_K= 0.035; s_K= 18.6%	$R^2 \approx 1$;
K_1= 163.6\$$trn.$; u_k= 1.782; ϑ_k= 0.038	Observed F-test values are greater than critical (significance level 95%)

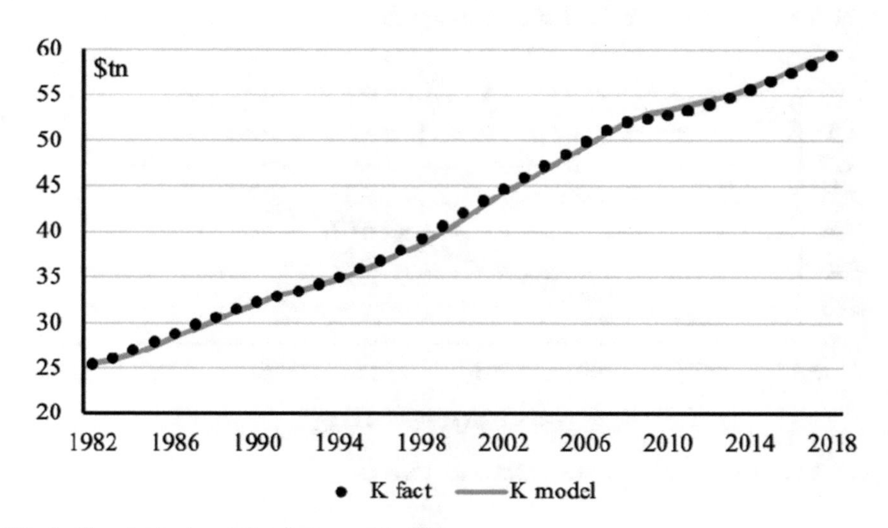

Fig. 1 The physical capital trajectory of the US economy

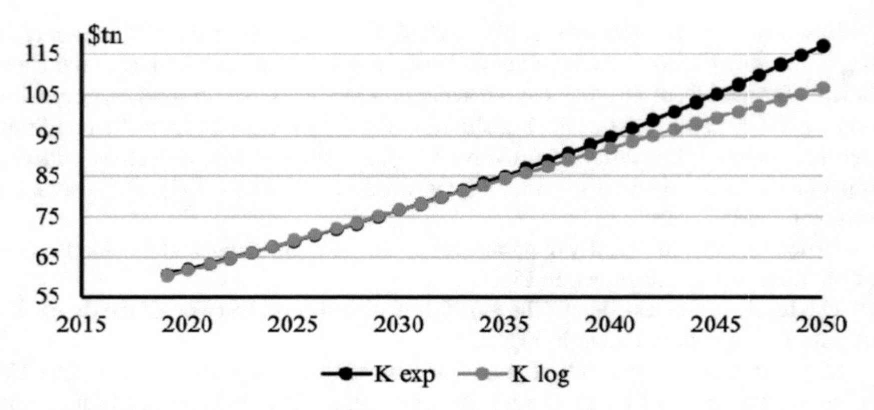

Fig. 2 Forecast dynamics of the physical capital of the US economy

Table 2 Calculation results for models of human capital accumulation

Parameters estimation	Accuracy assessment
$\sigma_H = 13.12$; $æ_H = 0.076$; $s_H = 29.5\%$; $\mu_H = 1.04\%$	$R^2 \approx 1$;
$H_1 = 567.5\$trn.$; $u_H = 1.399$; $\vartheta_H = 0.02$	Observed F-test values are greater than critical (significance level 95%)

The projected trajectory of physical capital is depicted in Fig. 3.

As a result, statistically significant estimates were obtained using the algorithm (Tables 1 and 2), on the basis of which the close to real values of the factors K and H and the predicted dynamics of physical and human capital for the United States in the first half of the twenty-first century are simulated.

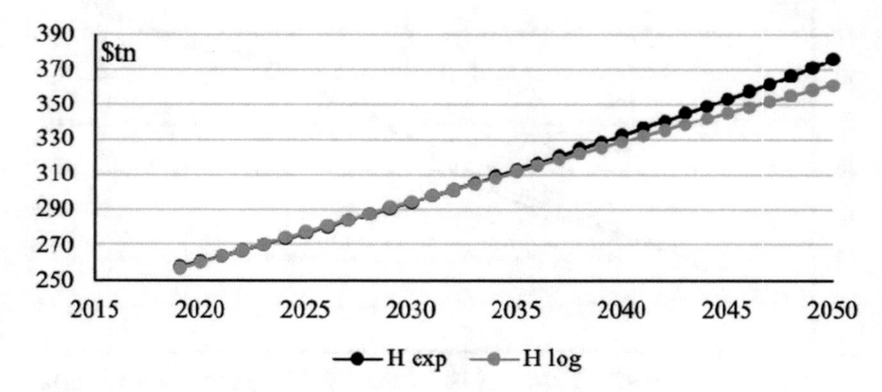

Fig. 3 Forecast dynamics of human capital of the US economy

4 Conclusions

In the context of the digital technologies rapid use, especially those aimed at replacing cognitive functions, intellectual capital embodied in the form of human capital grows at a higher rate than physical capital [24]. In this regard, our proposed forecasting algorithm, using capital models, corresponds to the ideas of Kaldor and Piketty and can be used both to determine the average annual growth rates of potential GDP and in related studies of the future dynamics of the capital resources of the economy.

Acknowledgements This article was prepared as part of the RFBR grant No. 20-010-00279 "An integrated system for assessing and forecasting the labor market at the stage of transition to a digital economy in developed and developing countries."

References

1. de Oliveira Aparecido, L., de Souza Rolim, G., da Silva Cabral De Moraes, J.R., et al.: Machine learning algorithms for forecasting the incidence of Coffea Arabica pests and diseases. Int. J. Biometeorol. **64**, 671–688 (2020). https://doi.org/10.1007/s00484-019-01856-1
2. Orfila, A., Ballester, J.L., Oliver, R., Alvarez, A., Tintoré, J.: Forecasting the solar cycle with genetic algorithms. A&A **386**(1), 313–318 (2002). https://doi.org/10.1051/0004-6361:20020246
3. Nagadevi, S., Ramesh, V., James, H.: Weather forecasting using machine learning algorithm. Int. J. Psychosoc. Rehabil. **24**(6), 4244–4250 (2020). https://doi.org/10.37200/IJPR/V24I6/PR260410
4. Moreira, M.W.L., Rodrigues, J.J.P.C., Kumar, N., Saleem, K., Illin, I.V.: Postpartum depression prediction through pregnancy data analysis for emotion-aware smart systems. Inf. Fus. **47**, 23–31 (2019)
5. Pabuçcu, H., Ongan, S., Ongan, A.: Forecasting the movements of Bitcoin prices: an application of machine learning algorithms. Quant. Financ. Econ. **4**(4), 679–692 (2020). https://doi.org/10.3934/QFE.2020031
6. Abramov, V., Popov, N., Istomin, E., Sokolov, A., Popova, A., Levina, A.: Blockchain and big data technologies within geo-information support for arctic projects (2019). In: Proceedings of the 33rd International Business Information Management Association Conference, IBIMA 2019: Education Excellence and Innovation Management through Vision, pp. 8575–8579 (2020)
7. Meng, F., Weng, K., Shallal, B., Chen, X., Mourshed, M.: Forecasting algorithms and optimization strategies for building energy management & demand response. In: Proceedings MDPI (2018). https://doi.org/10.3390/proceedings2151133
8. Vandeput, N.: Data Science for Supply Chain Forecast, 237 p. Independently Published (2018)
9. Ilin, I.V., Iliashenko, O.Y., Klimin, A.I., Makov, K.M.: Big data processing in Russian transport industry (2018). In: Proceedings of the 31st International Business Information Management Association Conference, IBIMA 2018: Innovation Management and Education Excellence through Vision, pp. 1967–1971 (2020)
10. Merrill, E., Ehrenhalt, S., Tay, A.: Forecasting in a Digital World. Deloitte Development LLC (2018)
11. Kuznetsov, V., Mohri, M.: Learning Theory and Algorithms for Forecasting Non-Stationary Time Series (2018). https://papers.nips.cc/paper/2015/file/41f1f19176d383480afa65d325c06ed0-Paper.pdf. Accessed 2021/02/01

12. Montero-Manso, P., Hyndman, R.J.: Principles and Algorithms for Forecasting Groups of Time Series: Locality and Globality. Working Paper 45/20, Monash University (2020). https://robjhyndman.com/publications/global-forecasting/. Accessed 2021/01/11
13. Makridakis, S., Spiliotis, E., Assimakopoulos, V.: Statistical and machine learning forecasting methods: concerns and ways forward. PLoS One **13**(3), e0194889 (2018). https://doi.org/10.1371/journal.pone.0194889
14. Parmar, D.: What Is Predictive Algorithmic Forecasting and Why Should You Care? (2019). https://www.nutanix.com/theforecastbynutanix/business/what-is-predictive-algorithmic-forecasting-and-why-should-you-care. Accessed 2020/12/20
15. Kaldor, N.: Capital Accumulation and Economic Growth/The Theory of Economic Growth, pp. 177–222. St. Martin is Press, New York (1961)
16. Piketty, T.: Capital in the Twenty-First Century. Harvard University Press, Cambridge and London (2014). https://doi.org/10.4159/9780674369542
17. Ilin, I.V., Levina, A.I., Dubgorn, A.S., Abran, A.: Investment models for enterprise architecture (Ea) and it architecture projects within the open innovation concept (2021). J. Open Innov. Technol. Market Complex. **7**(1), 1–18. статья № 69
18. Kurzenev, V., Matveenko, V.: Economic Growth [Jekonomicheskij rost]. SPb, Piter (2018)
19. Barro, R.J., Sala-i-Martin, X.I.: Economic Growth. The MIT Press (2003)
20. University of Groningen and University of California, Davis: Total Factor Productivity at Constant National Prices for United States [Data file]. Retrieved from FRED, Federal Reserve Bank of St. Louis. https://fred.stlouisfed.org/series/RTFPNAUSA632NRUG. Accessed 2021/01/22
21. U.S. Bureau of Economic Analysis, National Income and Product Accounts: Table 1.1.6. Real Gross Domestic Product, Chained Dollars [Data file]. https://apps.bea.gov/iTable/iTable.cfm?reqid=19&step=2#reqid=19&step=2&isuri=1&1921=survey. Accessed 2021/01/23
22. Christian, M.S.: Net Investment and Stocks of Human Capital in the United States, 1975–2013. International Productivity Monitor, Centre for the Study of Living Standards, vol. 33, pp. 128–149 (2017)
23. World Bank Data homepage. http://data.worldbank.org/. Accessed 2021/01/17
24. Egorov, D., Levina, A., Kalyazina, S., Schuur, P., Gerrits, B.: The challenges of the logistics industry in the era of digital transformation (2021). Lecture Notes in Networks and Systems, vol. 157, pp. 201–209 (2020)

Detection of Anomalies in IoT Systems by Neuroevolution Algorithms

Alexander Fatin, Evgeny Pavlenko, and Peter Zegzhda

Abstract The aim of the study is to detect anomalies in the operation of Internet of Things (IoT) systems, cyber-physical systems and Supervisory Control And Data Acquisition (SCADA). The approach is based on predictions and analysis of a multi-dimensional time series composed of data obtained from detectors and actuators of the system under consideration. The data in question is retrieved from a physical (low) level and a logical (high) level. Predictions of the future states of the system are performed by means of neuroevolutionary algorithms (symbiosis of neural networks and genetic algorithms) of the NEAT family, and the analysis is carried out by comparing the normalized data obtained with the predicted indicators of the system and then calculating the threshold value of the error. Testing is performed on a validation dataset. The results of the study are demonstration of the effectiveness of the method used with indicators of its correctness and accuracy.

Keywords Information security · Cyber-physical systems · IoT · Hypercube · Neuroevolution · Multidimensional time series

1 Introduction

Any cyber-physical system (CPS) operates on the basis of two types of flows—physical (low-level) and logical (high-level) [1–5]. The analysis of low-level components of the system (LLC), which consists in processing data received from measuring instruments, sensors and sensors, allows to assess the correctness of the execution of system processes [6]. It also allows you to study in real time the occurrence and manifestation of anomalous behavior at the earliest stages due to the absence of high-level abstractions and ease of access to the original data [7]. Similarly, high-level components of the system (HLC) analysis is performed in conjunction with the LLC

A. Fatin (✉) · E. Pavlenko · P. Zegzhda
Institute of Cybersecurity, Peter the Great St. Petersburg Polytechnic University, St. Petersburg 195251, Russia

C. Jahn et al. (eds.), *Algorithms and Solutions Based on Computer Technology*, Lecture Notes in Networks and Systems 387,
https://doi.org/10.1007/978-3-030-93872-7_5

analysis. It is due to the need to take into account the logic of operations, including the detection of anomalous behavior in logical space, when the physical parameters remain in the correct state [8].

Currently, many approaches and methods are used to detect cyber-physical attacks, logical errors and other anomalies in the state of systems [9]. The most effective are the methods of machine learning [10–15] and tools based on the identification of statistical indicators and assessment of the tendency of their change in the system [16–22].

Most of the existing work related to this topic is based on the use of statistical tools and the one dimensional time series, however, the most effective are the intelligent methods used for the prediction multidimensional time series. Most often, machine learning methods are used in conjunction with the multidimensional time series mathematical apparatus in order to achieve a higher response rate of the system, simplify work with data through aggregation and native work with continuously generated data in the dynamics of CPS behavior at the physical level, as well as reduce the magnitude of errors of the first and of the second kind. Additionally, the use of a multidimensional time series contributes to the creation of data propagation feedback dependencies by taking into account the sequence and cyclicality of system states [6, 9].

In this paper, we consider the application of the neuroevolutionary NEAT algorithm using a hypercube to analyze the multidimensional time series describing the state of the CPS in order to identify abnormal conditions.

The study of the described model is carried out on the verified dataset TON_IOT DATASETS [23]. The dataset used is primarily aggregated by the types of devices and the data they receive. The system topology is the structure of the Internet of Things (IoT).

A dataset consists of data sets transmitted by 7 network devices:

1. Workload
2. Current status value
3. Physical data measured by the device
4. Final recipient of the data.

The data was collected over 4 periods of operation—each of which amounted to 48 h of system operation. One of the periods is the reference one and includes the normal functioning of the system, and the remaining 3 contain various attacks and anomalies in the system, carried out discretely: DDoS, DoS, Backdoor, etc.

The sample includes data obtained from both the physical and logical levels of the system.

2 Materials and Methods

To predict the state of the CPS with the subsequent detection of anomalies based on the magnitude of the resulting error, a multidimensional time series is usually used due to the presence of a mechanism for storing feedbacks in the processed data. This approach makes it possible to ensure the presence of additional dependencies in the aggregated series and to improve the accuracy of predicting the state of the system using machine learning methods.

The classic multidimensional time series is a collection $X = \{X^{(1)}, X^{(2)}, \ldots, X^{(m)}\}$, where each value at time t_i is represented by a vector: $\{X^{(i)} = \{x_1^{(i)}, x_2^{(i)}, \ldots, x_n^{(i)}\}$.

To reduce the dimension of the problem, simplify and correctly process the data by the activation function of the neural network, the values of the multidimensional time series are normalized according to the following formula (1):

$$x_i = \frac{x_i - x_{min}}{x_{max} - x_{min}} \tag{1}$$

The dimension of the multidimensional time series is determined by the number of devices under consideration (7) and their bases (4), which amounted to 28. To preserve the promptness of response to incidents in the system, the time interval for obtaining the data sample was taken as Δ t = 1 s. Figures 1 and 2 show examples of the normal operation of the system during 48 h of its operation and the state of the system during the 48 h during which the systems were attacked.

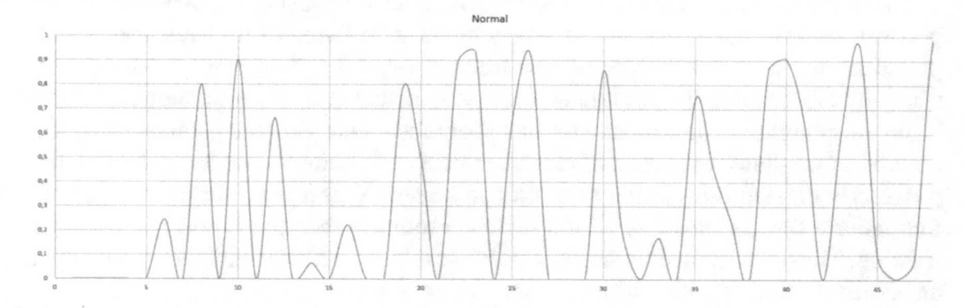

Fig. 1 An example of data changes during 48 h of system operation in normal state

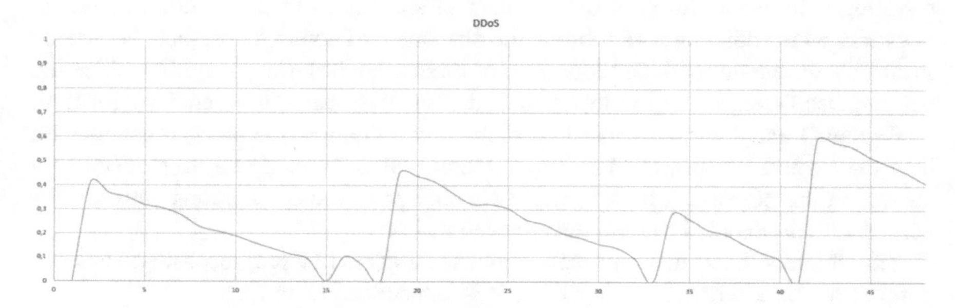

Fig. 2 An example of data changes within 48 h of working with DDoS attacks

3 Results

In general, a neuroevolutionary algorithm includes a neural network configuration optimized by means of a genetic algorithm.

In this study, the NEAT-Python library of the Python language was used to implement the network topology by means of the hypercube algorithm.

To configure the phenotype, a primary sandwich topology was implemented and used: two flat grids, one of which was a set of input neurons, and the other was a set of output neurons. Each of the layers can, by means of mutation, build connections from one to another.

Each point of the pattern, available for genetic operations and located in the hypercube, represents a connection between two nodes of the substrate graph. Based on this, the dimension of the hyperspace is twice the dimension of the underlying graph of the smallest dimension.

The primary topology of the "sandwich" type at the entrance takes a multidimensional time series from time t_i with dimension 28, obtained earlier. After the development of the topology of the system due to mutations on the output layer of the "sandwich", we obtain a multidimensional time series, consisting of the predicted data of the state of the system in time t_{i+1}.

The addition of any arbitrary gene for a node or link during the evolution of the network leads to the emergence of a new global dimension of the variation of link patterns, i.e., to the emergence of new traits through the phenotype substrate. The new way of changing the communication pattern comes down, ultimately, to modifying the genome of the structure of the hypercube—changing the parameters by the method of simulated evolution of rearrangement of links and/or nodes. Additionally, previously created links in the network can be reused as a basis for creating a new link pattern for the substrate with a higher resolution than the initial one used for training. Thus, this approach makes it possible to obtain a solution to the problem at any mesh resolution without limiting the vertex of the hypercube.

This approach makes it possible to simplify and limit the space for the development of large-scale neural networks, making it possible not to lose the efficiency of the resulting solution by leveling the problem of stagnation in the solution of neural networks by introducing variability in the placement of nodes. Thus, there is no need to strictly set the network configuration or modernize it during operation, since the network itself creates a new topology and adapts to it by means of a genetic algorithm.

Figures 3 and 4 show the used crossover and mutation operators, together with how they work. Since a change in the neural network graph can be reduced to a change in the placement of vertices and a change in the connections between the vertices, it was decided to limit ourselves to this set.

The obtained data, after normalization, is subjected to preprocessing: for each point of the time series, a predicted value is determined (Fig. 5).

For prediction of the subsequent value of the state of the model through the time series, you need to perform the operation according to the following formula (2):

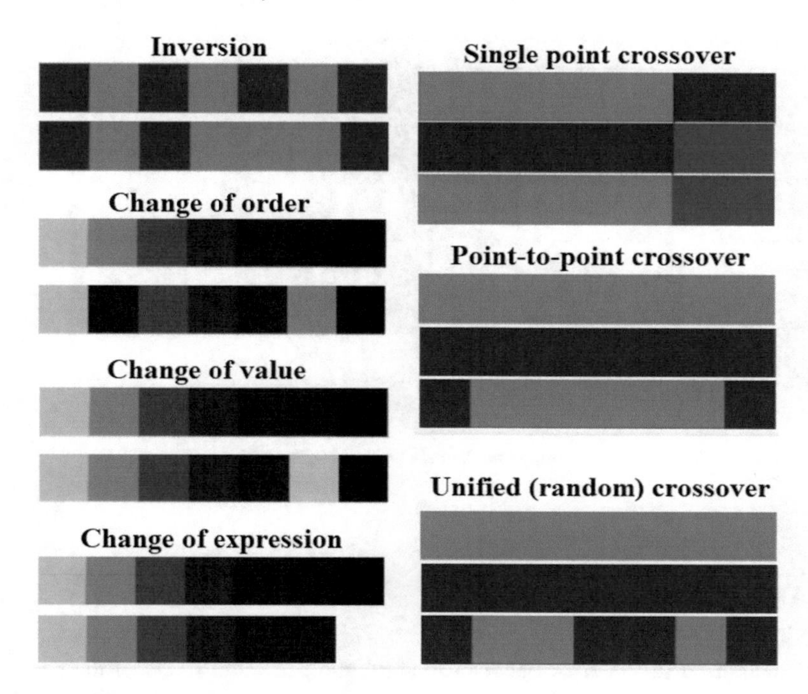

Fig. 3 Used crossover operators

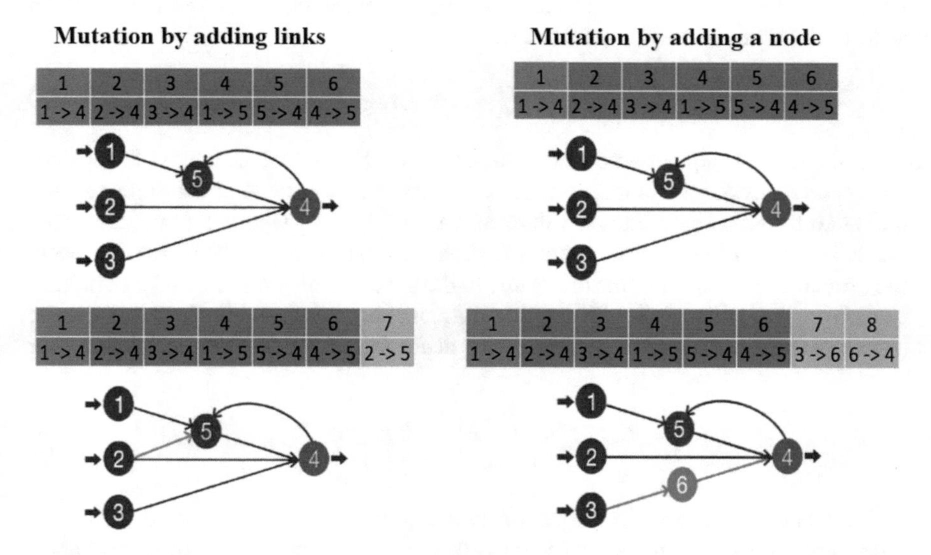

Fig. 4 Used mutation operators

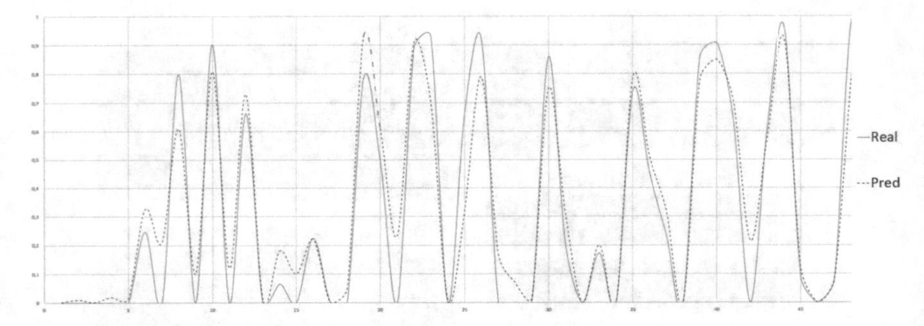

Fig. 5 An example of predicting the state of the system

$$\gamma_{pred} = pred(x_{t-1}, \ldots, x_{t-n}) \tag{2}$$

4 Discussion

The calculation of the error between the real and predicted states of the system is performed in 2 stages. At the first stage, it is necessary to calculate the difference between the real γ_{real} and the predicted γ_{pred} values of the state of the system according to the following formula (3):

$$err_t = \left|\gamma_{pred} - \gamma_{real}\right| \tag{3}$$

At the second stage, it is necessary to determine the anomalous behavior by calculating and analyzing the condition for exceeding the error value of the predicted and real state by more than a fixed value: $MAX(err_t) > T$, where T is the threshold value for the manifestation of anomalous behavior in the system. With this approach, the probability of false alarms arises due to the presence of "outliers" of large prediction errors in short periods of time; therefore, it is necessary to take into account the average error over a certain period of time according to the folowwing formula (4):

$$ERR_i = \sum_{t=i-k}^{i} MAX(err_t) > T \tag{4}$$

Figures 6 and 7 show examples of the magnitude of the normalized prediction error between the calculated on the basis of the neural network and the real state of the system in the presence and absence of attacks.

The threshold value of the prediction error is determined empirically and in this case is 0.4. With this value, the share of detected anomalies on the selected dataset was 91%. The Type I error on the studied set was 0.08, and the type II error was

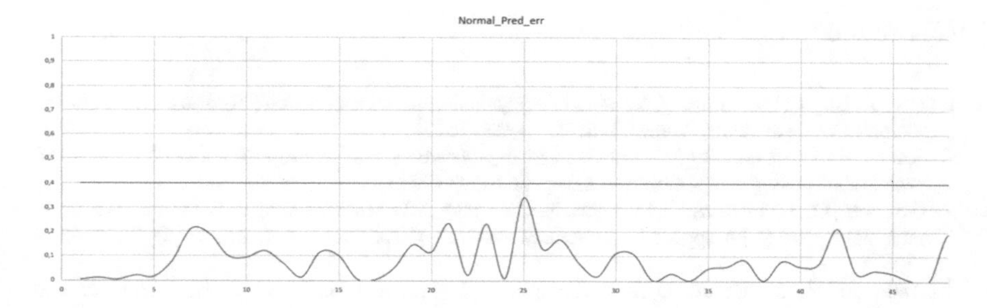

Fig. 6 An example of an error in predicting the state of the system in the absence of attacks

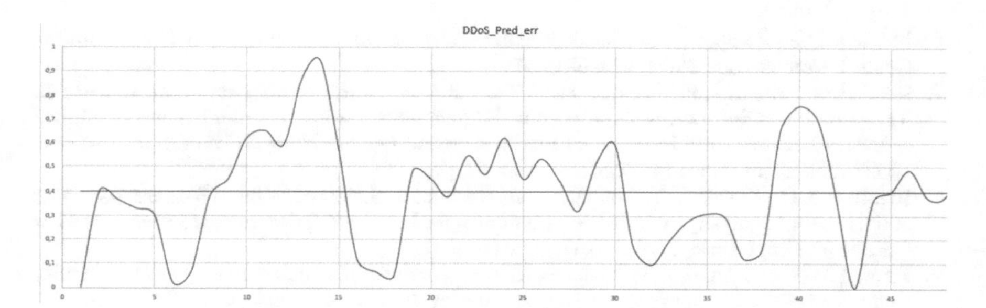

Fig. 7 An example of an error in predicting the state of the system in the case of a DDoS attack

0.12. These values are due to the choice of the boundary: when the threshold value decreases, the likelihood of type I error decreases, but the likelihood of type 2 error sharply increases. It was decided to focus on these values.

5 Conclusions

The use of the NEAT-hypercube mechanism based on the neuroevolutionary process has shown the possibility of successful and rapid detection of abnormal behavior in CPS. The described approach in terms of efficiency is not inferior to the classical approaches based on machine learning mechanisms and statistical tools. The main reasons for improving the results were: the possibility of restructuring the sandwich structure of the neuroevolutionary network in the process of stagnation in the search for a solution and the possibility of optimizing the topology of the neural network by means of a genetic algorithm.

Acknowledgements The study was carried out within the framework of the scholarship of the President of the Russian Federation for young scientists and graduate students SP-1689.2019.5.

References

1. Lee, J., Bagheri, B., Kao, H.A.: A cyber-physical systems architecture for industry 4.0-based manufacturing systems. Manuf. Lett. **3**, 18–23 (2015)
2. Huang, S., Zhou, C., Yang, S., Qin, Y.: Cyber-physical system security for networked industrial processes. Int. J. Autom. Comput. **12**(6), 567–578 (2015)
3. Givehchi, O., Landsdorf, K., Simoens, P., Colombo, A.W.: Interoperability for industrial cyber-physical systems: an approach for legacy systems. IEEE Trans. Ind. Inf. **13**(6), 3370–3378 (2017)
4. Ashibani, Y., Mahmoud, Q.H.: Cyber physical systems security: analysis, challenges and solutions. Comput. Secur. **68**, 81–97 (2017)
5. Ivanyo, Y.M., Krakovsky, Y.M., Luzgin, A.N.: Interval forecasting of cyber-attacks on industrial control systems. In: IOP Conference Series: Materials Science and Engineering, vol. 327, no. 2, pp. 1–7 (2018)
6. Lavrova, D.S.: An approach to developing the SIEM system for the Internet of Things. Autom. Control. Comput. Sci. **50**(8), 673–681 (2016)
7. Ilin, I., Borremans, A., Bakhaev, S.: The IoT and Big Data in the logistics development crude oil transportation in the arctic zone case study. In: Lecture Notes in Computer Science (including subseries Lecture Notes in Artificial Intelligence and Lecture Notes in Bioinformatics), vol. 12525, LNCS, pp. 148–154 (2020)
8. Ilyashenko, O., Kovaleva, Y., Burnatcev, D., Svetunkov, S.: Automation of business processes of the logistics company in the implementation of the IoT. In: IOP Conference Series: Materials Science and Engineering, vol. 940(1), no. 012006 (2020)
9. Fatin, A.D., Pavlenko, E.Y., Poltavtseva, M.A.: A survey of mathematical methods for security analysis of cyberphysical systems. Autom. Control Comput. Sci. **54**(8), 981–985 (2020)
10. Luo, T., Nagarajan, S.G.: Distributed anomaly detection using autoencoder neural networks in WSN for IoT. In: IEEE International Conference on Communications (ICC) (May 2018). https://arxiv.org/abs/1812.04872. Accessed 02/13/2021
11. Filonov, P., Lavrentyev, A., Vorontsov, A.: Multivariate industrial time series with cyber-attack simulation: fault detection using an LSTM-based predictive data model. NIPS Time Series Workshop (2016)
12. Nanduri, A., Sherry, L.: Anomaly detection in aircraft data using Recurrent Neural Networks (RNN). In: Integrated Communications Navigation and Surveillance (ICNS), pp. 5C2-1–5C2-8. IEEE (2016)
13. Yi, S., Ju, J., Yoon, M.-K., Choi, J.: Grouped Convolutional Neural Networks for Multivariate Time Series. https://arxiv.org/pdf/1703.09938.pdf. Accessed 01/12 /2021
14. Chen, S., He, H.: Stock prediction using convolutional neural network. In: IOP Conference Series: Materials Science and Engineering, vol. 435, no. 1, pp. 012026. IOP Publishing (2018)
15. Hundman, K., Constantinou, V., Laporte, C., Colwell, I., Soderstrom, T.: Detecting spacecraft anomalies using LSTMs and nonparametric dynamic thresholding. In: KDD '18: Proceedings of the 24th ACM SIGKDD International Conference on Knowledge Discovery & Data Mining, pp. 387–395 (2018)
16. Tulone, D., Madden, S.: PAQ: time series forecasting for approximate query answering in sensor networks. Wireless Sensor Networks: Third European Workshop, EWSN 2006, Zurich, Switzerland, pp. 21–37 (2006)
17. Wei, L., Kumar, N., Lolla, V., Keogh, E., Lonardi, S., Ratanamahatana, C.A.: Assumption-free anomaly detection in time series. In: SSDBM: Proceedings of the 17th International Conference on Scientific and Statistical Database Management, vol. 5, pp. 237–242 (2005)
18. Pincombe, B.: Anomaly detection in time series of graphs using ARMA processes. Asor Bull. **24**(4), 2–10 (2005)
19. Adams, R.P., MacKay, D.J.C.: Bayesian Online Changepoint Detection (2007). arXiv:0710.3742. Accessed 02/18/2021
20. Vyshemirsky, S.D.V., Macaulay, V.: Bayesian Changepoint Detection in Solar Activity Data, Glasgow, 52 p. (2014)

21. Kim, S.S., Reddy, A.L.N., Vannucci, M.: Detecting traffic anomalies using discrete wavelet transform. In: Proceedings of the International Conference on Information Networking, pp. 951–961 (2004)
22. Salagean, M., Firoiu, I.: Anomaly detection of network traffic based on Analytical Discrete Wavelet Transform. In: 8th International Conference on Communications. IEEE (2010)
23. TON_IOT DATASETS. https://ieee-dataport.org/documents/toniot-datasets. Accessed 01/12/2021

Implementing XGBoost Machine Learning Ensemble Algorithm to Predict Contact Pressure of Two 3D Bodies

Stepan Orlov, Kairzhan Aubekerov, and Stanislav Koptsev

Abstract The paper presents an ML-based approach to solution of contact mechanics problem. The problem itself is an example of Hertz' contact problem of two elastic bodies. The methodology consists of three steps. First-up, a numerical solution of a problem is retrieved by using Ansys FEM-package. Within the framework of this work, such results as contact pressure and contact status, depending on a displacement of the pin relatively to sheave, are taken into concern. Secondly, a medium-sized dataset of Ansys output results is gathered. The results consist of nodal coordinates of pin face and their contact status. Finally, an XGBoost ML-algorithm based on the above-mentioned dataset is trained. The expected model output includes accurate prediction of contact pressure in- and out-of-sample, depending on pin displacement relatively to sheave. A comparison between ML-based algorithm and Ansys Mechanical output is presented. Future plans of system development are outlined.

Keywords Ansys · XGBoost · Mechanics · Contact problem · Ensemble algorithms · Machine learning

1 Introduction

Nowadays, problems of designing structures and complex mechanisms lay in the field of mechanics of deformable structures and have quite a big scale and solution

S. Orlov (✉)
Institute of Machinery, Materials and Transport, Peter the Great St. Petersburg Polytechnic University, St. Petersburg, Russian Federation
e-mail: majorsteve@mail.ru

K. Aubekerov · S. Koptsev
Department of Transport Equipment and Logistic Systems, Karaganda Technical University, Karaganda, Kazakhstan
e-mail: kairzhanable@gmail.com

S. Koptsev
e-mail: stanislavrecovery@gmail.com

C. Jahn et al. (eds.), *Algorithms and Solutions Based on Computer Technology*, Lecture Notes in Networks and Systems 387,
https://doi.org/10.1007/978-3-030-93872-7_6

difficulty. Various methods and approaches to solving these problems, such as Finite Element Method, which is a numerical method that divides a complex structure into a discrete number of elements which have their own geometric and material properties [1]. Discretization of the whole, complex geometry into a finite number of elements is known as a process of generation of so-called computational mesh (further referred to simply as mesh). These meshes can reach big sizes and consist of millions of finite elements, thus increasing the difficulty of computational problem solution. Large meshes require more computing resources and time to reach a decent solution.

In this paper we address the two-body contact mechanics problem. Given geometry includes two elastic bodies—a pin which sticks by its face on a toroidal-shaped surface of sheave. Such mechanical characteristics as contact pressure distribution on a pin face, depending on a displacement of pin itself, relatively to the sheave surface are mainly considered. This problem can be solved by using above mentioned finite element method with the help of widely spread commercial finite element packages, such as Ansys. The main difficulty is a number of solution steps required to retrieve the desired result for any possible displacement of a pin. Both pin and sheave have quite a large mesh bringing computational difficulty and affecting time- and cost-effectiveness.

Our main goal is to offer an approach to speed up the solution of a two-body contact mechanics problem using an efficient deep-learning and ML-based Python library—XGBoost. The detailed description and overview of various ML-libraries and ensemble learning methods, such as Random Forest, Boosting, Extreme Boosting is provided in the number of publications [2, 3]. We took XGBoost for its ability to predict complex internal dependencies inside the dataset, fast performance and ease of use. XGBoost implements decision trees with boosted gradient, enhanced performance and speed. Meanwhile, because of the regular term, this method could be able to prevent common machine learning difficulties, such as over-fitting.

To give some clarity of our investigation, let us briefly describe the key steps we performed and overall description of the methodology of solving given problem.

Conceptually, the research consists of three steps. On the first step, the design geometry pin and sheave is built using Ansys Design modeler module. Then, ready-to-use geometry is exported to Ansys Mechanical module and a contact problem is set up. Initial settings include material properties of pin and sheave (in this paper we consider structural steel material imported from presets), target and contact surface behavior and all required constraints of two bodies.

The second step includes steps of gathering initial dataset by running several solution iterations with different value of pin displacement. The dataset consists of nodal coordinates of pin face and contact pressure on these nodes. After running a decent number of solution steps, we expect to have a dataset of a decent number of samples, ready to be fit into our XGBoost model.

The third and final step is fitting the XGBoost Python library and analysis of its output. This step includes model training on a given dataset, fine-tuning, performance check and work on common problems, such as overfitting. The final result of our work is a working ML-algorithm that takes arbitrary pin displacement relatively to sheave and is able to predict desired physical quantities in this region.

Next sections describe our methodology in detail. Further, the performance of the ML model is tested out-of-sample, when arbitrary pin displacement is given as the model input and the output result is then compared with numerical solution, provided by Ansys Mechanical.

2 Materials and Methods

2.1 Contact Problem Numerical Solution with Ansys Mechanical

The solution methodology of a standard contact problem using Ansys products has been thoroughly discussed and described in a number of works [4–7]. This section describes the geometry of a given problem, main contact settings in the Ansys Mechanical module and analysis parameters, such as initial constraints, displacement values, etc. that we used.

Figure 1 illustrates the geometry and basic parameters of a problem. The geometry is constituted by a toroidal sheave and a pin, whose body is obtained by extruding the cross-section.

It is worth noticing that final geometry consists of two physical bodies and the main point of interest is a contact between the pin and sheave faces. The material for both parts is set to be Structural Steel with Young's Modulus = 2e11 and Poisson's ratio = 0.3 and is assumed to be elastic.

After the construction, the designed geometry is then imported to the Ansys Mechanical module and the main contact parameters are set in this module.

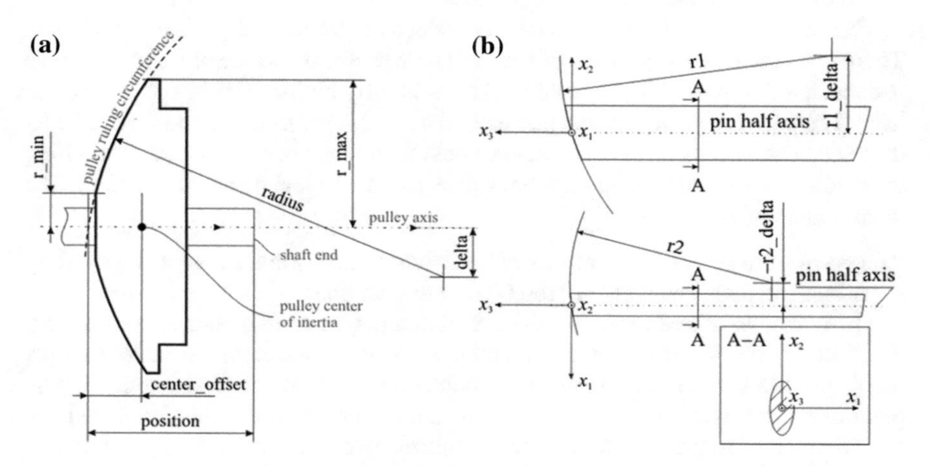

Fig. 1 The geometry of a contact problem. **a**: Sheave geometry. **b**: Pin geometry

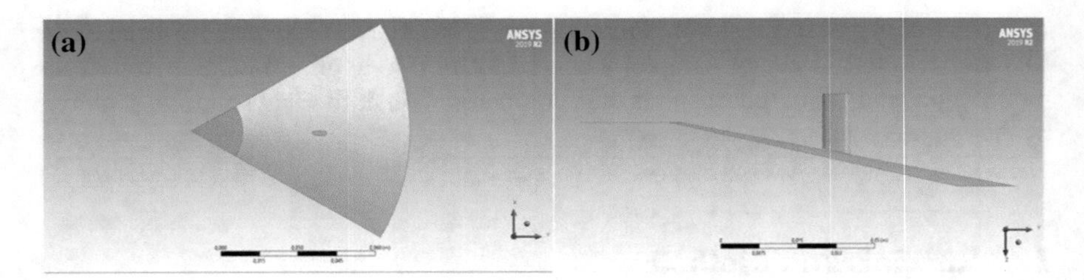

Fig. 2 Geometry of contact problem. **a**: a segment of toroidal sheave; **b**: lateral view of pin and sheave

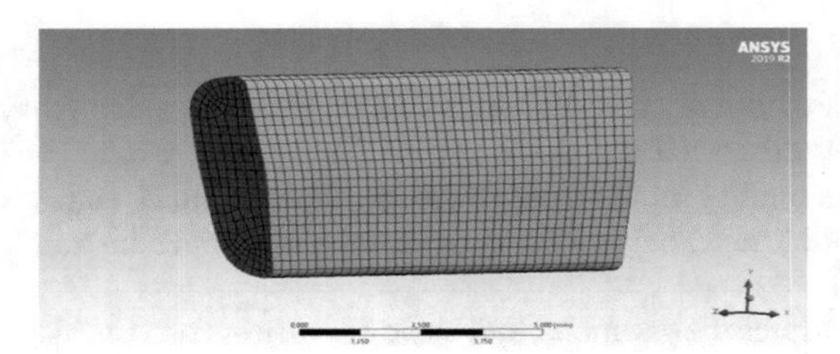

Fig. 3 Finite element mesh of the problem geometry. Computational mesh consists of 73 K nodes and 48 K elements. We expect that refining mesh and using the smaller element size will produce more accurate results but will cost more computational resources and time

Computational Mesh and initial problem constraints. Two simplified 3D models were used: (1) the segment of the toroidal sheave, and (2) the half of the pin divided by the symmetry axis. Figure 2 is composed of the frontal view, the lateral view of the whole structure and the view of pin itself.

Figure 3 represents computational mesh over a model geometry of pin and sheave. There is a contact between face of the pin and a uniform surface of sheave. Sheave is considered to have a fixed support; thus, it is fully constrained in space. The pin has relative to sheave displacements: ux and uy and is pressed into the sheave by uz displacement. Ux and uy displacement values are the subject to change, thus, running the solution with various displacement values and retrieving the results makes up the dataset generation process.

Ansys solution results. The results of numerous Ansys solution outputs according to various pin displacement are presented in this section.

Using the simplified models, a dataset of contact problem results is constructed. The dataset consists of four main parameters—x, y, z coordinates of nodes that lay on pin contact surface and the value of contact pressure in each of the nodes. These parameters are used to train an XGBoost model, as illustrated in Fig. 4 and the performance of trained model is then examined by comparing to Ansys numerical solution with the same input (pin face displacement value).

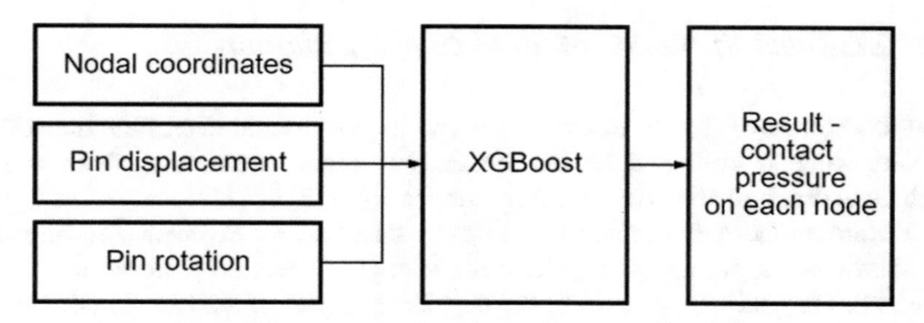

Fig. 4 Concept of XGBoost implementation in contact problem output prediction

In order to avoid changing pin displacement value by hand and start the solution process manually on each calculation iteration, built-in Ansys automatization tools, particularly ACT scripting mechanism was used.

The Ansys ACT Console is a special tool available in Workbench that uses Iron-Python 2.7* interpreter and specifically designed ExtAPI that allows to set up analysis parameters, solution steps, initial conditions and results export. For more information about Ansys Workbench scripting, ExtAPI usage and snippets, refer to [11, 18, 20].

With the help of automatization tools, usage of simplified model, we were able to reach a significant reduction in the computational time to gather a dataset of decent number of samples.

With this formulation, total number of 150 samples were retrieved. Table 1 shows a small subset of data, illustrating internal structure of features, upon which an XGBoost model is to be trained. Next section will cover up the process of XGBoost model train and validation.

Table 1 Ansys solution results—XGBoost train dataset sample

Pin displacement			Nodal coordinates		P, MPa
x	y	z	X	y	
0.3	0.2	0.338994	−0.33266	1.5968	1289.8
0.3	0.2	0.338994	−0.32382	1.3492	1804.1
0.3	0.2	0.338994	−0.31879	1.108	2252.4
0.3	0.2	0.338994	−0.3095	0.87	2610.6
0.3	0.2	0.338994	−0.29086	0.63065	2883.3
0.3	0.2	0.338994	−1.5817	0.58888	2949.6

2.2 *XGBoost Ensemble Method Implementation*

Brief description of the XGBoost library and its potential to determine the contact pressure value is evaluated in this section. For more information about various machine learning methods and their comparison, see [2] or [15].

XGBoost is one of the gradient boosting based tool [8]. It can be used for regression and classification problems [16]. Its basic idea is to generate multiple weak trees (learners), the goal of each is to fit the negative gradient of the loss function of the previous cumulative model. Gradient boosting finally produces a model in the form of an ensemble of weak learner, typically decision trees, called gradient boosting regression tree.

The form of XGBoost' objective function can be represented as [3]:

$$Obj = \sum_{i=1}^{n} L(\widehat{y}_i, y_i) + \sum_{t=1}^{k} \omega(f_t) \quad (1)$$

where L is the loss function for the model's bias and ω represents the regularization term which is used for decreasing the complexity of the model, thus, preventing the over-fitting problem.

In general, implementation the XGBoost ensemble learning method corresponds with traditional ML approaches. The backlog of our work is presented in Fig. 5 and

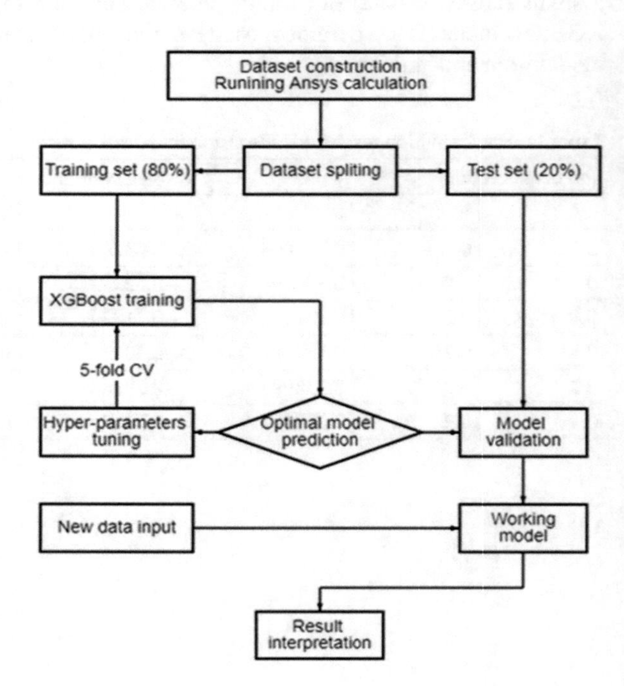

Fig. 5 Workflow of XGBoost model training and validation process

consists of four major steps: (1) collect the dataset to train the model and split it into training and testing sets; (2) train the model using train set. During this step, it is vital to pay attention to various model calibrating parameters (such as model complexity, regularization and depth of the decision tree). The goal of this step is to reach minimum predicting error, keeping in mind that we should avoid model overfitting, i.e., situation, when our model has very low predicting error on train set, but is failing to generalize the testing set. On the other hand, a situation of underfitting is also possible—in this case our model is fitting the training set poorly with the high error, but its error on the test set is barely higher than the training one. To know more about general concepts and approaches to train ML algorithms, refer to [9, 17]; (3) use the testing set to evaluate model prediction accuracy; (4) validate the model results by comparing it with any benchmark results. In this work we created the validation dataset which consists of input parameters that are not present in training and testing sets. We then run Ansys calculations upon parameters from the validation dataset and compare Ansys results with XGBoost model output. The results of validation step will be presented here after.

XGBoost model parameters. Once the required Ansys calculations are finished and thus, the dataset is constructed, the next step is to define XGBoost global parameters, as they are crucial for the model performance. We have chosen several key model parameters for optimization; they include (1) num_rounds: the number of rounds for boosting. This parameter is one of the main representations of the model complexity, hence the more rounds we run the boosting algorithm on train dataset, the more our model would be adjusted to given dataset structure. On the other hand, setting this parameter relatively high may lead to model overfit, and therefore—low performance out of train sample; (2) the learning rate (eta): an overfitting preventive term that shrinks each learning step size. After each boosting iteration, we get the weights of new features, and this parameter shrinks these weights to make the boosting process more conservative; (3) max_depth: the maximum depth of the tree. Increasing this value will make the model more complex and more likely to overfit, so it should be set carefully.

For the description of other available XGBoost tuning parameters, their affection to model performance and description, we advise seeing an official XGBoost documentation and references [12, 14].

Calibration of the hyper-parameters. The next step after collecting the dataset (the description of that process is available in previous sections), is to pick optimal XGBoost model parameters, as they are directly related with the model prediction ability and performance.

We have used grid search approach, realized in scikit-learn Python package [10] in combination with cross validation (CV) procedure. This is a common methodology to find an optimal set of model hyper-parameters.

Shortly said, grid search generates and runs model candidates from a grid of parameters, which are predefined in an array, which then is passed to GridSearchCV module of scikit-learn. The instance of GridSearchCV then iteratively generates regular XGBoost models and fits them on a dataset with all the possible combinations of hyper-parameters until the best combination is achieved.

However, when evaluating different hyperparameters for an ML model, there is a risk of overfitting, even using the train-test split because the parameters can be fitted to the particular samples of the test set and thus, bring poor general performance. One way to solve this problem is to have the dataset split on three parts—train set, test set and so-called "validation set", hence the model would be trained on training set; after that an evaluation would be done on the validation set and final model performance review is done on the test set. This practice may lead to success in model training and evaluation, nevertheless there can still be a problem when using this approach. By splitting the data into three subsets, we dramatically reduce the size of each subsample, designated for train and test purposes and the results may highly depend on a particular randomly taken samples from train and validation sets.

To solve this problem, we applied the procedure called cross-validation [15, 19]. During this procedure, final evaluation of the model still is done on test set, but there is no longer need for validation set. Instead, the training set is randomly split into k-smaller sets (this strategy is called k-fold CV). After splitting, for each of the k-folds, we train the model using k−1 folds of the data and perform validation on the remaining set [13]. The model performance is then calculated by the average value of k iterations. This approach may cost more computational resource, but it leads to shortage of data waste and eliminates the need of gathering more samples to the original dataset. In addition, the CV method can be effective when the dataset size is relatively small.

XGBoost performance evaluation metrics. Despite choosing suitable training and testing methodology, there is still need of quantitative metrics for evaluating model performance. We have chosen following metrics upon which our analysis and conclusions are performed.

Root Mean Squared Error. (RMSE) [3, 13]:

$$RMSE = \sqrt{\frac{\sum_{i=1}^{m}(P_i - T_i)^2}{m}} \tag{2}$$

where P_i is a predicted value by XGBoost, T_i—actual target value and m indicates the number of samples.

RMSE is a commonly used metrics to measure difference between actual target value and model prediction. By itself, this method does not give an unambiguous answer about the correspondence of the prediction to the reality, however, in combination with other criteria, it gives a clearer idea of the quality of the forecast. Intermediate steps of model training using grid search and CV with the corresponding RMSE value are shown in Fig. 6.

Geometrical compliance criterion. This criterion is based on minimizing the distance between the geometrical center of the contact spot obtained from the ML model and the geometrical center of the contact patch obtained from Ansys.

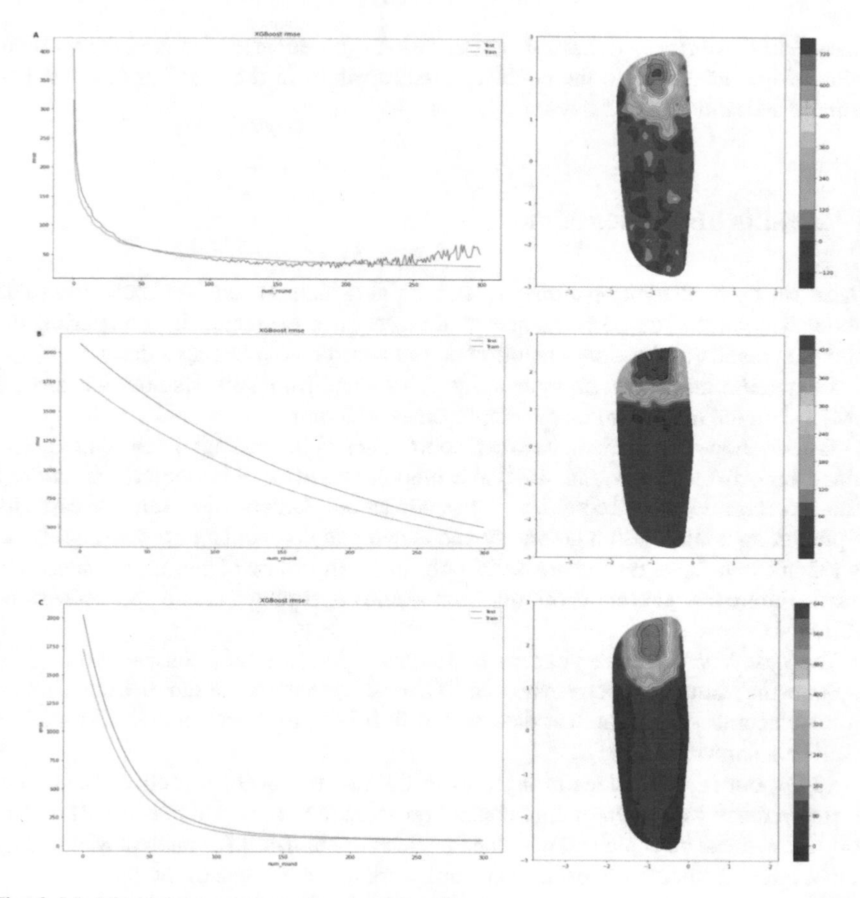

Fig. 6 Model training progress and various situations

Hence, the coordinates of the center of mass of a uniform flat figure are calculated by the formulas:

$$x_s = \frac{V_y}{2\pi S}, \quad y_s = \frac{V_x}{2\pi S} \tag{3}$$

where *Vx*, *Vy*—the volume of the body obtained by rotating the figure around the corresponding axis, S—figure area.

To assess the quality of the forecast using this criterion, the following operations are performed: (1) Calculation of the coordinates of the center of the contact spot, based on the contour of the contact spot obtained from Ansys; (2) Same calculation for the ML model; (3) Evaluation of distance between the obtained points.

Numerical Compliance Criterion. The criterion is based on minimizing the absolute value of the difference in the contact pressure values in the "key" nodes obtained from the ML model and Ansys.

3 Results and Discussion

Based on optimal hyper-parameters and training dataset, an XGBoost ensemble model was built. The performance of the model is evaluated by comparing the predicted results to the Ansys solution output based on the testing dataset.

The maximum contact pressure value determined from Ansys solution is around 3 MPa with the maximum uz pin displacement 0.3 mm.

Table 2 shows comparison between coordinates of the contact patch center determined both from Ansys and XGBoost model. Based on this output, we can say with a certain degree of confidence that our model demonstrates reliable capacity in predicting contact patch geometry and accompanying contact pressure field. Let us remark that these results are valid only in small orders of pin uz displacement values. Future thoughts about this question would be presented later in the discussions section.

The next key value we concern is a contact pressure between two bodies. To compare the results between Ansys and XGBoost, we have taken samples that contain pin displacement from the test dataset and fit it both to Ansys and XGBoost. The results are shown in Table 3.

So far, our investigations brought us to the fact that XGBoost ensemble model performs fairly well in predicting contact pressure in the interval between 1000 and 2000 MPa. It can be explained with the fact that our main point of interest was exactly in that interval since it represents standard working conditions of the pin.

Figure 6 represents several situations that may occur during training of an ML model. (a): Model with excessively high eta and excessive max_depth. Typical noises are observed on the RMSE graph, the test error begins to grow rapidly by the end

Table 2 Summary of contact patch geometrical compliance between Ansys and XGBoost results

Center coordinates from XGBoost	Center coordinates from Ansys	Error, mm
X: −0.95 Y: 2.1	X: −0.91 Y: 2.14	≈0.056
X: −0.94 Y: 1.73	X: −0.90 Y: 1.78	≈0.061
X: −0.94 Y: 1.64	X: −0.90 Y: 1.59	≈0.068
X: −0.93 Y: 1.34	X: −0.90 Y: 1.28	≈0.075
X: −0.93 Y: 1.04	X: −0.91 Y: 1.12	≈0.080
X: −0.92 Y: 0.46	X: −0.92 Y: 0.56	≈0.095
X: −0.92 Y: 0.1	X: −0.92 Y: 0.23	≈0.121
X: −0.91 Y: 0.14	X: −0.91 Y: −0.01	≈0.159

Table 3 Contact pressure determined from Ansys solution and XGBoost prediction. We have taken the results from Ansys as the "actual" results, that conform with reality and XGBoost output as prediction

$P_{predicted}$ (MPa)	P_{actual} (MPa)	Error (%)
456	523	12.8
640	720	12.5
946	1063	11.1
1288	1382	6.8
1689	1780	5.1
1845	2012	8.3
2098	2345	10.5
2468	2831	12,8
2750	3154	12.8

of training, and the noise is persistent. The pattern of contact pressure distribution is blurred, the model does not define areas in which there is no contact, chaotic foci of high pressure are observed, which do not exist in reality. The maximum contact pressure in the model is significantly higher than the actual; (b): 2—undertrained model, low number of iterations, low model depth, low eta. The RMSE graph shows that the model has not yet reduced to the optimum, the error is high both on train and test samples. There is a significant gap in the error value between Train and Test by the end of training. The contact pressure picture is still blurred, the model already qualitatively defines the geometry of the high contact pressure distribution. The maximum contact pressure in the model is significantly less than the actual contact pressure; (c): Model with optimal parameters. The RMSE plot shows a smooth convergence of the model to the optimum, no noise, and a minimal gap in error between train and test by the end of training. The pattern of the contact pressure distribution quite clearly follows the actual geometry of the high contact pressure distribution. The slip zone and the contact patch are visible, as well as the model determines the areas in which there is no contact. The maximum contact pressure in the model is close to the value of the actual contact pressure.

4 Conclusions

This paper presents fundamental steps of implementing a supervised learning task on a prebuilt dataset of results of a contact problem solution.

The methodology of predicting the results of contact problem using XGBoost ensemble algorithm based on the dataset gathered by running Ansys numerous solution iterations is presented. The hyper-parameters of the model were determined using grid search method and five-fold CV procedure. This process, however, is not detailed within this work due to the page limitation. The results demonstrate satisfying performance of the trained model in prediction of contact pressure value and contact field of the pin face. For the analyzed contact problem, the training time can be considered irrelevant compared to the time to solve a 3-dimensional contact problem in Ansys, for example.

It should be noted that our model training was ran on a relatively small dataset due to technical and computing time limitations. It is expected to get more accurate results with the dataset of bigger size. Our goal was to present an algorithm of solving mechanical problems with the help of an ML approaches. We have chosen the XGBoost library for its convenience, ease of use and certain adoptability to data of various shape. To make fair comparisons, future work can be done, such as comparison between several ML algorithms or ANNs. These remarks make a decent field of further development and investigations.

References

1. Mueller, D.W.: An introduction to the finite element method using MATLAB. Int. J. Mech. Eng. Educ. **33**, 260–277 (2005). https://doi.org/10.7227/IJMEE.33.3.8
2. Zhang, W., Zhang, R., Wu, C., Goh, A.T.C., Lacasse, S., Liu, Z., Liu, H.: State-of-the-art review of soft computing applications in underground excavations. Geosci. Front. **11**, 1095–1106 (2020). https://doi.org/10.1016/j.gsf.2019.12.003
3. Feng, D.C., Wang, W.J., Mangalathu, S., Hu, G., Wu, T.: Implementing ensemble learning methods to predict the shear strength of RC deep beams with/without web reinforcements. Eng. Struct. **235**, 111979 (2021). https://doi.org/10.1016/j.engstruct.2021.111979
4. ANSYS: ANSYS Contact Technology Guide.
5. Stolarski, T., Nakasone, Y., Yoshimoto, S.: Application of ANSYS to contact between machine elements. In: Engineering Analysis with ANSYS Software, pp. 375–509. Elsevier (2018). https://doi.org/10.1016/b978-0-08-102164-4.00007-8.
6. Елисеев, К.В., Кузин А.К., Орлов.С.Г.: Вычислительная механика. Вычислительный практикум в системе ANSYS: учебное пособие. Изд-во Политехн. ун-та, Санкт-Петербург (2004). https://doi.org/10.18720/SPBPU/2/si20-389.
7. Hattori, G., Serpa, A.L.: Contact stiffness estimation in ANSYS using simplified models and artificial neural networks. Finite Elem. Anal. Des. **97**, 43–53 (2015). https://doi.org/10.1016/j.finel.2015.01.003
8. Konstantinov, A. V., Utkin, L. V.: Interpretable machine learning with an ensemble of gradient boosting machines. Knowledge-Based Syst. 222, (2021).https://doi.org/10.1016/j.knosys.2021.106993
9. Ng, T.S.: Machine learning. Stud. Syst. Decis. Control. **65**, 121–151 (2016). https://doi.org/10.1007/978-981-10-1509-0_9
10. 3.2. Tuning the hyper-parameters of an estimator—scikit-learn 0.24.1 documentation. https://scikit-learn.org/stable/modules/grid_search.html. Accessed 23 Apr 2021
11. ANSYS: ANSYS ACT Developer' s Guide (2017)
12. XGBoost Parameters—XGboost 1.5.0-SNAPSHOT documentation. https://xgboost.readthedocs.io/en/latest/parameter.html. Accessed 23 Apr 2021
13. Utkin, L.: An imprecise extension of SVM-based machine learning models. Neurocomputing **331**, 18–32 (2019). https://doi.org/10.1016/j.neucom.2018.11.053
14. Notes on Parameter Tuning—XGboost 1.5.0-SNAPSHOT documentation. https://xgboost.readthedocs.io/en/latest/tutorials/param_tuning.html. Accessed 23 Apr 2021

15. Fisher, W.D., Camp, T.K., Krzhizhanovskaya, V.V.: Crack detection in earth dam and levee passive seismic data using support vector machines. Procedia Comput. Sci. 577–586. Elsevier B.V. (2016). https://doi.org/10.1016/j.procs.2016.05.339
16. Chen, T., Guestrin, C.: XGBoost: a scalable tree boosting system. In: Proceedings of the ACM SIGKDD International Conference on Knowledge Discovery and Data Mining, pp. 785–794. Association for Computing Machinery, New York, NY, USA (2016). https://doi.org/10.1145/2939672.2939785
17. Introduction to Boosted Trees—XGboost 1.5.0-SNAPSHOT documentation. https://xgboost.readthedocs.io/en/latest/tutorials/model.html. Accessed 23 Apr 2021
18. ANSYS Customization Suite-18.0 Release Reference Guide for Mechanical. https://storage.ansys.com/corp/ACT_Reference_Guide_doc_v180/Mechanical/index.html. Accessed 23 Apr 2021
19. 3.1. Cross-validation: evaluating estimator performance—scikit-learn 0.24.1 documentation. https://scikit-learn.org/stable/modules/cross_validation.html#cross-validation. Accessed 23 Apr 2021
20. ANSYS: ANSYS ACT Customization Guide for Mechanical (2019)

On Cloud Solution to Improve Business Performance for Product Deliveries

Alex Tejada, Konstantin Frolov, and Eduard Overes

Abstract In this paper, we introduce a case study where we analyze a late product deliveries problem that a large Russian wholesaler has been struggling with, since the beginning of its operations. From the problem identification, we address modern on cloud technology alternatives and then propose a solution to support this company's logistics operations in its central warehouse to enhance the delivery time. The result explains how to effectively send imported goods to final customers using SAP and Azure Integration for SAP Workloads to support Enterprise Information System processes. With such integration, business data will be consolidated in a reliable technology platform. Thus, the challenge is to deal with both systems and offer the smartest solution to meet the customers' expectations. On the one hand, SAP will support business processes management. On the other hand, Azure will support the cloud infrastructure resources and the interconnection with the LAN gateway. Regarding the scope of our research, the delivery of goods will be limited to Russian cities inside the urban perimeter. Besides, we use a cloud integration approach to allow: (i) Simplifying prebuilt integration. (ii) Learning best practices based on successful integrations. (iii) Leveraging SAP SaaS applications into Azure SaaS and PaaS applications. (iv) Redefining the user experience. (v) Integrating deployment flexibility. Consequently, in the end, we will appreciate the ease of combining this inter-cloud solution with ML, BI, and Big Data Analytics to fulfill the delivery requirements and optimize the whole service.

Keywords Logistics · Last-mile delivery · Integration · Cloud infrastructure · Cloud applications · Technology platform · Inter-cloud

A. Tejada (✉) · K. Frolov
Peter the Great St. Petersburg Polytechnic University, 195251 St. Petersburg, Russia
e-mail: tehadaparedes.a@edu.spbstu.ru

K. Frolov
e-mail: k.frolov@sap.com

E. Overes
ZUYD University, 6200 AP Maastricht, The Netherlands
e-mail: ed.overes@zuyd.nl

C. Jahn et al. (eds.), *Algorithms and Solutions Based on Computer Technology*, Lecture Notes in Networks and Systems 387,
https://doi.org/10.1007/978-3-030-93872-7_7

1 Introduction

One primary problem with our wholesaler logistics is that there is an internal communication neglect. This problem is usually derived from an absence of effective coordination, some warehouse management errors, and a lack of connection with end customers. They want their orders to arrive as fast as possible, which requires more local storage and delivery difficulty from the wholesaler. Certainly, all those previous facts alter punctuality. Besides, what we see behind the scenes are sequential steps that must be synchronized to accomplish the offered times. Thus, it is the goal of this work to conclude with a delivery optimization proposal combining business and technology strategies. Consequently, with our analysis, we pretend to give a valid alternative that combines cloud computing for business through SAP and cloud computing for virtual infrastructure through Azure.

To overcome the problem, we will include statistics as a crucial player when it was time to identify late delivery factors. Analyzing and classifying those statistics allow to concentrate resolving efforts on the problems' origin. As a result, the final model will help the wholesaler to cover not only the delivery problems but also the right platform operation that our solution demands with SAP and Azure integration. Also, with the achieved optimization the wholesaler will face the following logistics challenges more efficiently: (i) Reducing Transportation Costs. (ii) Improving Business Processes. (iii) Enhancing Customer Service. (iv) Improving Supply Chain Visibility. (v) Supply Chain Finance. (vi) Impact of the Economy. (vii) Driver Shortage. (viii) Government Regulations. (ix) Sustainability. (x) Technology Advancements. In broad terms, we address the problem in our case study using SAP and Azure's cloud resources.

2 Materials and Methods

2.1 Related Literature

Having a WBS as a start point of this study (Fig. 1), we have searched for some specific material that complements each part of such WBS. The components of this WBS are: (i) Initiation. (ii) Proof of Concept. (iii) Design. (iv) Implementation. (v) Testing. (vi) Deployment. Thus, considering this structure, some paper resources that we have found are: (i) Definition of cloud project requirements. (ii) Analysis of technology feasibility for a cloud project. (iii) Design of a cloud solution. (iv) Implementation considerations for a cloud solution. (v) Testing of a cloud computing system. (vi) Deployment of a cloud project.

To analyze the cloud requirements in the initiation stage, we found in the work of Alghamdi et al. [1] some advantages of using cloud infrastructure to overcome any business requirement. These advantages are mainly related with reducing costs, improving agility and flexibility, and controlling security [2]. Hence, in our case

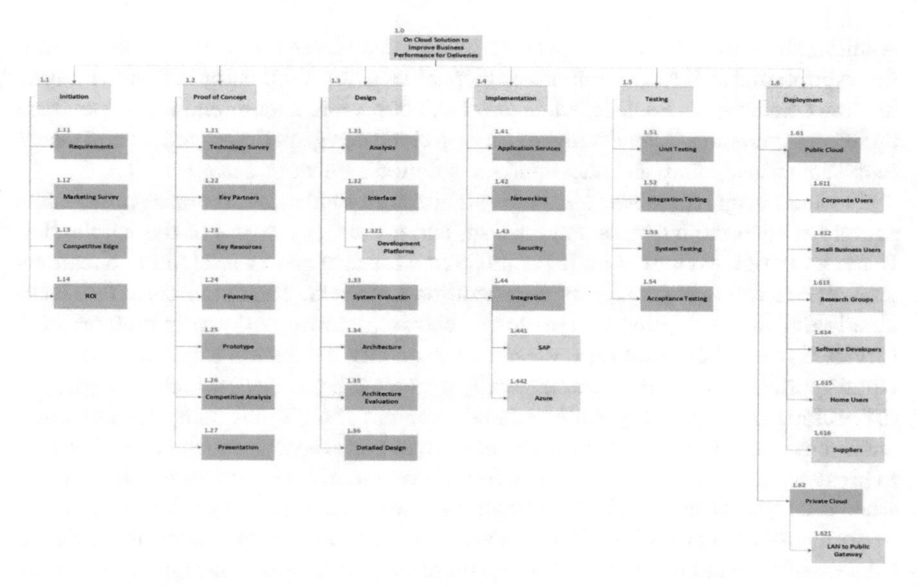

Fig. 1 Work breakdown structure

study we identify the following requirements: (i) Service management. (ii) Platform management tools integration. (iii) Security, reliability, visibility, and reporting. (iv) Interfaces for users, developers, and administrators. First, service management applications will help create policies for data and workflows to ensure business functions are fully efficient and resources are provided to systems in the cloud. Then, our solution platform should work with SAP and Azure management services, and open APIs to integrate existing OAM&P systems. Third, to guarantee SLAs, OLAs, security, and general service compliance, the final solution should offer robust real-time reporting and visibility, managing system performance, supervising customer service, and monitoring all processes. Finally, SSO feature for SAP and Azure integrated accounts will make interfaces available to ease business functions for users and complex cloud computing duties for administrators, automating regular deployment tasks, reducing operation costs, and progressively enhancing product deliveries to customers.

Among the techniques that have been developed to analyze technology feasibility and key resources in the proof-of-concept stage, Ben Halima et al. [3] discuss in their work how cloud solutions are increasingly popular computing paradigm that provides on-demand services to organizations for deploying their business processes over the Internet as it reduces their needs to plan for provisioning resources. For instance, one of the major challenges for the wholesaler as an SAP customer considering migration to SAP S/4HANA is finding a proof-point [4] to determine whether a system conversion will work well for their existing landscape. A proof of concept can be a useful way to gain more experience with the new technology within the context of the business, without the need to execute a full migration immediately.

Additionally, from the Azure perspective we can cover the starting points, from the aspirational DX to the important tactical steps of administration and resource naming conventions. Besides, when we check out Azure's requirements, we analyze architecture features related with security, on cloud design patterns, and several visual representations to help us understand the solution's network design.

For the design stage, we obtain an architecture model that represents the interaction between the business, application, and technology layers of the wholesaler. In the work of Gesvindr et al. [5] is illustrated for instance PaaS cloud applications architecture following a prototype generation approach. The authors use this PaaS cloud domain considering its benefits of elastic platform with many prefabricated services. Also, they propose a design-time quality evaluation approach for PaaS cloud applications based on automatically generated prototypes, which are deployed to the cloud and repeatedly evaluated in the context of multiple quality attributes and environment configurations. Looking back to our case study, we take those insights to break up Azure applications into hundreds of underlying services that have value when used by SAP applications in our inter-cloud schema. Thereby, releasing optimal performance along the cloud infrastructure will support user processes and improve subsequently customer experience. Furthermore, with a right design we consider around how application components communicate to include overall performance as well [6]. We achieve this goal understanding how the integrated delivery application will scale under an increasing load. Lastly but crucially, it is the security element that our model includes. As a cloud-based solution, we leverage IAM. The wholesaler administration will develop mature IAM capabilities to become significantly more agile at configuring security for supplementary cloud-based applications.

Moving on to the implementation stage we couple the architecture model with apps, communication, security, and hybrid cloud integration. Thus, in the research developed by Mansouri et al. [7] the authors examine how to implement hybrid cloud for performance evaluation of distributed databases. In like manner, they coincide with us when this type of implementation enables users to horizontally scale their on-premises infrastructure up to public clouds to improve performance and cut up-front investment costs. Most important, they emphasize the reachable automation of cloud bursting for open or closed source apps. In our case study, the hybrid cloud integration [8] must be done between the wholesaler private cloud and the acquired SAP and Azure services. Adjusting this prior approach to our cloud deployment model for applications and data, the final solution will align IT and business needs considering flexibility, security, speed, automation, cost, locality, service levels, and system interdependencies. For the integration, the wholesaler must understand the impact of these connections and employ best practices to address the existing enterprise systems. They will inevitably be challenged to blend their existing in-house IT investment with their newly adopted cloud services. At the same time, with the right networking settings our solution will ensure that appropriate connectivity is available to support overall business processes and even disaster recovery requirements. Another point to consider is the governance policies and service agreements developing because the cloud services combined to create a hybrid cloud computing environment require an overall governance framework maintained by the wholesaler that should consider various cloud service agreements established with SAP and Azure. As we stated in the previous stage above, security is a paramount requirement in our model. For this reason, in the implementation, we handle interfaces between the different environments, the movement of applications and data between the environments, and the organized control of assets across these environments. Also, the wholesaler—as a consumer of public cloud resources and services—must enable management of the complete hybrid cloud system and adapt APIs and integration points for management capabilities. And equally important, for the implementation, our model includes a careful and automated planning of backup, data archive, and disaster recovery mechanisms—with the ability of monitoring the frequency of such operations and determining what resiliency and backup capabilities are provided out-of-the-box for the cloud services portion of the hybrid cloud deployment.

Once implementation is done, it is vital to conduct different functional verifications. To analyze all the instances, we carry out: (i) Unit Testing. (ii) Integration testing. (iii) System Testing. (iv) Acceptance Testing. Certainly, with each type of examination, we pretend to find any functional failure that could affect the entire system performance or even represent a security breach. As part of QA and QC procedures, it is preferable to localize and realize system failures in a timely fashion before the solution works in production. Acceptance testing, for example, it is discussed in the paper of Smara et al. [9] where the authors exhaust a component-based cloud computing system, so the fault detection capacity can alert about errors. They highlight how important system-monitoring is, despite the extreme difficulty of keeping track of all cloud components. Applying this approach to our case study, we could spot specific issues and fix them during the testing. Another work about testing effectiveness is the analysis developed by Song [10] where he stresses his cloud system, controlling security and concluding that reliability should be assessed through environment testing, technology evaluation, and service appraisal. In our case, extra verification could be demanded [11], with functional tests (e.g., System Verification Testing, and Interoperability Testing) and not functional tests (e.g., Availability Testing, Multi-Tenancy Testing, Performance Testing, Security Testing, Disaster Recovery Testing, and Scalability Testing). Not least, we acknowledge that performing the testing operations, our solution obtains benefits like: (i) Dynamic availability of testing environment. (ii) Low cost. (iii) Easily customizable. (iv) Scalability.

Finally, the deployment merges the private and public clouds in a single hybrid cloud. Most of the deployment configuration has to be done in the public cloud—SAP and Azure—and the rest in the private cloud—wholesaler's on-premises network. To support business functions, SAP environment will be configured using S/4HANA and CRM modules for corporate users, small business users, research groups, software developers, home users, and suppliers. While to support logical infrastructure functions, Azure services are configured using Azure Data Lake, Azure Data Factory, Azure Logic Apps, Power BI, and Cognitive Services. From the private cloud, there will be the connection interface setup to the public cloud—LAN to public gateway. Bringing from the SAP implementation stage—the preparation, sizing, functional development, cutover, and going live—we will continue with SAP cloud deployment options [12]. For instance, we leave the environment ready through the sizing of high availability and disaster recovery, and with functional development we make available the change management, operations management, and testing. Next, for the SAP cloud deployment option we use Azure. In Azure, we will keep our SAP applications available and redirect traffic from troubled instances to healthy ones that are running smoothly. Once our applications are deployed, Azure will take care of the rest—from provisioning to load balancing [13]. Our applications will be backed by 99.99% SLA. Moreover, we will use Azure Resource Manager–based deployment model—provided by Azure Cloud Services (extended support)—to increase regional resiliency. With this deployment model, we will use capabilities such as RBAC, tags, policy, and support for deployment templates.

2.2 Case Study

The study object for this work is a large importer and distributor company in Russia. The company distributes various categories of products including computer, digital and home appliances, office accessories, office furniture and business gifts. At present, the organization is facing many challenges to improve its "delivery in time" processes. For this reason, the Board of Directors wants to bring a global modern solution that fixes the common delivery problems. Certainly, most of those problems are related to its systems reliability. Looking for modern technology alternatives, the company's research team found in SAP S4/HANA the feasibility of implementing Embedded Predictive Models, and with its Data for Model Training the possibility of obtaining historic sales order, including customer, product, employee, shipping, etc.

For a better comprehension of what the company needs and expects from an integral solution, we start analyzing its job in the market. Thus, we have the corporation acquires goods and services produced in a foreign country and resell those goods or services in Russia. Then, the company sells the imported products for rates that cover all the expenses associated with the initial purchase, any applicable tariffs, and the transportation costs involved to ship the goods into the country. One part of those rates is reserved as the profit. Additionally, the company works as an agent for foreign firms that wish to enter in the Russian retail market. At the end of the distribution process (Fig. 2), the company delivers the various categories of products. As a complementary fact, we identify the following pains the company is suffering: (i) Not have spearheaded the local market. (ii) Not have become a broad-line distributor on the local market. (iii) Not have modern logistics solutions. (iv) Not have a reliable system to fix common delivery problems. (v) Not have embedded Predictive Models. (vi) Not have classified data for Model Training. In the same way, we identify the following gains the company is looking for: (i) Have a valid solution which permits to interact the delivery process components in an efficient way. (ii) Monitor

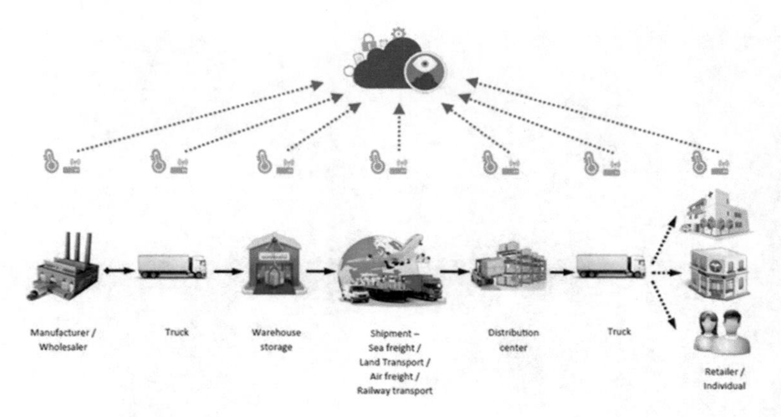

Fig. 2 Distribution process

the whole process until final customers receive their orders. (iii) Measure the satisfaction grade of each order. (iv) Permanently learn hits and setbacks of the process. (v) Apply corrections in future deliveries. (vi) Avoid issues and propose possible actions or automatic decisions. (vii) Predict possible delays before they occur. (viii) Automatize prefill during order capturing. (ix) Run Smart Variant Configuration. (x) Predict order probability of quotations.

This is how the problem can be tackled. We are going to define the customer profile with the customer jobs, and the gains and pains described above. Then, in a value map and customer profile canvas (Fig. 3), we define the gain creators and pain relievers with the products and services that we will implement in a cloud solution.

Next, to prioritize our work, we sort out what the gain creators are (Fig. 4). Also, we sort out what the pain relievers are. Afterward, to continue with the solution design we take the Needed Gain Creators and Needed Pain Relievers as the most significative conditions to satisfy the main requirements.

From the list of gain creators, we establish that the Needed Gain Creators are: (i) Full control (Central Management). (ii) Organization and standardization. (iii) Services integration. (iv) Security. (v) Simplify orders process. (vi) Punctuality. (vii) Track events. (viii) Process evaluation. And, from the list of pain relievers, we determine that the Needed Pain Relievers are: (i) Enable a supply chain. (ii) Setup an

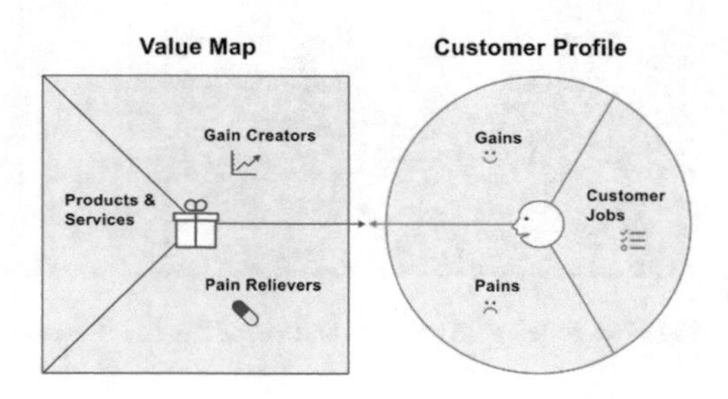

Fig. 3 Value map and customer profile

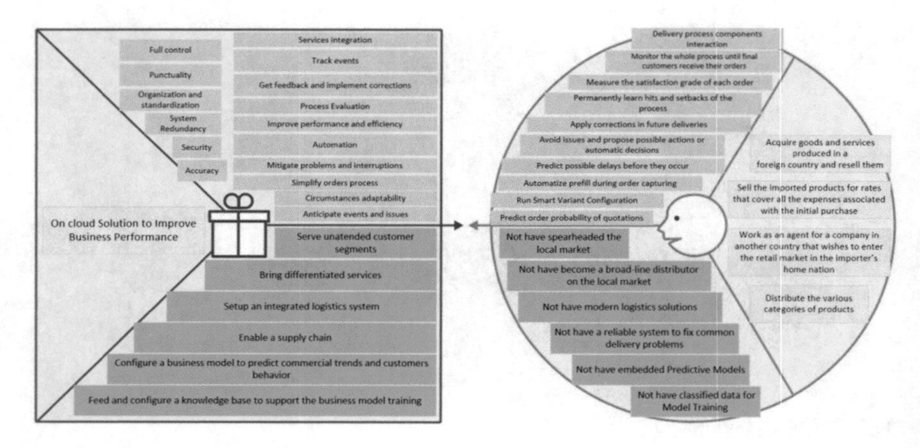

Fig. 4 Gain creators and pain relievers for the identified gains and pains

integrated logistics system. (iii) Configure a business model to predict commercial trends and customers' behavior. (iv) Feed and configure a knowledge base to support the business model training. Afterward, to make sense the needed features from the value map, we match them (Fig. 5) with the gains and pains of the customer profile that we are going to solve with our proposal. Therefore, during the implementation and deployment, the Needed Gain Creators and Needed Pain Relievers must be the first enabled services.

The other gain creators and pain relievers can be considered as a complement of the needed ones. In our analysis, we call them Preferred Gain Creators and Preferred Pain Relievers and they can be enabled in a second instance because the needed ones are the most important to support the main problems of the company. From the

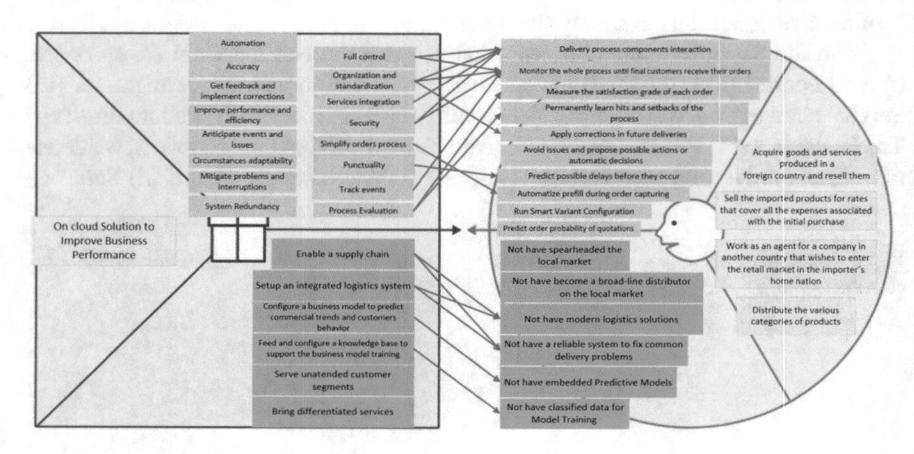

Fig. 5 Needed gain creators and needed pain relievers matching

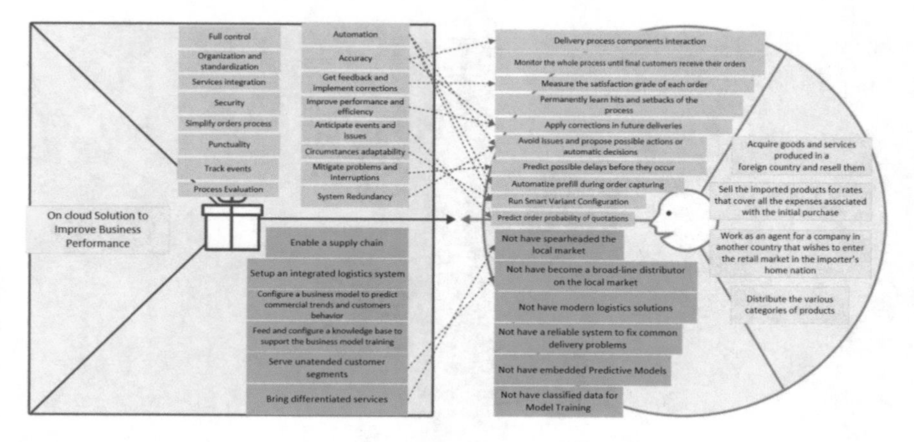

Fig. 6 Preferred gain creators and preferred pain relievers matching

list of gain creators, we establish that the Preferred Gain Creators are: (i) Automation. (ii) Accuracy. (iii) Get feedback and implementation corrections. (iv) Improve performance and efficiency. (v) Anticipate events and issues. (vi) Circumstances' adaptability. (vii) Mitigate problems and interruptions. (viii) System redundancy. And, from the list of pain relievers, we determine that the Preferred Pain Relievers are: (i) Bring differentiated services. (ii) Serve unattended customer segments. Next, to make sense the preferred features from the value map, we match them (Fig. 6) with the gains and pains of the customer profile that we are going to solve with our proposal. As we stated above, during the implementation and deployment, the Preferred Gain Creators and Preferred Pain Relievers can be enabled once the needed ones are completely functional and operative. Of course, the preferred ones' deployment will ensure that all the desires and expectancies of the company have been covered in the final solution.

Advancing with our study, we must know what occurs in the last-mile delivery (Fig. 7) so we can orientate our solution to bring the right applications to support the punctuality problem and all the logistics issues that surround it. Measuring the last-mile road signs—through its KPIs—provide us insights into the cost drivers and capabilities of the transportation network [14]. Using these insights, we will be able to recognize the missing resources to fulfill customer orders efficiently and cost effectively in the last-mile journey. Thus, the road signs that we have identified are: (i) Vehicle Capacity Utilization in Kgs. (ii) Stops Planned vs Actual. (iii) Cost per Km. (iv) Service Time. (v) On-Time Deliveries. Each road sign indicates the KPIs that need to be considered.

Firstly, examining the Vehicle Capacity Utilization in Kgs, we obtain the capacity of a delivery vehicle measured against the used capacity. Thus, the KPI is the total used capacity divided by the total available capacity. Applying this concept to our case study in 5 different deliveries, we obtain the results expressed in Table 1. As part of this capacity analysis, we obtain the chart depicted in Fig. 8, where we can appreciate

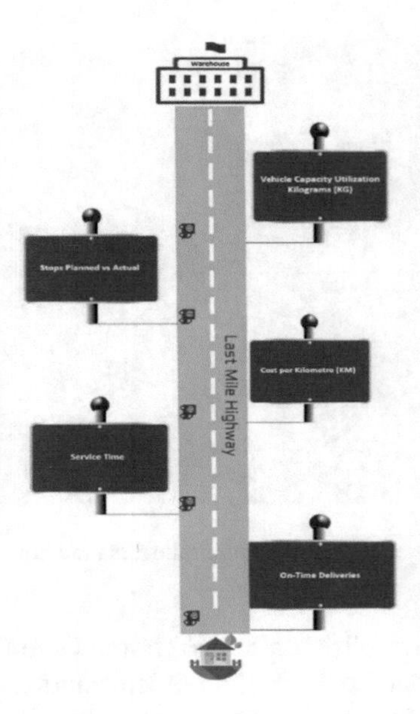

Fig. 7 Last mile delivery

Table 1 Vehicle capacity utilization in kilograms

Vehicle	Capacity	Used	Utilization (%)
1	4100	3650	89.02
2	4200	3800	90.48
3	4100	3500	85.37
4	4500	4200	93.33
5	4000	3750	93.75

the vehicles never carry the 100% of their capacity. In 5 different deliveries and vehicles, the highest utilization was 93.75%, and the lowest utilization was 85.37%. To leverage the full capacity of a vehicle, our solution should alert about available room, so we can optimize the current vehicle's journey and more customers obtain their products in the same delivery without more delays, otherwise their products will be sent in the next coming vehicle.

Secondly, examining the Stops Planned versus Actual, we obtain the number of planned stops compared to the actual number of stops. Thus, the KPI is the total number of planned stops divided by the actual number of stops. Applying this concept to our case study in 5 different deliveries, we obtain the results expressed in Table 2. Not all the evaluated delivery vehicles have the same characteristics, but we obtain the average of the planned and actual deliveries, and the percentage performance as well. With those average values, and as part of this stops' analysis, we obtain the chart depicted in Fig. 9, where we can appreciate the vehicles always make more stops than planned. In 5 different deliveries and vehicles, the highest number of stops

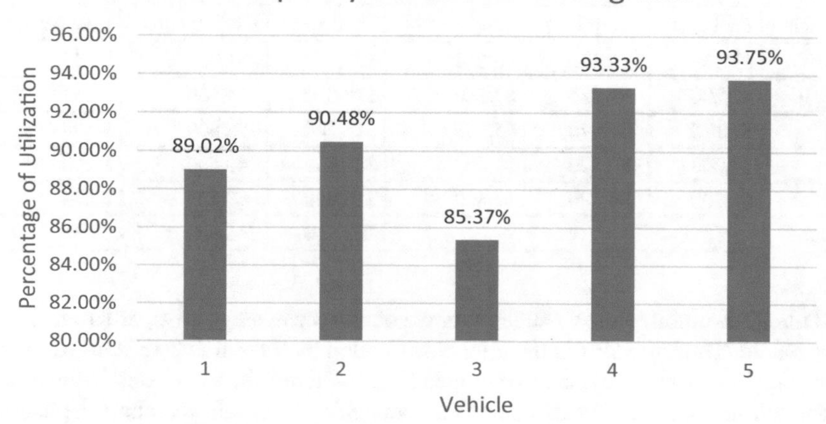

Fig. 8 Vehicle capacity utilization in kilograms

Table 2 Stops planned versus actual

Vehicle	Planned	Actual	Performance (%)
1	11	17	64.7
2	10	16	62.5
3	13	18	72.2
4	12	19	63.2
5	14	20	70.0
Average	12	18	66.5

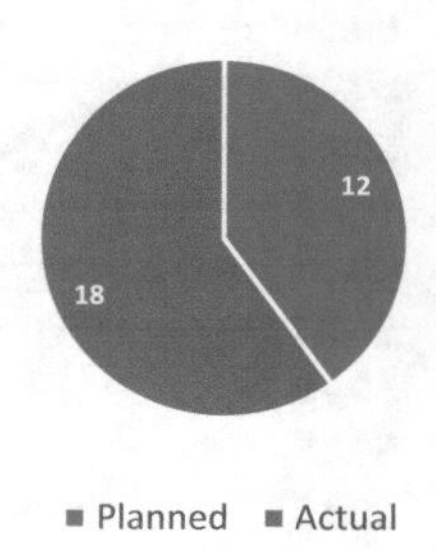

Fig. 9 Stops planned versus actual

was 19, and the lowest number of stops was 15. To optimize a vehicle's journey, our solution should recommend the right route to get the customers' addresses and avoid unnecessary stops, so we can optimize the current vehicle's journey and more customers obtain their products without delays.

Table 3 Cost per Km

Vehicle	Fuel cost	Labor cost	Vehicle cost	Total cost	Km travelled	Cost per Km
1	$3,330	$10,212	$8,658	$22,200	51,000	$0.44
2	$2,200	$8,280	$7,520	$18,000	54,700	$0.33
3	$2,175	$6,670	$5,655	$14,500	38,000	$0.38
4	$3,000	$9,500	$8,300	$20,800	49,500	$0.42
5	$2,700	$6,100	$7,800	$16,600	48,600	$0.34
Total				$92,100	241,800	$0.38

Thirdly, examining the Cost per Km, we obtain the average cost per Km travelled per vehicle. Thus, the KPI is the total cost divided by the number of Kms travelled. Applying this concept to our case study in 5 different deliveries, we obtain the results expressed in Table 3. As part of this cost analysis, we obtain the chart depicted in Fig. 10, where we can appreciate that along a year the vehicles travel more Kilometers depending on their cost. In 5 different delivery vehicles, the highest cost per Km was $0.44, and the lowest cost per Km was $0.33. To leverage the total cost of a vehicle, and then reduce the average cost per Km, our solution should alert about its entire mechanical conditions, so we can assign it more deliveries according to its performance and more customers can be served in shorter periods of time.

Fourthly, examining the Service Time, we obtain the length of service time from in store to delivery. Thus, the KPI is the total length of time divided by the number of parcels delivered. Applying this concept to our case study in 11 different deliveries, we obtain the results expressed in Table 4. As part of this time analysis, we also obtain the chart depicted in Fig. 11. Looking at the table results we can appreciate the

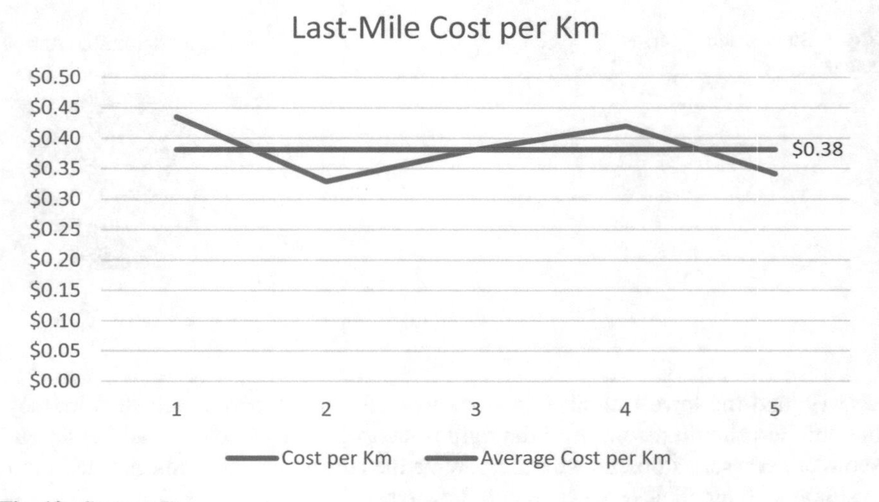

Fig. 10 Cost per Km

Table 4 Service time

Parcel	Preparation time (mins)	Delivery time (mins)	Total time (mins)
1	22	35	57
2	15	35	50
3	16	40	56
4	15	40	55
5	21	40	61
6	15	36	51
7	17	30	47
8	20	41	61
9	21	35	56
10	16	38	54
11	18	42	60
Total service time	196	412	608
Average service time per parcel			55

Fig. 11 Service time

parcels that have to be sent to the customers spent a minor time during the preparation stage, and then the delivery stage takes usually the double. The total service time, consequently, is high due to the delivery inconveniences—in other words it is still the bottleneck of the whole process. In those evaluated deliveries, the highest total service time was 61 min, and the lowest total service time was 47 min. The average time, as result of this evaluation is 55 min. To optimize the global service time, our solution should support not only the orders preparation process but also the delivery process as it was discussed in the previous statistics, so we can reach the Service Time in a value near the 30–45 min order fulfillment productivity [15]. Therefore, the applications to be used must support the warehouse and delivery automation.

Fifthly, examining the On-Time Deliveries, we obtain the number of parcels delivered on-time. Thus, the KPI is the total number of on-time deliveries divided by the total number of deliveries. Applying this concept to our case study along the last year deliveries, we obtain the results expressed in Table 5. As part of this on-time delivery analysis, we obtain the chart depicted in Fig. 12, where we can appreciate the total on-time deliveries in a year are barely 15.2%. Monthly evaluation shows that the highest rate of compliance was 16.4% in May, and the lowest one was 14.1% in September. To improve such compliance, our solution should alert the business

Table 5 On-time deliveries

Month	On-time	Late	Total	Compliance (%)
1	11,200	61,400	72,600	15.4
2	10,300	60,500	70,800	14.5
3	13,100	72,300	85,400	15.3
4	12,700	73,500	86,200	14.7
5	15,200	77,600	92,800	16.4
6	15,100	78,300	93,400	16.2
7	14,300	79,200	93,500	15.3
8	14,100	80,200	94,300	15.0
9	12,800	78,300	91,100	14.1
10	14,400	76,800	91,200	15.8
11	14,200	82,900	97,100	14.6
12	15,700	85,600	101,300	15.5
Total	163,100	906,600	1,069,700	
Average				15.2

Fig. 12 On-time deliveries

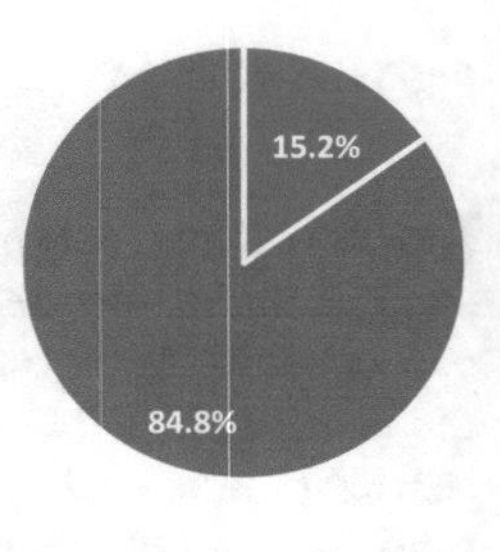

about all the delivery problems with key insights, so any immediate action can be taken by the responsible areas. Those insights have to be oriented to apply best practices [16] for optimizing last-mile delivery like: (i) Having a proper plan in place. (ii) Leveraging modern technologies that fit the business needs. (iii) Analyzing and assess data. (iv) Establishing standard procedures. (v) Monitoring driver, not just the vehicle. (vi) Monitoring the customers. (vii) Inventory tracking. (viii) Managing third-party drivers. (ix) Using real-world data to help optimize routes in real-time. (x) Getting proof of delivery.

3 Results

The contribution of this work is completed with a solution schema—considering that we successfully passed the initial and intermediate phases established for the WBS—where we incorporate the business variants into the technology deployment (Fig. 13) and tracking core cloud KPIs like performance, cost, security, and data backup/recovery. For the Technology Layer deployment, we orientate our work to develop the Azure approach with: (i) Cloud Support. (ii) Infrastructure. (iii) Virtual Network. For the Application Layer deployment, we orientate our work to develop the SAP approach with: (i) Business. (ii) Analytics. (iii) ML. Next, the solution involves the platform integration with functional components. Such integration is possible using Azure Data Factory which incorporates connectors to SAP BW, SAP Cloud for Customer, SAP ECC, and SAP HANA.

While deploying the Technology Layer [17], Azure Active Directory (AAD) acts as a security control plane providing centralized identity store with seamless SSO for SAP applications deployed across on-prem, in-Azure, SAP SaaS solution, as

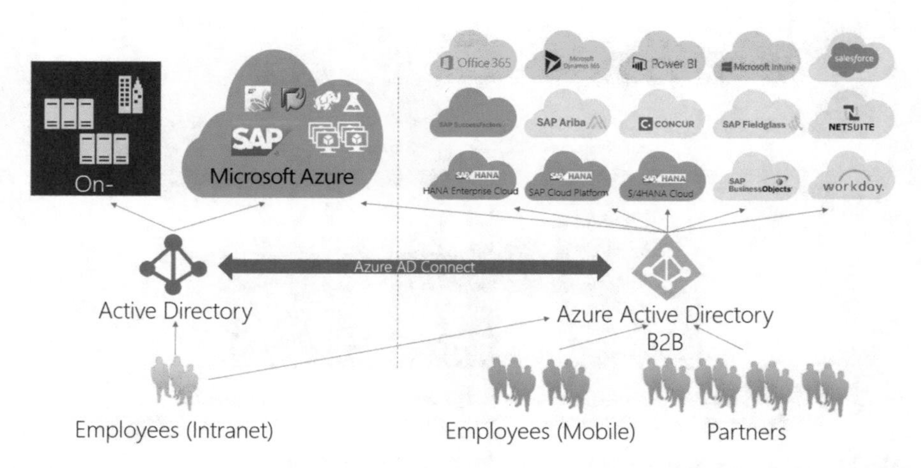

Fig. 13 Solution schema

well as SAP Cloud Platform services. Indeed, Azure AD integrates with SAP cloud IAS (Identity Authentication Service) and IPS (Identity Provisioning Service) to provide a more seamless, secure, and enhanced application access experience across SAP Cloud Platform services, enabling cross-cloud consumption of content and service at a higher scale. In addition, some of the key Azure security controls that are leveraged in our solution to create a strong security foundation are: (i) Identity and access management. (ii) Network security. (iii) Data protection. (iv) Security management.

Going on in the Technology Layer deployment, we work with Azure Data Factory (ADF) to integrate SAP data sources with Azure [18]. In our case study, ADF ingests data from those data sources using data driven workflows and then processes and transforms data using tools such as HDInsight and Spark (Fig. 14). The results can then be published to Power BI for consumption by Intelligent Solutions. Also, by using ADF's automated pipelines, our wholesaler company can use, transform, and analyze SAP data to gain insights on customer behavior such as customer churn and other risk factors. Similarly, when combined with other analytics services or ML services, ADF can provide powerful insights hitherto unavailable to customers.

When reaching the Technology Layer stability, it is time for the Application Layer deployment [19]. This requires an architecture depicted in Fig. 15—an overall frame-

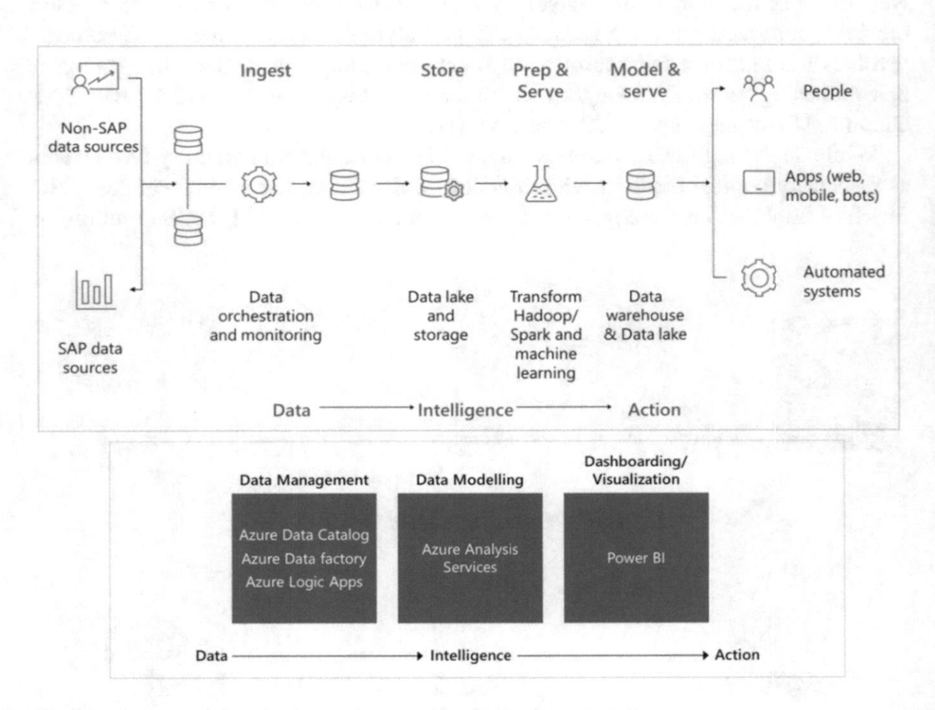

Fig. 14 Technology layer—Azure data services for business insights

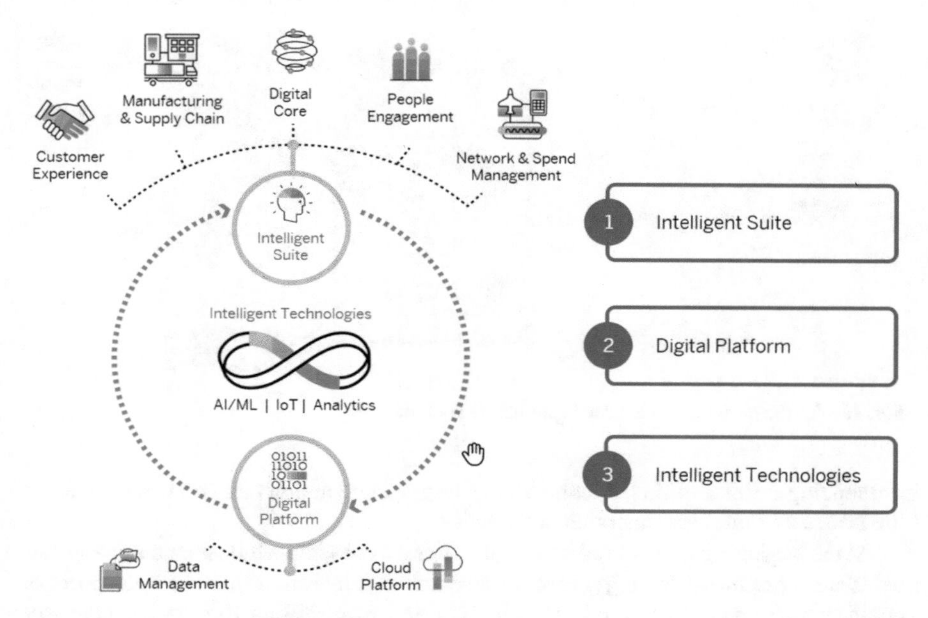

Fig. 15 Intelligent enterprise framework

work—consisting of three components: (i) The intelligent suite: To enable automation of day-to-day business processes and better interaction with customers, suppliers, and employees through applications that have intelligence embedded. (ii) The digital platform: To facilitate the collection, connection, and orchestration of data as well as the integration and extension of processes in integrated applications. (iii) Intelligent technologies: To enable leveraging data to detect patterns, predict outcomes, and suggest actions. Applying this framework to our case study and by considering the influencing criteria of the probable delivery delays, appropriate actions can be taken to avoid the issues and increase customer satisfaction.

To continue with the Application Layer deployment there is the feasibility of SAP S/4HANA Embedded Predictive models, and as data requirements the data for model training like historic sales order including customer, product, employee, shipping, etc. Thus, Delivery Performance is a broadly used standard KPI measurement in supply chains to measure the fulfillment of a customers demand to the requested date [20]. With the SAP Fiori App Delivery Performance, a sales manager is enabled to monitor his current delivery performance situation, and instantly recognizes the effect of delivered as requested ratio of sales orders, because a critical delay of delivered goods endangers customer satisfaction and retention in the future. Currently, the determination of the planned delivery date of a sales order item relies on the result of a planning or scheduling engine, e.g., ATP (Available to Promise). So, it cannot be considered probable deviations that arise in the future. Consequently, with SAP ML/Predictive Analytics (Fig. 16), it will be possible to learn from historic data to predict how likely it is that a sales order item will be delayed. By considering the

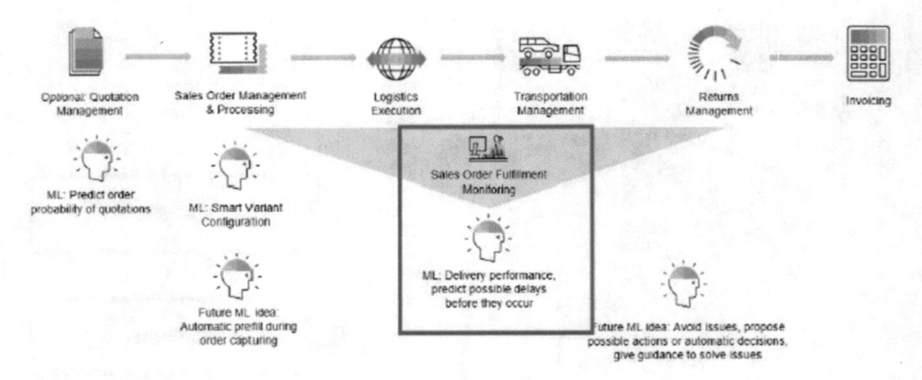

Fig. 16 Application layer—SAP ML/predictive analytics

influencing criteria of the probable delay, appropriate actions can be taken to avoid the issues and increase customer satisfaction.

At the beginning of this Predictive Analytics process, the wholesaler company can use ***Sales Quotation Management*** for a seamless integration into the sales process chain, which spans from quote to shipment and from billing to booking revenue. Also, the company can represent presales business processes using sales inquiries and sales quotations. Customers can respond to a sales quotation with a purchase order, which in turn triggers a sales order. Next, ***Sales Order Management and Processing*** allows the company to execute business transactions based on sales documents, such as inquiry, quotation, and sales order, defined in the system. Within sales order management and processing, the system users enter a sales document based on customer requirements. When they create or change sales documents, the system confirms dates and quantities. They can also display and change the sales documents to respond to customer questions. Afterward, the ***Logistics Execution System*** (LES) allows an administrator to manage the information and processes involved in all stages of the supply chain, from procuring raw materials to distributing finished products. Moreover, LES connects SCM processes involved in procurement, order processing, production, storage, inventory management, shipping, and sales. Soon, ***Transportation Management*** (TM) supports the company in all activities connected with the physical transportation of goods from one location to another. In this solution, the company use TM to transfer orders and deliveries from SAP S/4HANA, dispatch and monitor the transportation, and calculate the transportation charges. Furthermore, the company uses TM to create and monitor an efficient transportation plan that fulfills the relevant constraints (e.g., service level, costs, and resource availability). Administrators determine options to save costs and to optimize the use of available resources. Also, they react to transportation events and find solutions to possible deviations from the original transportation plan. Simultaneously to the two previous stages, the ***Sales Order Fulfillment*** app allows the company to resolve issues that impede sales orders from being fulfilled. The cockpit offers the users a list of all sales orders that cannot be completed for one or more reasons. The cockpit highlights impediments and provides supporting information and specific options to resolve issues. This allows administrators to keep track of sales orders in critical stages, collaborate with internal and external contacts, and efficiently address issues to ensure that sales orders in critical stages are fulfilled as quickly as possible. Then, if due to any complaint there are customer returns, ***Advanced Returns Management*** (ARM) enables the company to manage such returns. Customer returns involve the return of goods from a customer to a company location. With ARM, the company can handle all returns scenarios, such as returns made at the counter, returns that involve a shipment to the warehouse—including subsequent reverse logistics—and direct shipments from customer to vendor. At different times during the process, the company can refund the customer with a credit memo or compensate them with a replacement material. Later, at the ***Customer Invoicing*** stage, an invoice run converts the invoice requests into invoicing documents and releases the invoices for sending to the customer. Also, the released invoices are automatically posted as receivables in financial accounting and an open item is created in the customer account. This is tracked until the corresponding payment is received. The payment is posted as a cash receipt in financial accounting.

Along the whole Predictive Analytics process, ML will be used for constant improvement in the different stages. For instance, when Sales Quotation Management is performed, ML predicts order probability of quotations. During the Sales Order Management and Processing, ML runs Smart Variant Configuration to optimize time, and automatic prefill is executed during order capturing. One of the major aims of this work is achieved when Delivery Performance predicts possible delays before they occur through ML—cduring the Logistics Execution, Transportation Management, and Sales Order Fulfillment Monitoring. Furthermore, to reduce not only delays but also customer returns, ML shows how to avoid issues, proposes possible actions or automatic decisions, and gives guidance to solve issues. When running the integral solution, the expected business value will be shown as: (i) Manual effort decreasing to monitor and resolve issues. (ii) Better delivery performance achieving, and hence increasing customer satisfaction. It is clear then, with this solution the company has gained control, improved reliability, increased effectiveness, and supported business continuous improvement.

4 Conclusions

This case study made it possible to specifically analyze the logistics and supply process to determine the causes that may delay distribution and their impact on sales. Indeed, the operating factors in the logistics and warehouse area impact the sales operation. Therefore, in today's data-driven economy, organizations are mandated with building intelligent applications and making decisions that are based on hard facts. Looking for modern technology options, our organization—a wholesale company—chose SAP to support the logistics functions. This business platform, however, needs a robust infrastructure. Hence, Azure adjusted as a perfect complement. Our solution design and deployment, then, was split into two approaches: the technology layer with Azure, and the application layer with SAP.

Our company's customers increasingly want and expect faster and more frequent deliveries throughout the Russian territory. Rather than considering this demand as a problem, the company decided to treat it as a significant improvement challenge. Nevertheless, meeting that rising demand without a business strategy does not play out in terms of margin. Thus, the company meets this need by joining the technology and business approaches in one solution. In this way, the company seizes the top-line opportunity while managing margin risk. It is demonstrated then, SAP and Azure as smart technologies can run the right operating model for delivery, innovative methods to manpower, and a willingness to assist consumers. With these functions in place, the company increases its profitability at the core of last-mile delivery.

To sum up, performing well in the last-mile delivery means going the business the extra mile. Last-mile delivery is becoming an increasingly important part of providing the services that customers expect. Therefore, with our solution, their requirements are always first. Being a customer-driven business is a necessity of the company today and 24/7 customer care, website, mobile app, and social networks play a vital role in boosting its logistics and supply chain business. Confronting those facts with SAP and Azure tools, the company will simply look at its business with a fresh eye and start applying the proposed strategies one by one to improve its delivery service and lift its customers' satisfaction ratio. For data analytics purposes and permanent feedback, the company, from now, will always look at the last-mile journey as the most important source of KPIs.

References

1. Alghamdi, B., Ellen Potter, L., Drew, S.: Validation of architectural requirements for tackling cloud computing barriers: cloud provider perspective (2021). https://doi.org/10.1016/j.procs.2021.01.193
2. Cleo: Cloud infrastructure types, requirements and benefits (2021). https://bit.ly/3wKQzJp
3. Ben Halima, R., Kallel, S., Gaaloul, W., Maamar, Z., Jmaiel, M.: Toward a correct and optimal time-aware cloud resource allocation to business processes (2020). https://doi.org/10.1016/j.future.2020.06.018
4. NTT Data: Why it makes sense to do an SAP S/4HANA conversion proof of concept (2019). https://bit.ly/321trYY
5. Gesvindr, D., Gasior, O., Drew, S., Buhnova, B.: Architecture design evaluation of PaaS cloud applications using generated prototypes: PaaSArch Cloud Prototyper tool (2020). https://doi.org/10.1016/j.jss.2020.110701
6. TechBeacon: 5 Steps to building a cloud-ready application architecture (2021). https://bit.ly/3tloJ4x
7. Mansouri, Y., Prokhorenko, V., Ali Babar, M.: An automated implementation of hybrid cloud for performance evaluation of distributed databases (2020). https://doi.org/10.1016/j.jnca.2020.102740
8. IBM: Hybrid cloud integration in 7 easy steps (2016). https://ibm.co/3dbvUqh
9. Smara, M., Aliouat, M., Khan Pathan, A., Aliouat, Z.: Acceptance test for fault detection in component-based cloud computing and systems (2017). https://doi.org/10.1016/j.future.2016.06.030

10. Song, H.: Testing and evaluation system for cloud computing information security products (2020). https://doi.org/10.1016/j.procs.2020.02.023
11. Software Testing Help: Getting started with cloud testing (2021). https://bit.ly/3a6tfvS
12. NetApp: SAP deployment: a 5-step process and 3 cloud deployment options (2021). https://bit.ly/3ahylpn
13. Microsoft: Azure cloud services (2021). https://bit.ly/3x7TLPF
14. Perry Lavergne Consultant: KPIs—"The Last-Mile" (2020). https://bit.ly/3gsTGQm
15. Capgemini: The last-mile delivery challenge (2021). https://bit.ly/32CBagw
16. Peerbits: How last mile delivery optimization can boost your logistics & supply chain business (2021). https://bit.ly/3vdliNs
17. Microsoft: Top 5 reasons to deploy SAP on Azure (2020). https://bit.ly/3esJbKl
18. Microsoft: Integrating SAP application data with Azure (2018). https://bit.ly/3vg8FkX
19. Digitalist Magazine: What is the intelligent enterprise and why does it matter? (2019) https://bit.ly/3tRunvd
20. SAP Jam: Delivery performance/delivery in time (2020). https://bit.ly/3sOOGbt

Algorithmizing the DDM Model for Predictability of Stock Returns

Yuri Ichkitidze, Daria Pelogeiko, Anastasia Sudakova, and László Ungvári

Abstract The purpose of the research is to develop algorithms for predicting stock returns and to test them in various ways. The conducted forecasts show how well the estimates of the tested model reflect the real estimates in comparison with the generally accepted "market" estimates by CAPM. As a result, the possibility of predicting returns was scientifically proved, moreover the developed models gave correct estimates of returns.

Keywords Stock returns predictability · Nonlinear forecasting models · Structural breaks · Dividend discount model · Efficient markets

1 Introduction

Various studies have attempted to improve and correct the formulation of the market efficiency hypothesis [1–3]. In this paper, we rely on the definition given by Timmermann and Granger [4], the main difference of which is that here market efficiency is presented as a relative concept that depends on several parameters: the set of information, the chosen forecasting model, and the search technology used. In this case, the identification of the predictability of profitability does not contradict the effectiveness of the market as a whole.

Y. Ichkitidze (✉) · D. Pelogeiko · A. Sudakova
National Research University Higher School of Economics, Kantemirovskaya 3A, Saint-Petersburg, Russia
e-mail: yichkitidze@hse.ru

D. Pelogeiko
e-mail: daria.pelogeiko@mail.ru

A. Sudakova
e-mail: avsudakova_1@edu.hse.ru

L. Ungvári
Development in Relations of Industrial and Education Management GmbH, Karl-Marx-Str. 118, 15745 Wildau, Germany

C. Jahn et al. (eds.), *Algorithms and Solutions Based on Computer Technology*, Lecture Notes in Networks and Systems 387,
https://doi.org/10.1007/978-3-030-93872-7_8

Alternative theories about asset pricing have also been formulated in the scientific discussion, in particular theories about overreaction [5–7] and underreaction [8–10]. In addition, many studies in recent decades have shown that there is publicly available information that can be used to predict returns—for example, dividend yield, interest rates, inflation, and others.

In this study, we want to take a more comprehensive approach to the assessment of profitability and analyze it from the point of view of fundamental analysis. For this purpose, discounted cash flow models were used, in particular, the Dividend Discount Model (DDM) [11, 12] based on overall market assessment offered on the Damodaran website [13]. Using the analytical form of the model, we predict the rate of internal rate of return (IRR).

We formulate the basic hypothesis, which is tested in this paper, as follows: estimates of returns from forecast models based on a modified discounted cash flow model will be at least 50% more accurate than estimates of forecast models based on CAPM. In this case, the alternative hypothesis is formulated as follows: estimates based on the CAPM forecast are more often than half of the time more accurate than the estimates of the alternative model. That is, we expect to develop a model that can, on a par with CAPM, reflect "market" estimates of returns and will be at least as consistent. We test this hypothesis on two samples: 62 companies from the NASDAQ-100 index and 50 companies from the S&P500 index.

2 Materials and Methods

2.1 DDM-Based Equity Returns Estimation Model

In this study, the forecast variables are discounted, so the DDM model was used to estimate the returns. The classic DM model was transformed using EPS and the price-to-revenue ratio multiplier developed by Fisher et al. [14]. When NPV = 0 (net present value of the investment), the value of the rate $r_e \stackrel{\text{def}}{=} IRR$ and the fair value formula takes the form:

$$P_{\text{рын}} = \sum_{t=1}^{T} \frac{k_{dt} \times ROS_t \times SPS_t}{(1+IRR)^t} + \frac{SPS_T \times (P/S)_T}{(1+IRR)^T} \quad (1)$$

where IRR (Internal Rate of Return). By calculating it using this formula, we get the value of the current yield at a certain point in time at a certain market price.

Using this specification allows you to increase the dependence of the model on the revenue per share (SPS) indicator, the dynamics of which can be traced on historical data, reduce the impact of the dividend forecast and get rid of the need to predict the discount rate. The inclusion of the simplest factor relationships in the model—management benchmarks (k_d, ROS, P/S) in the SPS forecast—allows us to improve the final quality of the forecast of the yield indicator.

2.2 *Forecasting Models for Key Financial Reporting Indicators*

For the study, three methods of forecasting the parameters of financial statements were selected. The first method is based on a linear time series forecasting model of the ARIMA class. The following model specifications were selected for the study (Table 1).

The reason for the logarithm is the fact that the indicator theoretically cannot become negative. The transition to the first differences for revenue is justified by the stationarity of this variable in 95% of cases at the 1% level of significance and in 99% of cases—by 5%. In cases of non-stationary ROS and P/S indicators, the ARIMA model (0, 0, 0) was used. In addition, an important criterion is the principle of economy (compactness of the specification), especially when the depth of historical data for evaluation is not enough. As an example, Figs. 1 and 2 show forecasts of financial indicators for Electronic Arts from the position of year 1999.

The second method of forecasting, in contrast to the first, takes into account the non-linearity in the revenue forecast. The theoretical basis is the process of innovation diffusion described by Rogers [15]. Taking into account the strong dependence of the final results on the SPS forecast, the inclusion of non-linearity is particularly relevant. A non-linear revenue forecast is obtained using a combination of three models:

Table 1 Model specifications for forecasting

Indicator	Forecast model
SPS, sales per share	ARIMA(1,1,0) with logarithms
ROS, return on sales	ARIMA(1,0,0)
P/S, Fisher multiplier	ARIMA(1,0,0) with logarithms

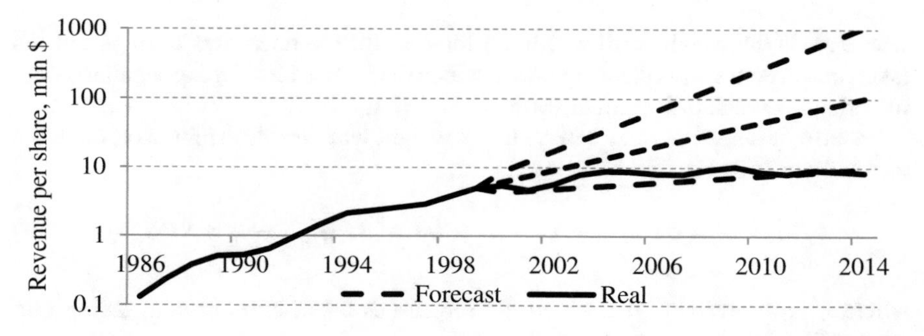

Fig. 1 SPS forecast for Electronic Arts in 1999 using a linear model (*Source* Authors' calculations)

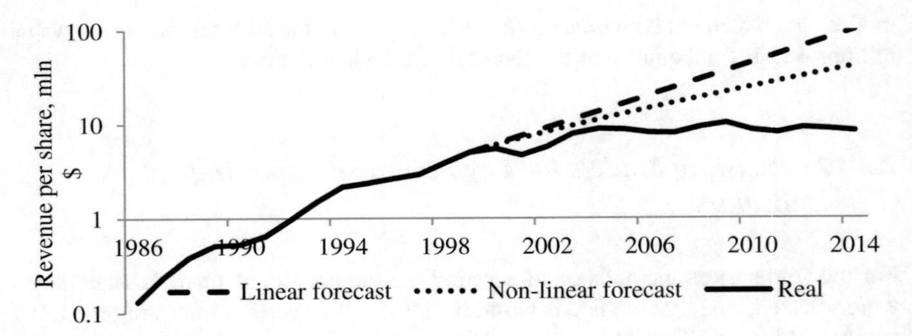

Fig. 2 SPS forecast for Electronic Arts in 1999 based on the model taking into account non-linearity (*Source* Authors' calculations)

1. linear model ARIMA (1, 1, 0) with logarithms:

$$d1SPS_{tlin} = \alpha + \beta_1 \cdot sd1SPS_{t-1} + \varepsilon_t \tag{2}$$

2. autoregression models taking into account the exponential trend:

$$d1SPS_{texp} = \alpha_1 \cdot se^{\alpha_2 t} + \beta_1 + \beta_2 \cdot s\left(d1SPS_{t-1} - \alpha_1 \cdot se^{\alpha_2 t-1}\right) + \varepsilon_t \tag{3}$$

   where the multiplier $\alpha_1 \cdot e^{\alpha_2 t}$ evaluates the exponential trend parameter. The negativity of α_2 indicates a decrease in the SPS differences.

3. models of structural discontinuity, the presence of which is tested using the Chow test [16].

The selection of weights with which the models described above will be included in the final one was selected taking into account deviations based on the RMSE indicator:

$$SPS_t = m_t \times SPS_{tlin} + n_t \times SPS_{texp} \tag{4}$$

where m_t is the weight with which the forecast for the linear model in period t is taken into account; n_t is the weight with which the forecast for the exponential model in period t is taken into account, with $m_t + n_t = 1$.

In case of detection of structural break weight was also determined on the basis of RMSE as follows:

$$SPS_t = l_t \times SPS_{tSB} + (1 - l_t) \times (m_t \times SPS_{tlin} + n_t \times SPS_{texp}) \tag{5}$$

where l_t is the weight with which the forecast is taken into account, taking into account the structural gap in period t.

In addition, it was decided to adjust the time series for inflation in order to eliminate the distortion associated with changes in the level of inflation in the past. An example is shown in Fig. 2.

As a third method for predicting financial variables, ARIMA linear forecasting models are used with the specification selection using the auto.arima function.

2.3 Sampling and Data

The models described above are tested on two samples. The first is the shares included in the calculation of the NASDAQ-100 index in 1999. The general population consists of 113 companies. Some of the companies ceased to exist at the time of the study, so the final NASDAQ sample included 62 companies. The second sample consists of companies included in the S&P500 index in 1994. From the general population consisting of 483 companies, 50 companies were randomly selected. For each of the companies, data on financial reporting indicators were collected for each year (from 1980 to 2019): Sales (S), Net Income (NI), Dividend per share (DPS), Shares Outstanding (SO). To build market estimates using CAPM models, daily values were collected.

The data sources for financial indicators were the Thomson Reuters Eikon database, for daily stock quotes and indices—the Yahoo Finance resource [17], for CAPM factors—the Kenneth R. French resource [18], the annual values of the consumer Price Index (CPI) were taken from the Federal Reserve Economic Data portal [19], the values of the Implied Premium (DDM), the values of dividends for the S&P500 index and the risk—free rate (T. Bond Rate)—from Damodaran's website [13].

3 Results

3.1 Forecasted Returns

Based on the forecast values for each of the companies, the internal rate of return (auto_IRR, lin_IRR, model_IRR) was calculated for the period from 1991 to 2019 (or for the maximum possible period, taking into account the available data). Forecast returns were also calculated for three CAPM models (CAPM 1, CAPM 2, CAPM 3). Based on real data using DM, the values of actual returns (real_IRR) for each year from 1991 to 2012 (or for the maximum possible period, taking into account the available data) are calculated.

Figures 3 and 4 show the results of calculating returns based on the dividend discount model for Adobe, which was included in the NASDAQ sample, and for IBM from S&P500 sample.

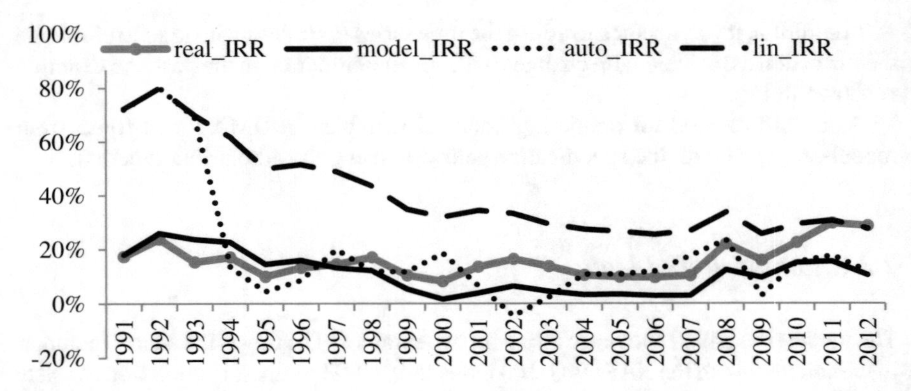

Fig. 3 Dynamics of predicted returns on the DDM model and real realized returns for Adobe, NASDAQ sample (*Source* Authors' calculations)

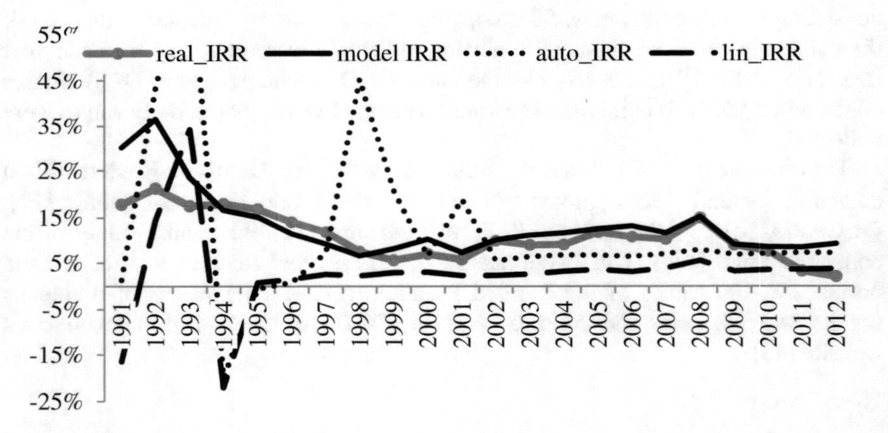

Fig. 4 Dynamics of predicted returns on the DDM model and real realized returns for IBM, S&P500 sample (*Source* Authors' calculations)

The graphs show that for these companies, the fair value model estimate, taking into account non-linearity in revenue forecasts, generally corresponds to the values of real returns for both companies, while models based on linear forecasts (lin_irr, auto_irr) tend to overestimate or underestimate the yield estimates, especially for the period up to 2000. However, with this approach to analyzing the results obtained, we can draw conclusions about the quality of the model for predicting profitability for an individual company, but not for the sample as a whole, and therefore it is necessary to analyze the predictive accuracy of the models.

Table 2 Summary statistics of the best regression models: 2 samples and 6 estimates of predictive models

		NASDAQ			S&P500		
		α	β	Adj.R^2	α	β	Adj.R^2
Model	Time RE	0.11***	0.10**	0.01***	0.11***	0.18***	0.31
Auto_model	Pooling	0.14***	0.0021	−0.0012	0.11***	0.20***	0.11
Lin_model	Pooling	0.15***	−0.01	−0.0006	0.08***	0.56***	0.48
CAPM	Time FE		−1.70***	0.02		−2.23***	0.03
CAPM-3	Pooling	0.13***	0.44***	0.02	0.18***	−0.21***	0.02
CAPM-5	Pooling	0.11***	0.75***	0.14	0.17***	−0.15***	0.01***

Source Authors' calculations

3.2 Predictive Accuracy Analysis

Panel data regression analysis. One of the options to check the correctness of the developed models is to build regressions, where the dependent variable will be a number of values of actual returns, and the explanatory variable will be a number of values obtained from one of the six forecast models (Table 2).

For each of the 6 pairs of variables in each sample, estimates were obtained using the end-to-end regression model, the model with random effects on time/objects (OLS method), and the model with fixed effects (LSDV method). To select the best model, the Hausman (FE vs. RE) and Broich-Pagan (REvs. Pooled) and F-test (FE vs. Pooled). Since the purpose of regression analysis is to check how well the models considered in this study explain a number of real return values, it is sufficient to analyze the significance of the coefficients and the model's correspondence to the data (Table 5). For the combined model and CAPM, the presence of random and fixed effects was revealed, respectively. The coefficients are significant for almost all models, but a relatively high R^2 is observed only in the S&P500 sample for the combined (Model) and linear (Lin_model) models. This means that these models can be used to predict the stock returns of a portfolio of companies from the S&P500 sample. The end-to-end regression estimates for these models are shown in Figs. 5 and 6, respectively.

RMSE results analysis. Another way is to analyze the deviations of the predicted values from the real ones using the Root Mean Square Error (RMSE):

$$RMSE_{forecast} = \sqrt{\frac{1}{n} \times \sum_{j=1}^{n} (r_{forecast\_i} - r_{real\_i})^2} \quad (6)$$

where n is the number of companies in the sample; $r_{forecast}$ is the value of the forecast return; r_{real} is the real realized return.

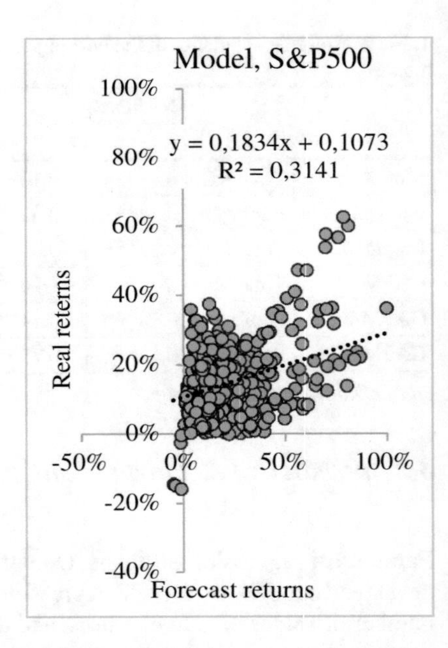

Fig. 5 Dependence of real returns on returns based on the combined model estimates, S&P500 sample (*Source* Authors' calculations)

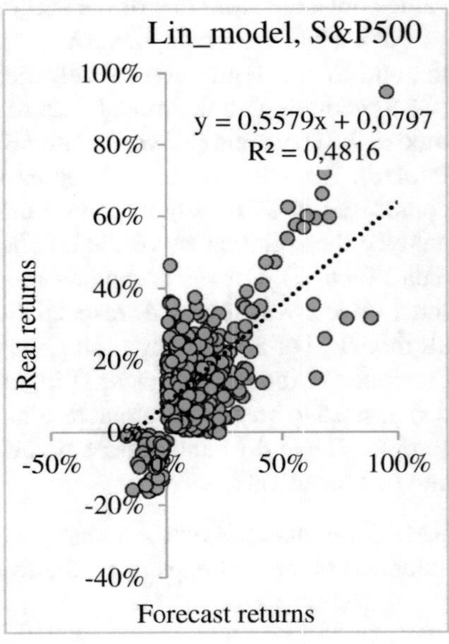

Fig. 6 Dependence of real returns on returns based on linear model estimates, S&P500 sample (*Source* Authors' calculations)

This approach will allow you to see which of the models gave the best estimate in each of the periods. The averaged values for all periods (RMSE_ind) will give the aggregate characteristic of each of the models. Then, for the NASDAQ sample, the best models are—CAPM, CAPM3 and Model, for the S&P500—models of the CAPM class.

Stock portfolio analysis. Above, we analyzed the behavior of the model in the case of predicting the return on shares of an individual company or some average company, but in reality, in the stock market, investors form portfolios to diversify their risks. Moreover, many studies on financial markets are conducted on portfolios. Let's analyze the predictive accuracy of the models when predicting the profitability of a portfolio in which all the companies included in the sample in a certain year have equal weight. Then the return on the portfolio is equal to the average value of the return on all companies (Figs. 7 and 8).

Figures 9 and 10 show that the Model, taking into account the non-linearity of forecasts (Model), systematically overestimates the portfolio return regardless of the sample, while the CAPM model systematically underestimates the return.

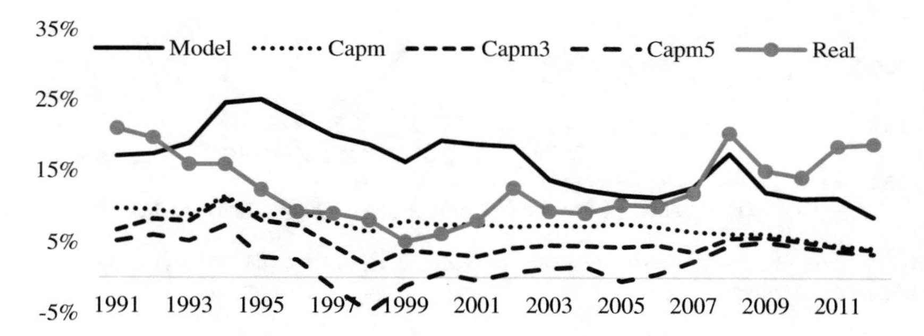

Fig. 7 Real and CAPM models dynamics of portfolio profitability based on the NASDAQ sample (*Source* Authors' calculations)

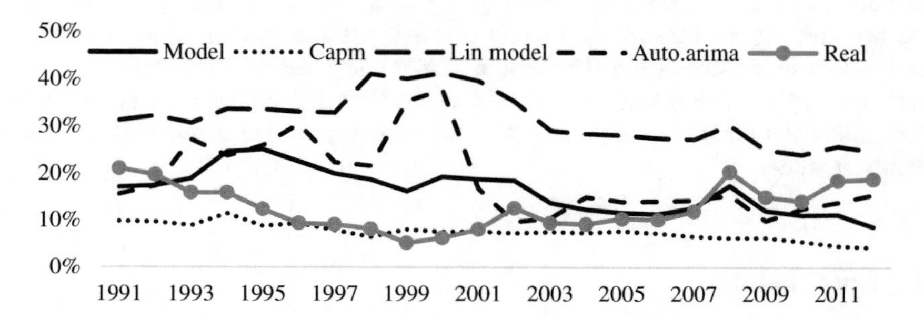

Fig. 8 Real and DDM models dynamics of portfolio profitability based on the NASDAQ sample (*Source* Authors' calculations)

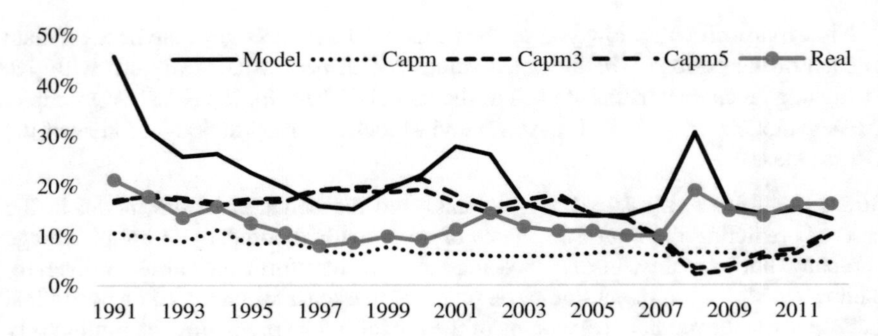

Fig. 9 Real and CAPM models dynamics of portfolio returns for the S&P500 sample (*Source* Authors' calculations)

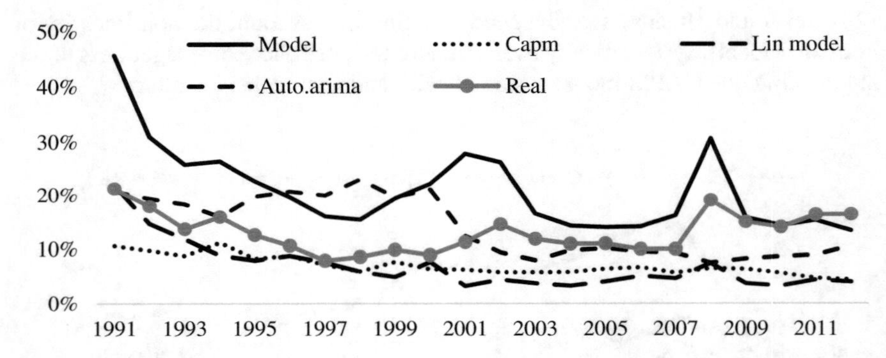

Fig. 10 Real and DDM models dynamics of portfolio returns for the S&P500 sample (*Source* Authors' calculations)

One reason may be that the market as a whole is underreacting, meaning that the market price of a stock is systematically below its fair value. And the fundamental models (in particular, the DDM model), especially the model taking into account the non-linearity of forecasts, correctly assess the future dynamics of profitability and include an underreaction. This assumption is consistent with the results of Hong and Stein [8], which show that as long as fundamental information is gradually disseminated among market participants, the market prices of assets will be below their fair value.

4 Conclusions

This paper provides a theoretical justification for the ability to predict returns. To identify the properties of profitability predictability, a model based on the dividend discounting model was used, presented as a decomposition into SPS (revenue per share) and management benchmarks (ROS–return on sales, k_d–dividend payout ratio, P/S–price-to-sales ratio). To calculate the forecast values of profitability, forecasts of financial indicators were constructed in three ways: using the linear first-order autoregression model AR(1), using the linear model using the auto-order autoregression function (auto.arima), and using a model that takes into account non-linearity in revenue forecasts (the presence of structural gaps and exponential trend). The inclusion of non-linearity in the revenue forecast is consistent with the theory of "diffusion of innovation" [15]. Capital asset pricing models (one-, three-, and five-factor CAPM) were considered as alternative valuation models.

Based on two samples of NASDAQ (62 companies) and S&P500 (50 companies), real and six variants of forecast values of returns for the period from 1991 to 2012 were calculated. The predictive accuracy of the models was analyzed by analyzing individual averaged deviations, portfolio deviations, and regression analysis. In the course of the study, it was revealed that the models of discounting dividends are able to predict the assessment of profitability.

References

1. Fama, E.F.: Efficient capital markets: II. J. Financ. **46**(5), 1575–1617 (1991)
2. Malkiel, B.G.: The cost of capital, institutional arrangements and business fixed investment: an international comparison (1992)
3. Fama, E.F., French, K.R.: Permanent and temporary components of stock prices. J. Polit. Econ. **96**(2), 246–273 (1988)
4. Timmermann, A., Granger, C.W.: Efficient market hypothesis and forecasting. Int. J. Forecast. **20**(1), 15–27 (2004)
5. Kahneman, D., Slovic, S.P., Slovic, P., Tversky, A. (eds.) Judgment Under Uncertainty: Heuristics and Biases. Cambridge University Press (1982)
6. De Bondt, W.F., Thaler, R.: Does the stock market overreact? J. Financ. **40**(3), 793–805 (1985)
7. Abarbanell, J.S., Bernard, V.L.: Tests of analysts' overreaction/underreaction to earnings information as an explanation for anomalous stock price behavior. J. Financ. **47**(3), 1181–1207 (1992)
8. Hong, H., Stein, J.C.: A unified theory of underreaction, momentum trading, and overreaction in asset markets. J. Financ. **54**(6), 2143–2184 (1999)
9. Frazzini, A.: The disposition effect and underreaction to news. J. Financ. **61**(4), 2017–2046 (2006)
10. Ichkitidze, Y.: Temporary price trends in the stock market with rational agents. Q. Rev. Econ. Finance **68**, 103–117 (2018)
11. Damodaran, A.: Applied Corporate Finance. Wiley (2010)
12. Damodaran, A.: Investment Valuation: Tools and Techniques for Determining the Value of Any Asset, vol. 666. Wiley (2012)
13. Damodaran's website Homepage, http://pages.stern.nyu.edu/~adamodar/New_Home_Page/datafile/histimpl.html. Accessed 28 Apr 2021
14. Fisher, J., Geltner, D., Pollakowski, H.: A quarterly transactions-based index of institutional real estate investment performance and movements in supply and demand. J. Real Estate Financ. Econ. **34**(1), 5–33 (2007)
15. Rogers, E.M.: Diffusion of Innovations. Simon and Schuster (2010)
16. Gujarati, D.N.: Basic Econometrics. Tata McGraw-Hill Education (2009)
17. Yahoo Finance Homepage. https://finance.yahoo.com/. Accessed 28 Apr 2021

18. The Kenneth R. French Homepage. http://mba.tuck.dartmouth.edu/pages/faculty/ken.french/index.html. Accessed 28 Apr 2021
19. The Federal Reserve Economic Data Portal Homepage. https://fred.stlouisfed.org. Accessed 28 Apr 2021

Mathematical Foundations Intellectually Coordination of Data for Group Expert Innovative Processes Evaluation Within the Framework of Scientific and Industrial Cooperation

Arthur Zaenchkovsky, Elena Kirillova, and Zoltan Zeman

Abstract In the article the technique and algorithm for group expert innovative processes evaluation based on analityc hierarchy process (AHP) is suggested. It is reasonable for support decision-making in the innovative processes execution within the framework of scientific and industrial cooperation and taking into account the specifics of the initial stages innovative processes implementation. The key elements of this methodology will be an innovative evaluation of different directions, the result of which is a quantitative and qualitative innovative processes assessment. The application of the developed methodology will improve the objectivity, efficiency and effectiveness of decision-making in group expert innovative processes evaluation, will avoid parametrically and organizationally uncertain stages of expert's negotiations in the conditions of incompleteness and unreliability of relevant information.

Keywords Innovative project · Group expert evaluation · Analityc hierarchy process · Scientific and Industrial Cooperation

1 Introduction

Successful implementation of innovative processes plays an important role in the modern Russian economy according to the strategy of transiting to an innovative development model and increasing the competitiveness. The careful attention to the scientific analysis and evaluation of the innovation process is due to its considerable

A. Zaenchkovsky · E. Kirillova (✉)
Moscow Power Engineering Institute, Smolensk Branch of the National Research University, Smolensk, Russia
e-mail: kirillova.el.al@yandex.ru

A. Zaenchkovsky
e-mail: no@sbmpei.ru

Z. Zeman
Szent Istvan University, Godollo, Hungary
e-mail: zeman.zoltan@gtk.szie.hu

C. Jahn et al. (eds.), *Algorithms and Solutions Based on Computer Technology*, Lecture Notes in Networks and Systems 387,
https://doi.org/10.1007/978-3-030-93872-7_9

importance for making operational and, above all, strategic decisions in economics [1]. Moreover, technological progress and making the competition process more intensive and global cut life cycles of products, force industrial enterprises to implement innovations quicker and develop products and services more effectively [2]. The growing integration of different technologies makes innovations more expensive and risky, which in turn emphasizes the role of the assessment and analysis stage in innovation management itself. In this regard, the responsibility in decision making management on the selection and effectiveness evaluation of such projects increases [3–5]. To date, these issues are presented as not quite developed. However, almost all existing methods and models are based on the same universal methodological provisions, which significantly reduces the practical possibility of taking into account the specifics of projects [6, 7]. Also, most of the existing methods involve more or less modeling with respect to which the decision is made [8, 9]. In the case of innovation, this is practically impossible. In addition, innovative projects are characterized by specific uncertainties due to the technical and market novelty of innovations, the complexity and ambiguity of their impact on the object, as well as its social environment [10]. That is why the main method used for the selection and evaluation of innovative projects is the method making it by the experts [11–13].

The initial stage of innovation process, which is a set of searching, selection and preparation for innovative ideas, is one of the most poorly structured [14]. So it has the prerequisites for system optimization. Systematic search, transformation and promotion of innovative ideas should be carried out centrally within the selected area [15–17], in order to systematize and unify this process, as well as to achieve a synergetic effect. One of the priority problems in this stage are problems with expert decision-making, experts selection and coordination of their assessments.

Making management decisions with limited resources provides the specialization in the framework of only one stage of the innovation process, rarely two. However, most of the existing methods and tools of its assessment [18, 19] practically do not take into account such aspects of industrial enterprise activity as scientific-industrial cooperation. At the same time, the level of such cooperation is not considered seriously most likely due to the existing objective problems in its decomposition. In this case, the analysis of a significant number of qualitative characteristics, subjective and relative variables with a probabilistic entity of its realization are assumed. At the same time, it is highly demanded to search for methods and ways to assess the current level of this cooperation, as the degree of development in the industry can ensure a significant contribution to the value of its innovation potential. This experience can be realized through a synergistic relationship between industrial enterprises and research organizations through the formation of stable connect on the basis of experience exchange, which will be the core for a rising of the innovative potential value [20]. It is also necessary to emphasize the positive impact on the innovative potential of an industrial enterprise—its inclusion in innovative scientific and industrial clusters, where the infrastructure objects being formed a key contribution to a closer exchange of information which results in a cluster of innovative processes aimed at coordinating all participants of the value chain when creating new technologies and expanding the range of high-value-added knowledge-intensive products. This would allow, among other things, to reduce the gap in patenting results between domestic and foreign enterprises, for example, in non-ferrous metallurgy (e.g., the American multinational corporation Alcoa has a patent portfolio of more than 4500 intellectual property objects, with a patenting rate of about 30 patents a year, while the largest Russian aluminium producer RUSAL is the rights holder of only about 100 patents).

Taking into account these features, the technique and algorithm for group expert innovative project evaluation on the initial stage of the project in regional scientific and industrial cooperation using the AHP, taking into account the internal and external consistency of expert data and heterogeneity of the expert staff is suggested.

The cooperation in the innovation sector is of particular importance for the effective running of the national economy [21]. The actual problem of modern innovative development of domestic economy enterprises is "lack of coordination in terms of the innovation policy articulation and implementation" [22].

Forcing of integration and cooperation processes and uprising tends in the international specialization of regions, as well as enhancing the role of the regional level in decision-making on the development of territories and shifting the emphasis to decentralization, contribute to the creation of conditions when large-scale and high-tech production spread through the boundaries of one region or state. Individual business elements and region authorities interact more actively, creating large-scale integrative scientific and technological complexes, which gives us such the main consequences of the world economy globalization as synergy effect [23]. Under the influence of such processes, the barrier feature of borders decreases, while the contact function of borders promotes the activation of foreign economic activity of closed regions and reinforces the innovative component in the social and economic development of neighbouring territories.

At the new stage of technical and economic development, there are a number of qualitative changes that require a fundamental rethinking of approaches to the assessment of internal factors that determine the success of an organization, whether it is an industrial enterprise, a financial institution, a scientific or social organization. Firstly, the composition of organizational resources is becoming significantly more complex, with intangible resources and the problems of their reproduction beginning to play the leading role [24]. Dynamic capabilities of such cooperation's are aimed at farming new resources. In particular, organizational procedures and practices in new knowledge generation, especially important in the pharmaceutical, IT, and other knowledge-intensive industries [25].

This is confirmed by statistical evidence of the practice of implementation of inter-organizational interaction at the current time. As it is visible in practice realization of inter-organizational interaction in the Russian Federation now is carried out weakly and fragmented.

In 2019, the level of innovation activity of organizations was 9.1%, which is lower than in previous years (in 2018—12.8%, in 2017—14.6%). According to the surveys, we should not expect significant growth of this indicator in the coming years either. Only every 10th organization reported that it has an intention to implement innovations in 2020–2022. One of the factors hindering innovation is the underdevelopment of cooperative relationships, and the rating of this indicator has remained at the same level for the last 10 years.

In 2019, the share of government funds in the total domestic spending on research and development in Russia was 66.3%. This is the highest value of the indicator among the countries which lead on the scale of spending on science. Despite this, the share of expenditures on innovation activity in the total volume of shipped goods, performed works and services is 2.1%, which is much lower in comparison with other countries (including the indicators of 2018). The share of funds of the entrepreneurial sector amounted to 30.2%, which is closer to the bottom positions among the leading countries on the scale of spending on science. It is worth noting that the largest amount of funds is oriented, in turn, to the entrepreneurial sector (60.7%), and the smallest to the higher education sector (10.6%) [26].

Overall, 18.2% of innovatively active enterprises were involved in joint R&D projects in the Russian Federation in 2019. One of the negative moments seems to be the decrease of organizations that participated in joint projects for research and development share comparing with the total number of organizations that carried out technological innovations in the period 2000–2018. In particular, this indicator for this period decreased by 25%. It is worth noting that cooperation is carried out mostly within a single project (79.1%), rather than permanently (52.2%). Evaluating technological independence in the development of innovations, it can be noted, a low percentage of joint development with other organizations (25.5%). Among the sources of information for technological innovation we can see a decrease in the share of institutional sources, and a slight increase of internal sources, namely organizations within the larger association to which this organization belongs. Also the greatest interest in participation in joint projects on carrying out research and development by scientific organizations is identified. However, it is worth noting that in the last three years more than a quarter of innovative organizations realized innovative goods, works, services by orders of users, which is caused by the presence of demand for such goods, works and services.

The effectiveness of the results of innovation activities is confirmed by the statistics, namely, among the organizations that had ready innovations in 2016–2018 there is more improvement in quality (40.4%) and expansion of the range of goods, works, services (37.5%), as well as preservation of traditional markets (36.4%). Much less noted is the expansion of markets (25.1%) and improvement of information links within the organization or with other organizations (20.2%), which respectively require the development of inter-organizational interaction to improve results and stimulate innovation activities.

The volume of innovative goods, works, services under state and municipal contracts in 2018 was 7.3% ratio of the total volume of innovative goods, works, services.

2 Materials and Methods

Set of alternatives $X = \{x_1, x_2,..., x_n\}$ will be described by a set of criteria (metrics) $K = \{K_1, K_2,..., K_N\}$. Each alternative will be evaluated by an expert or a group of experts $E = \{E_1, E_2,..., E_m\}$ for each of the criteria, and then on the basis of the obtained expert estimates, a generalized estimate for each alternative is formed. Using this the decision maker (DM) chooses the best alternatives. Each criterion *K*—has its own weight, which determines the degree of importance (significance) of evaluation on this criterion, and each expert *E*, characterized by the competence coefficient—*c*, which allow to take into account the importance of his opinion in the formation of a generalized assessment.

Experts evaluate alternatives not only by certain criteria, but by the whole set of criteria, which is possible within the framework of a pairwise method of comparing alternatives with the subsequent formation of pairwise comparison matrix.

3 Results

The algorithm for choosing the best alternative is based on the formation of a generalized evaluation by the particular criteria (expert's) assessments aggregation or individual experts preferences matrices. In cases where the set of alternative solutions has a large dimension, it is important to choose promising alternatives for further quantitative analysis, i.e. reducing the dimension. This choice can be made on the basis of approximate estimation models.

There are three main strategies for reducing individual expert assessments:

- generalized evaluation cannot be better than the worst of the particular estimates;
- overall assessment due to the best of the private evaluations;
- the generalized estimate is intermediate between the private estimates involved in the aggregation.

Consider the operator of the aggregation in general form:

$$\alpha(x) = \mathrm{Agg}(W, A) \tag{1}$$

where *x*—an alternative of a given set, $\alpha(x)$—is it's generalized estimate, $A = (a_1,..., a_n)$—vector of partial estimates of the alternative, $W = (w_1,.., w_n)$—vector of weight coefficients (w_i determines the degree of aggregates influence on the generalized estimate). If *A*—the evaluation vector of the alternative by criteria (indicators), *W*—the set of criteria weights, then $\alpha(x)$—a multi-criteria evaluation of the alternative. If *W*—a set of experts' competence coefficients and *A* is a set of estimates obtained respectively from these experts, then $\alpha(x)$ is a group assessment of the alternative.

In the formation of a generalized evaluation, two aggregation schemes are possible:

1. Agg1 $(W, A) =$ Agg1 $(g(w_1, a_1), g(w_2, a_2),\ldots, g(w_n, a_n)) = \alpha$,
2. Agg2 $(W, A) =$ Agg2 $(g1(W), g2\ (A)) = (w, \alpha)$.

In the first case, aggregates $g(w_i,\ a_i)$ are first constructed for all i from 1 to n, which are then folded into a generalized evaluation of α. In the second case, the aggregation of weights and partial alternatives estimates is carried out separately, with the actual assessment of the alternative being α, and w being considered as the degree of confidence in this assessment.

We consider a model for the generalized estimate formation, assuming that the partial estimates of the alternatives and the weighting coefficients are numerical from [3, 4]. For generalized assessment formation can be used the operator of aggregation ordered weighted averaging (OWA): n—local OWA-operator associated with a vector of weights $W = \{w_1,\ldots,w_n\}$ satisfying the conditions $W_i \in [0, 1]$ and $\sum_{i=1}^{n} w_i = 1$, there is a map F: $[0, 1]^{n+1} \Rightarrow [0, 1]$ such that:

$$F(W, A) = \sum_{i=1}^{n} w_i \cdot b_i \tag{2}$$

where B $= (b_1,\ldots,b_n)$ is a vector obtained from $A = (a_1\ldots,a_n)$ the ordering of the elements in non-increasing order.

Note that $F \cdot (W, A)$ implements disjunctive aggregation and $F \cdot (W, A)$ implements conjunctive aggregation. For an arbitrary OWA operator, the inequality holds

$$F \cdot (W, A) \Leftarrow F(W, A) \Leftarrow F \cdot (W, A) \tag{3}$$

which means that in the general case $F(W, A)$ is the averaging operator.

Special values are introduced to classify OWA operators with respect to bundles *AND* and *OR*:

$$orness(W) = i \tag{4}$$

characterizes the proximity to the disjunction operators, and

$$andness(W) = 1 - orness(W) \tag{5}$$

characterizes the proximity to the conjunction operators.

It follows from the definition that $orness(W) \in [0, 1]$, for *max* $orness(W^*) = 1$, for *min* $orness(W^*) = 0$. For an arbitrary OWA-operator, if $orness(W) > 0.5$, then the corresponding operator will be called a quasi-conjunction, if $orness(W) < 0.5$, then a quasi-conjunction. Operators of quasi-union and quasi-intersection are relevant for those cases when the expert is difficult to identify with full confidence the type of operation underlying the formation of a generalized estimate.

Some evaluation models are focused on the compensation properties presence of the aggregation operator, when small values of alternative estimates for one criterion (indicator) are compensated by large values of estimates for another or other criteria (indicators). Given the characteristics of the operators $orness(W)$ and $andness(W)$, we can assume that for an arbitrary operator $F(W, A)$ the closer the value of $orness(W)$ to 1, the more this operator has compensatory properties.

By changing the vector of weights, we can increase or decrease the value of $orness(W)$, thereby improving or worsening the compensation properties of the corresponding operators. This gives the opportunity for purposeful construction of quasidisks and quasicanonical with different intensity compensation properties.

In order for the operator to have the best compensation properties, we need to make sure that the value of the *orness* value is the maximum for this set of weights. Therefore, in one of the method stages, we should add the sorting of the weight vector by non-growth. This guarantees a value of *orness* not <0.5, which means that the aggregation operator will be a quasi-conjunction.

In the process of agreeing on expert views, it is important to take into account the availability of repeated access to experts and their competence with respect to each other. Only after the final structure is approved we should come to the set of priorities.

The entire solution process is reviewed and rethought at each stage, allowing for an assessment of the solution quality. If the number of experts is more than one, then the consistency of these opinions is determined after the opinions on each of the hierarchy levels elements. Verification is performed by calculating the concordance coefficient. When evaluations are discordant we should determine the connection, which consist of experts: the confrontation or the coalition. In confrontation, the Delphi method [21] is used as a method of reconciliation insofar because expert's opinions are independent and everyone is interested in their own goals.

The interaction scheme presented in Fig. 1 consists of two parts. In the first part, the preliminary preparation for the evaluation procedure is carried out, and the evaluation itself is carried out with the help of AHP [27]. The second part reflects the process of matching expert data by OWA-operators.

Features of the group decision making process on this scheme:

- selection of experts and specialists can be carried out in parallel and independently;
- evaluation is conducted at the same time without negotiations between experts;
- the main time spent are for only: experts and specialists recruitment, evaluation, analysis of results.

Among the advantages of described algorithm a special place is taken by reducing the time for the examination from the beginning (a set of experts) to the end (output of results). This can be said, judging by the reduction in the number of blocks of iterative processes.

The responsibilities of DM are in the area of experts and criteria selection. This situation does not allow experts to select specialists and evaluation criteria based on their own taste, as this can lead to the fact that negotiations will be held for a long time and experts will select based on their preferences rather than professional skills.

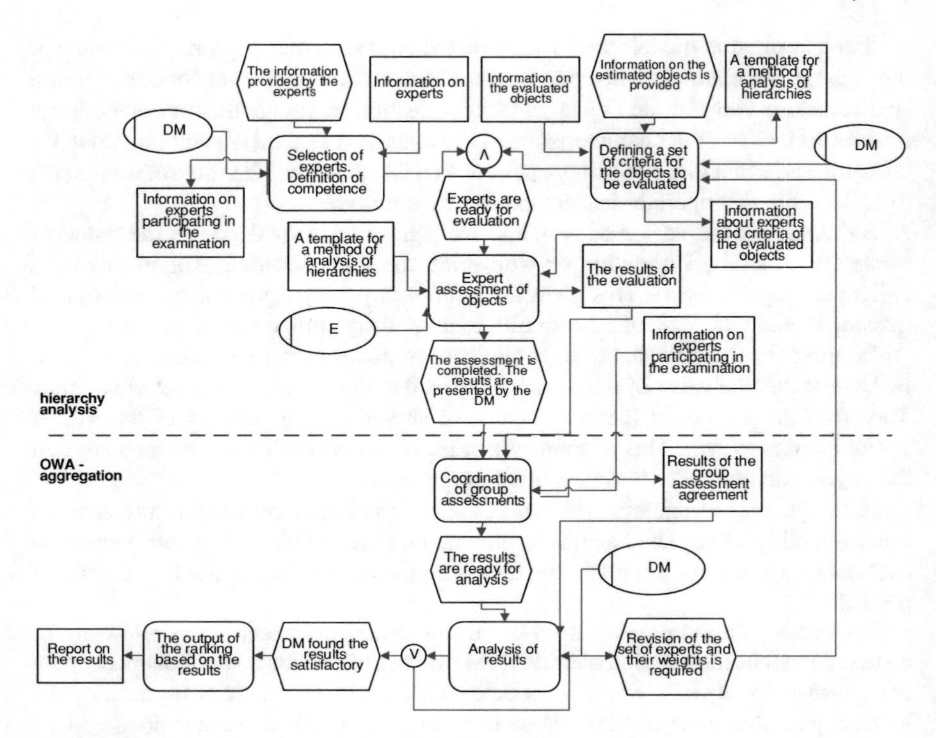

Fig. 1 Interaction scheme for group expert decision making

During the evaluation process, each of the experts is informed about the presence of individual assessments inconsistency. If an expert receives such an information message, it means that he made a mistake in the arrangement of estimates in the matrix of paired comparisons. In this case, the expert can use the system advice and change the estimates so that they are consistent with each other.

The scheme implies that each expert has the right to be informed about the consistency of his/her assessments and at the end of the evaluation process each expert will have no inconsistency in individual assessments.

If we remove the ability of experts to be informed about the presence of inconsistency, then the DM will be transferred to expert data, which can be both coherent and inconsist. This requires that the DM will be able to verify the consistency of individual assessments and then decide on further action.

Figure 2 shows the scheme of interaction for making a group decision, taking into account the lack of informing experts about the inconsistency of individual assessments.

This scheme has all the features of the previous one, except that experts cannot learn about the consistency of their individual assessments. Using the expert data consistency checker, the DM analyses the results and identifies those experts who have mismatched estimates.

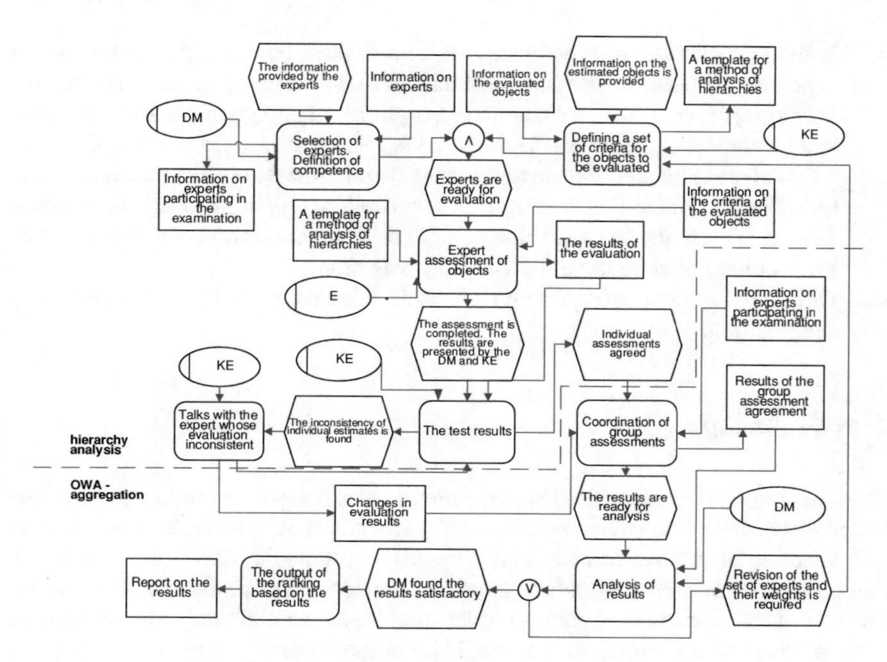

Fig. 2 Interaction scheme for group decision making without informing about inconsistency

The decision is made by the DM on the basis of its own knowledge and experience, because sometimes to save time and money it will be advantageous to choose another method of solving the problem of inconsistency. The advantage of this scheme is to reduce the time of the evaluation process by experts due to the fact that they are not informed about the presence of inconsistency, and therefore do not spend time on revaluation. It also allows the DM to obtain "clean" results that have not been subjected to a process of changing estimates to achieve greater consistency.

Taking into account the above features, authors propose a method and algorithm of group decision making based on the method of analysis of hierarchies for decision-making within the boundaries of inventory filters, taking into account the internal and external consistency of expert data and heterogeneity of the expert staff. The methodology identifies three types of actors: decision maker (DM), knowledge engineer (KE), expert (E).

Step-by-step single examination procedure includes the following steps:

1. KE choose the experts, the DM defines the evaluating criteria and also preliminary estimates of experts;
2. KE within the AHP forming a hierarchy for innovative projects or choose a typical pattern for multiple comparison;
3. E assess the project according to the selected criteria, in parallel with this KE make a technical assessment of experts;

4. Individual matrices of E estimates are coordinated, then group coordination of expert data is performed using linear and OWA-operators, $orness(W_i)$ indicators for aggregation vectors are calculated, a priority aggregation vector is selected and a complex estimate is calculated taking into account $\Delta orness(W_i)$;
5. KE analyzes obtained information, when he identificate critical inconsistencies he make the decision on the assessment procedure repetition or the inconsistent data exclusion, the identification of critical compensation vector properties of the specified final aggregation vector aggregation;
6. DM draws up the final report and makes decision on the results of the collective examination.

4 Conclusions

The most poorly formalized and having prerequisites for system optimization are the initial stages of the innovation process. This stages are the set of searching, initializing and selecting of innovative ideas and proposals. Systematic search, transformation and promotion of innovative ideas should be carried out centrally within the inter-sectoral macro-formation region, in order to systematize and unify this process, as well as to achieve a synergetic effect [23]. On the basis of these aspects, authors propose to formalize and improve the process of data coordination in the group innovative projects evaluation at the initial stages within the framework of scientific and industrial cooperation. The specificity of innovative projects selected for implementation in the framework of such clusters, which determines the general requirements for the system of expert evaluation of such projects: the impossibility of a complete numerical-parametric description of the process, a significant number and heterogeneity of subjects engaged in filtering interests, the lack of relevant information sources for the work; as a consequence of the large amount of heterogeneous social units, the difficulty of determining their dimension, scaling and quantification, interdependence of decision criteria. Taking into account these features, the method and algorithm of group innovative projects evaluation based on AHP is suggested 28. It is taking into account the internal and external consistency of expert data and heterogeneity of the expert staff.

The stage single procedure of evaluation is offered, and it is proved that application of the specified technique will increase efficiency and adaptation of group expert decisions in the innovative stage, will allow to avoid parametrically and organizationally uncertain stages of experts negotiations in the conditions of incompleteness and unreliability of relevant information.

Acknowledgements The research was carried out with the financial support of the «Council for grants of the President of the Russian Federation for state support of young Russian scientists—candidates of science» according to the project MK-4087.2021.2.

References

1. Bril, A., Kalinina, O., Ilin, I.: Small innovative company's valuation within venture capital financing of projects in the construction industry. MATEC Web of Conferences, 106, статья № 08010 (2017)
2. Ilin, I.V., Levina, A.I., Dubgorn, A.S., Abran, A.: Investment models for enterprise architecture (EA) and it architecture projects within the open innovation concept. J. Open Innov.: Technol. Market Complex. **7**(1) (69), 1–18 (2021)
3. Kalimullina, A.M., Yungbludb, V.T., Khodyrevab, E.A.: Characteristic features of innovation project management aimed at university human resource development. Int. J. Environ. Sci. Educ. **11**(9), 2237–2253 (2016)
4. Okorokov, V.R., Kalchenko, O.A.: An innovative project evaluation technique under conditions of information uncertainty. Int. J. Bus. Inform. **10**(2), 180–197 (2015)
5. Maghsoudi, S., Duffield, C., Wilson, D.I.: Innovation evaluation: past and current models and a framework for infrastructure projects. Int. J. Innov. Sci. **7**(4), 281–298 (2015)
6. Zakharov, P.N., Nazvanova, K.V., Posazhennikov, A.A.: Improvement of methodological approaches to the effectiveness assessment of innovation development of regional economy. Lect. Notes Netw. Syst. **57**, 1155–1168 (2019)
7. Alesinskaya, T.V., Arutyunova, D.V., Orlova, V.G., Ilin, I.V., Shirokova, S.V.: Conception BSC for investment support of port and industrial complexes. Acad. Strat. Manag. J. **16**(Special issue 1), 10–20 (2017)
8. Nikolova, L.V., Kuporov, J.Ju, Rodionov, D.G.: Risk management of innovation projects in the context of globalization. Int. J. Econ. Financ. Iss. **5**, 73–79 (2015)
9. Ilin, I., Levina, A., Borremans, A., Kalyazina, S.: Enterprise architecture modeling in digital transformation era. Adv. Intell. Syst. Comput. (AISC) **1259**, 124–142 (2021)
10. Danelyan, Y.: Formal methods of expert assessments. Econ. Stat. Inform. **1**, 183–187 (2015)
11. Santalova, M.S., Lesnikova, E.P., Chudakova, E.A.: Expert models for the evaluation of innovative entrepreneurial projects. Asian Soc. Sci. **11**(20), 119–126 (2015)
12. Pimonov, A., Gorbachev, T.F., Raevskaya, E.: Expert evaluation of innovation projects of mining enterprises on the basis of methods of system analysis and fuzzy logics. In: E3S Web of Conferences The 1st Scientific Practical Conference "International Innovative Mining Symposium (in memory of Prof. Vladimir Pronoza)", no. 15, p. 01021 (2017)
13. Chertina, E.V., Aminul, L.B., Eremenko, O.O.: Decision-making on investment of IT-innovation based on fuzzy expert information. Vestnik Astrakhan State Tech. Univ. **1**, 103–111 (2018)
14. Stevanović, M., Marjanović, D., Štorga, M.: Idea assessment and selection in product innovation—the empirical research results. Tehnički vjesnik. **23**(6), 1707–1716 (2016)
15. Zinovyeva, I.S., Guzeeva, O.G., Sibiryatkina, I.V., Kolesnichenko, E.A.: Methodology of identifying most appealing start-ups for stakeholders. In: 32nd International Business Information Management Association Conference, IBIMA 2018—Vision 2020: Sustainable Economic Development and Application of Innovation Management from Regional expansion to Global Growth, pp. 1787–1796 (2019)
16. Batraga, A., Salkovska, J., Braslina, L., Legzdina, A., Kalkis, H.: New innovation identification approach development matrix. Adv. Intell. Syst. Comput. **783**, 261–273 (2019)
17. Plaskova, N.S., Prodanova, N.A., Oteshova, A.K., Rodionova, L.N., Arlanova, O.I.: Innovative activity of the organization: improving methodological and accounting and analytical support. J. Crit. Rev. **7**(5), 693–696 (2020)
18. Zaytsev, A., Dmitriev, N., Asaturova, Y.: Developing innovative activity management tools as a way to increase the market capitalization of an industrial enterprise. In: 15th European Conference on Innovation and Entrepreneurship, pp. 702–712 (2020)
19. Science, Technology and Innovation Outlook 2021: Times of Crisis and Opportunity. OECD Publishing, Paris (2021)
20. Izotov, A., Rostova, O., Dubgorn, A.: The application of the real options method for the evaluation of high-rise construction projects. E3S Web Conf. **33**, 03008 (2018)

21. Ilin, I.V., Bolobonov, D.D., Frolov, A.K.: Innovative business model as a factor in the successful implementation of IIoT in logistics enterprises. In: Proceedings of the 33rd International Business Information Management Association Conference, IBIMA 2019: Education Excellence and Innovation Management through Vision 2020, pp. 5092–5102 (2019)
22. Statovsky, D.A., Platonov, V.V.: Research of innovation microsystems as an element of innovation activity regulation at the regional level. **11**(217), 44–50 (2016)
23. Ilin, I.V., Izotov, A.V., Shirokova, S.V., Rostova, O.V., Levina, A.I.: Method of decision making support for it market analysis. In: Proceedings of 2017 20th IEEE International Conference on Soft Computing and Measurements, SCM 2017, 7970732, pp. 812–814 (2017)
24. Henderson, R., Cockburn, I.: Measuring competence? Exploring firm effects in pharmaceutical research. Strateg. Manag. J. **15**(8), 63–84 (1994)
25. Gokhberg, L.M., Gracheva, G.A., Ditkovsky, K.A., et al.: Indicators of innovative activity: 2021: statistical collection. National Research. Higher School of Economics, Moscow, Higher School of Economics (2021)
26. Zartha, J.W., Montes, J.M., Vargas, E.E., Palaci, J.C., Hernández, R., Hoyos, J.L.: Methods and techniques in studies related to the Delphi method, innovation strategy, and innovation management models. Int. J. Appl. Eng. Res. **13**, 9207–9214 (2018)
27. Abu-Sarhan, Z.: Application of analytic hierarchy process (AHP) in the evaluation and selection of an information system reengineering projects. Int. J. Comput. Sci. Netw. Secur. **11**(1), 172–176 (2011)
28. Gunduz, M., Alfar, M.: Integration of innovation through analytical hierarchy process (AHP) in project management and planning. Technol. Econ. Dev. Econ. **25**(2), 258–276 (2019)

Threshold Isogeny-Based Group Authentication Scheme

Elena Aleksandrova, Olga Pendrikova, Anna Shtyrkina, Elena Shkorkina, Anastasya Yarmak, and József Tick

Abstract Due to the active evolution of information technologies, there is a need to develop authentication algorithms taking into account the needs of modern multi-user distributed systems of the digital industry and new trends in mathematical problems that ensure their security. One of the perspective structures in post-quantum mathematical era are isogenies of supersingular elliptic curves. The analysis of approaches to group authentication is presented. The dynamic interactive threshold digital signature scheme based on isogenies is proposed, which provides full anonymity and unforgeability. Some security issues of the scheme, including compromising the identity of signature initiator and generating a signature by a user who is not a member of the group, are considered. The scheme can be applied in distributed systems of the digital industry that require authentication, provided that the number of members and interaction structure changes dynamically.

Keywords Supersingular elliptic curves · Isogenies · Binomial coefficients · Threshold signature

E. Aleksandrova (✉) · O. Pendrikova · A. Shtyrkina · E. Shkorkina · A. Yarmak
Peter the Great St. Petersburg Polytechnic University, St. Petersburg, Russia
e-mail: helen@ibks.spbstu.ru

O. Pendrikova
e-mail: pendryak@mail.ru

A. Shtyrkina
e-mail: anna_sh@ibks.spbstu.ru

E. Shkorkina
e-mail: shkorkina.en@edu.spbstu.ru

A. Yarmak
e-mail: yarmak.av@ibks.spbstu.ru

J. Tick
Óbuda University, Budapest, Hungary
e-mail: Tick@uni-obuda.hu

C. Jahn et al. (eds.), *Algorithms and Solutions Based on Computer Technology*,
Lecture Notes in Networks and Systems 387,
https://doi.org/10.1007/978-3-030-93872-7_10

1 Introduction

At present, there is a widespread digitalization of life-forming areas of human activity, including industry. This trend leads to the development of new types of systems characterized by a distributed infrastructure [1–3]. A growing number of devices imposes a number of restrictions on classical approaches to ensuring the cybersecurity of such objects, as a result of which fundamentally new approaches appear, including interdisciplinary ones [4–7]. The use of standard authentication methods is problematic because they do not take into account the distributed architecture that is typical for the objects of the digital industry. Taking into account the above, one of the promising approaches to data authentication is the approach focused on group interaction of nodes [8].

Progress in the field of quantum computing threatens the functioning of public key cryptographic algorithms, the security of most of which is based on the problems of number factorization and discrete logarithm in the cyclic group of prime order. There is a need in fundamentally new algorithms that are resistant to quantum computer attacks [9]. One of the problems, the complexity of solving which on a quantum computer is currently exponential, is the problem of elliptic curves isogeny computation.

A lot of cryptographic schemes based on isogenies of elliptic curves are proposed nowadays: key exchange [10, 11], public key encryption [12], zero-knowledge proofs, different signature schemes (group, undeniable, blind, directed), hierarchical authentication scheme, etc. [13–16]. But there are no threshold isogeny-based schemes. We propose a protocol of group authentication, taking into account the requirements of the minimum acceptable threshold of group members.

2 Materials and Methods

Digital signature is one of the most popular tools that ensure data integrity and authentication, confidentiality, untraceability, non-repudiation, anonymity, and other properties.

A situation may occur in modern digital systems, such as electronic voting systems, blockchain technologies, e-wallets, IoT systems [17], etc., when a message is to be signed by a group of participants, and some verifier would have the opportunity to check it. Group authentication seems to be the best way to solve this problem. The most known are group, ring, multisignature and threshold schemes (Table 1).

These concepts can be combined or supplemented with different properties depending on the requirements of specific practical applications. For example, there are undeniable group and ring signatures, blind threshold signature, ring threshold signature, multisignature, etc.

Group and ring signatures are generated by only one group member, so the consent of the others is not required (they do not even know that they participated in signing

Table 1 The properties of different signature schemes

Signature type	Properties		
	The number of signers	Trusted party	Anonymity of the signer
Group signature	One group member	Yes	No
Ring signature	One group member	No	Yes
Multisignature	All members of the group	No	No
Threshold signature	Any number of group members	Optionally	Yes

procedure). Multisignature is formed by all members of the group [18]. To form a threshold signature, the consent of the threshold number of participants t is sufficient.

In the group signature [19] scheme trusted manager is involved in key generation and distribution, while in the ring and multiple schemes no trusted party is needed. In the classic threshold scheme the trusted party can participate in key distribution and recovery, but you can also do without any dealer at all.

A group signature implies opening manager; therefore, it allows revealing the identity of the signer. For multisignature, this property comes from the definition and structure itself. Ring and threshold signatures allow you to ensure complete anonymity of the signer without making additional changes to the scheme.

For all these types of signatures, there are provably secure algorithms and schemes, so the choice between them is justified by the properties provided and the ease of use. The threshold signature [20] was chosen as the structure under study, since it provides all the necessary properties: the minimum threshold number of participants who agree to the sign a message, full-anonymity of the signers, and the ability to deal without any trusted party.

3 Results

To provide group authentication, a threshold digital signature scheme is proposed, the security of which is due to the complexity of supersingular elliptic curves isogeny problem. The scheme is based on secret sharing protocol [21] and involves four algorithms:

- Parameter generation: given security parameter, public parameters of the scheme are initialized;
- Key generation: given public parameters of the scheme, group public and private keys are formed, as well as a key distribution table based on the values of the

threshold t and the number of group members n, and the group private key is shared between the members;
- Sign: given a group member private key and a part of group public key, the secret is recovered and signature σ of the message m is generated. Algorithm doesn't allow correct signing by the group of less than t users;
- Verify: given group public key, public parameters of the scheme, the message m and digital signature σ, algorithm returns $True$ for the correct signature, and $False$ otherwise.

3.1 Parameter Generation

Public parameters of the scheme are generated as follows:

1. For security parameter λ generate the field characteristic $p = l_A^{e_A} \cdot l_B^{e_B} \cdot l_S^{e_S} \cdot f \pm 1$, where l_A, l_B, l_S are small primes and f is cofactor.
2. Generate supersingular elliptic curve $E(\mathbb{F}_{p^2})$ with the number of points equal to $\left(l_A^{e_A} \cdot l_B^{e_B} \cdot l_S^{e_S} \cdot f\right)^2$.
3. Find a couple of points $\{P_S, Q_S\}$, which are generators of torsion group $E\left[l_S^{e_S}\right]$.
4. Find a couple of points $\{P_{ver}, Q_{ver}\}$, which are generators of torsion group $E\left[l_A^{e_A}\right]$.
5. Generate a point $P_M \in E\left[l_S^{e_S}\right]$ at random.
6. Define hash-function $H : \{0, 1\}^* \to \mathbb{Z}$.

Public parameters are $p, E, \{P_S, Q_S\}, \{P_{ver}, Q_{ver}\}, P_M, H$.

3.2 Key Generation and Distribution

For the group of n members, public and private keys are generated and the private key is shared between all the members to provide a threshold of t participants as follows:

1. Form the key distribution table from the number n og the members and a threshold t.

    a. Compute maximum forbidden subsets $S_0, S_1, \ldots, S_{k-1}$ for the group of n members, where $k = C_n^{t-1}$, and a threshold t as in key sharing schemes.
    b. For i from 0 to $k - 1$
        (1) Set table[i] $= 2^n - 1$, where *table* is key distribution table.
        (2) For j from 0 to $n - 1$ set *table* [i] $=$ *table* [i] $\wedge (1 \ll j)$, if $u_j \in S_i$.

2. Choose random $m_S, n_S \in \mathbb{Z}/l_S^{e_S}\mathbb{Z}, l_S \nmid m_S, n_S$, and generate private isogeny $\varphi_s : E \to E_s$ using the kernel $< m_S P_S + n_S Q_S >$.
3. Choose random $m_{ver}, n_{ver} \in \mathbb{Z}/l_A^{e_A}\mathbb{Z}, l_A \nmid m_{ver}, n_{ver}$, and generate test isogeny $\varphi_{ver} : E \to E_{ver}$ using the kernel $< m_{ver} P_{ver} + n_{ver} Q_{ver} >$.

4. Construct isogeny chain $\varphi_0 \to \varphi_1 \to \cdots \to \varphi_{k-1}$, where $\varphi_i : E_{i-1} \to E_i, E_{-1} = E_{ver}$, choosing generators $P_i, Q_i \in E_i\left[l_r^{e_r}\right], i = 0, \ldots, k, r = \{A, B\}$, and kernel coefficients $m_i, n_i \in \mathbb{Z}/l_r^{e_r}\mathbb{Z}$, non-divisible by l_r, at random.
5. Compute isogeny composition $\psi(P_M) = \varphi_{k-1} \circ \varphi_{k-2} \circ \ldots \circ \varphi_1 \circ \varphi_0 \circ \varphi_{ver}(P_M)$. Group public key is $gpk = \{E_S, (E_i, P_i, Q_i | i = 0, \ldots, k-1), E_{ver}, \psi(P_M), deg\psi\}$.
6. For each group member set user private key $sk_i = \{('sec' : m_S, n_S), ('ver' : m_{ver}, n_{ver}), (w : m_w, n_w)\}$, where $w \in \{0, \ldots, k-1\}$ are such that $table\ [w] \& (1 \ll i) = 1$.

In fact, key distribution is performed in such a way that the recovery of the entire isogeny chain requires the agreed participation of at least t users. This is achieved by constructing distribution table $table$, where bit representation of the element $table\ [i]$ indicates whether the jth user receives isogeny φ_i (if the jth bit of $table\ [i]$ is 1) or not (otherwise).

3.3 *Sign*

To sign a message M on behalf of the group of n members, the user S, initiating the sign process, performs as follows (see Fig. 1).

1. Chooses at least $t-1$ members, who agree to sign this message.
2. Computes the hash of the message M: $h = H(M)$.
3. Generates point $Q \neq P_M$, $Q \in E\left[l_S^{es}\right]$, at random and sets $Q_M = hQ$.
4. From the part of private key $('ver' : m_{ver}, n_{ver})$ and the part of group public key (E, P_{ver}, Q_{ver}) generates isogeny $\varphi_{ver} : E \to E_{ver}$ using the kernel $\langle m_{ver} P_{ver} + n_{ver} Q_{ver} \rangle$ and computes $P_{-1,S} = \varphi_{ver}(P_S), Q_{-1,S} = \varphi_{ver}(Q_S), Q_{-1,M} = \varphi_{ver}(Q_M)$. $E_{-1} = E_{ver}$.

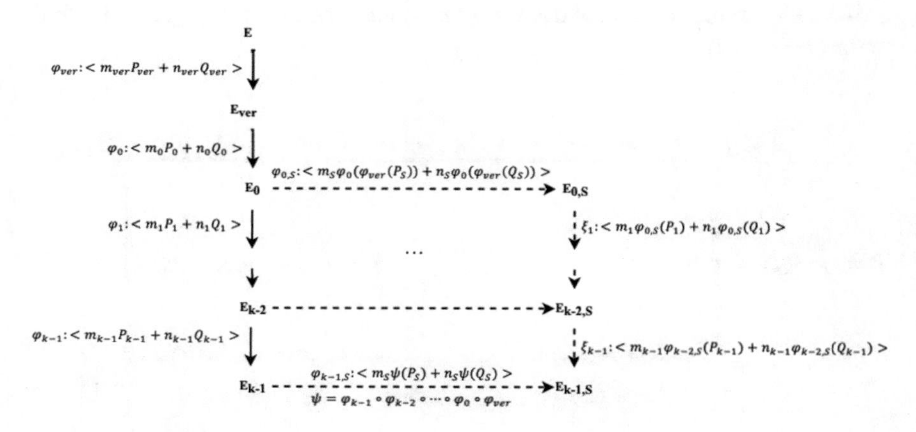

Fig. 1 Generation of threshold signature

5. For i from 0 to $k-1$:

   a. If the user S possesses data for computing isogeny φ_i, then he generates isogeny $\varphi_i : E_{i-1} \to E_i$ using the kernel $\langle m_i P_i + n_i Q_i \rangle$ and computes $\varphi_i(P_{i-1,S}), \varphi_i(Q_{i-1,S}), \varphi_i(Q_{i-1,M})$ from the part of private key $(i : m_i, n_i)$ and the part of group public key (E_{i-1}, P_i, Q_i).
   b. Otherwise, the user S chooses any user u, who possesses data for computing isogeny φ_i, from the key distribution table $table$ and sends him the values $i, P_{i-1,S}, Q_{i-1,S}, Q_{i-1,M}$.
   c. The user u generates isogeny $\varphi_i : E_{i-1} \to E_i$ using the kernel $\langle m_i P_i + n_i Q_i \rangle$ and computes $\varphi_i(P_{i-1,S}), \varphi_i(Q_{i-1,S}), \varphi_i(Q_{i-1,M})$ from the part of private key $(i : m_i, n_i)$ and the part of group public key (E_{i-1}, P_i, Q_i).
   d. The user u generates secret isogeny $\varphi_{i-1,S} : E_{i-1} \to E_{i-1,S}$ using the kernel $\langle m_S P_{i-1,S} + n_S Q_{i-1,S} \rangle$ and computes $\varphi_{i-1,S}(P_i), \varphi_{i-1,S}(Q_i)$ from the part of private key $('sec' : m_S, n_S)$ and the part of group public key $(E_{i-1}, P_{i-1,S}, Q_{i-1,S})$.
   e. The user u generates isogeny $\xi_i : E_{i-1,S} \to E_{i,S}$ using the kernel $\langle m_i \varphi_{i-1,S}(P_i) + n_i \varphi_{i-1,S}(Q_i) \rangle$ from the part of private key $(i : m_i, n_i)$ and the values $\varphi_{i-1,S}(P_i), \varphi_{i-1,S}(Q_i)$.
   f. The user u sets $P_{i,S} = \varphi_i(P_{i-1,S}), Q_{i,S} = \varphi_i(Q_{i-1,S}), Q_{i,M} = \varphi_i(Q_{i-1,M})$.
   g. The user u sends the tuple $\sigma_i = \{E_{i,S}, P_{i,S}, Q_{i,S}, Q_{i,M}\}$ to the user S.
   h. After receiving the tuple σ_i, the user S, using the values $(E_{i,S}, P_{i,S}, Q_{i,S})$ verifies if the user u knows secret isogeny φ_S. Namely, he computes isogeny $\varphi_{i,S} : E_i \to E'_{i,S}$ using the kernel $\langle m_S P_{i,S} + n_S Q_{i,S} \rangle$ and checks the equality $E'_{i,S} = E_{i,S}$.

6. The tuple $\sigma = \{Q, Q_{k-1,M}\}$ is the signature for the message M.

The fact that all the signers are indeed members of the group is provided by a zero-knowledge proof protocol based on a commutative isogeny diagram of elliptic curves (see Fig. 2).

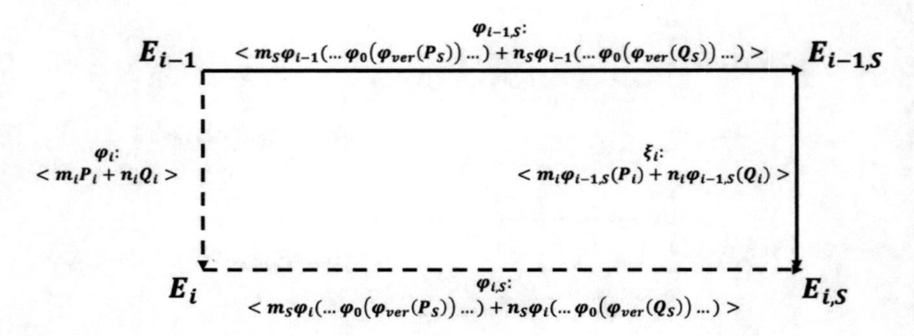

Fig. 2 Commutative diagram

3.4 Verify

In order to verify the signature $\sigma = \{Q, Q_{k-1,M}\}$ for the message M verifier checks the equality for bilinear pairings

$$e_{l_S^{e_S}}\left(\psi(P_M), Q_{k-1,M}\right) \stackrel{?}{=} e_{l_S^{e_S}}(P_M, hQ)^{deg\psi},$$

using a part of group public key $\{\psi(P_M), deg\psi\}$. If it holds then he concludes that the signature is valid, otherwise the signature is invalid.

The verification procedure is based on the following property of bilinear maps:

$$\begin{aligned} e_{l_S^{e_S}}\left(\psi(P_M), Q_{k-1,M}\right) &= e_{l_S^{e_S}}(\psi(P_M), \psi(Q_M)) = e_{l_S^{e_S}}(\psi(P_M), \psi(hQ)) \\ &= e_{l_S^{e_S}}(P_M, hQ)^{deg\psi}. \end{aligned}$$

4 Discussion

There is an isomorphism between the group of isogeny classes and the group of ideal classes of elliptic curves. It is known for non-supersingular elliptic curves, that the ring of endomorphisms of the curve is commutative and, accordingly, the group of ideal classes is abelian. That's why, it became possible to attack non-supersingular elliptic curve cryptosystems with an algorithm that solves Computational SuperSingular Isogeny problem (CSSI) on quantum computer subexponentially. In turn, the ring of supersingular elliptic curve endomorphisms is not commutative, and a set of ideal classes is not a group. So, this attack is not possible, and the problem of isogeny computation between supersingular elliptic curves is resistant to the attacks on quantum computer.

Note, that for $t = 1$ the scheme proposed is just a group signature, where each user can sign a message on behalf of the group; at the same time, this scheme provides anonymity. But in this case, it cannot be used for group signature generation, as, what is unacceptable, all the group members have the same secret key.

Compromising the identity of signature initiator. Given public parameters, group public key and the signature $\sigma = \left(Q, Q_{k-1,M}\right)$ it is impossible to disclose the identity of signature initiator, as the secret chain of isogenies is completely restored while signing, regardless not only of initiator, but also of the identities of other users. The known parameters do not give any information that would identify the group members. So, this scheme gives full-anonymity, without any compromising the identity of signature initiator and the other users.

User's private key recovery. User's private key is a part of secret isogeny chain. While signing, the initiator selects the participant at random and transmits him the current information. This participant, in turn, forms the next partial signature.

So, attacker is to find all C_{n-1}^{t-1} isogenies to compromise the user. Given public parameters $(p, E, \{P_S, Q_S\}, \{P_{ver}, Q_{ver}\}P_M, H)$ and group public key $(gpk = \{E_S, (E_i, P_i, Q_i | i = 0, \dots, k-1), E_{ver}, \psi(P_M), deg\psi\})$ private key recovery is reduced to the problem of finding kernels of isogeny φ_i. At the same time, if an attacker intercepts data transmitted between the initiator and the signer, he can only get images $\varphi_i(Q_{i-1,M}), \varphi_i(P_{i-1,S}), \varphi_i(Q_{i-1,S})$ and corresponding preimages. This problem is computationally hard, so, it is impossible to recover from public and intercepted information even a part of the user's private key.

Generating a signature by a user who is not a member of the group. The concept of threshold signature allows signing a message by more than t users. Theoretically, the signature initiator, even without being a member of the group, can try to form a correct signature, choosing not $t-1$ but t users, who can restore the secret chain of isogenies without his participation. To prevent that, one more isogeny, $\varphi_{ver} : E \to E_{ver}$, was included into isogeny chain, which kernels are known only to the group members. Illegal user, wishing to initiate signing, is to solve the same problem as in the attack above, since there is also the problem of finding isogeny between known curves, given public parameters and the corresponding points images. Thus, it is impossible to form a signature by a participant who is not a member of the group.

Integrating the intruder's signature into the chain. Suppose, illegal user tries to integrate his own partial signature into the chain before the ith step. He receives data from the previous partial signature $\sigma_{i-1} = \left(Q_{i-1,M}, P_{i-1,S}, Q_{i-1,S}\right)$ from the signature initiator. To form a signature, it is necessary that the list of elliptic curves in group public key contains information about the elliptic curve E_{int}, moreover, this curve must have the same properties as E_{i-1}, because the next user, who gets partial signature, is to compute the images $\varphi_i(Q_{i-1,M}), \varphi_i(P_{i-1,S}), \varphi_i(Q_{i-1,S})$ as if isogeny φ_i maps E_{int} into E_i (when $E_i = E_{k-1}$, this requirement is not mandatory). Also, this user must prove the knowledge of secret isogeny φ_s to the signature initiator, by computing isogeny $\varphi_{int,S} : E_{int} \to E_{int,S}$ ant sending elliptic curve $E_{int,S}$ to the initiator. To do this, he must know the kernel of given isogeny, in particular the values of (m_S, n_S), which are known only to the legitimate users. If the attacker is a group member and has these values in his private key, he will be able to prove knowledge of the secret isogeny to the initiator and form a partial signature, which he will pass to the initiator for further signing and verification. But integrating the additional isogeny into the chain forms composition

$$\psi^* = \varphi_{k-1} \circ \varphi_{k-2} \circ \dots \circ \varphi_{int} \circ \dots \circ \varphi_1 \circ \varphi_0 \circ \varphi_{ver},$$

where there are $k+2$ isogenies (because of the attacker's one) instead of $k+1$. So, if the degree of the initial composition is $deg\psi$, the degree of the resulting composition is $deg\psi^* = deg\psi * deg\varphi_{int}$, and even if the signature is successfully generated, the equality

$$e_{l_S^{e_S}}\left(\psi(P_M), \psi^*(hQ)\right) \stackrel{?}{=} e_{l_S^{e_S}}(P_M, hQ)^{deg\psi}$$

does not hold during the verification process, because the group public key includes the image of legitimate chain $\psi(P_M)$ and the degrees of compositions are different: $deg\psi \neq deg\psi^*$.

In addition to the fact that signing and verifying algorithms make it impossible to carry out this attack, it also does not make practical use, since the signature is formed on behalf of entire group, provided full anonymity of all the signing participants, and integrating additional isogeny into the chain will not give the attacker any useful information.

Message forgery. For the attacker who wants to spoof the message, the problem of message forgery is reduced to the problem of finding hash-function collision. Let an attacker wants to replace the message M, with $\sigma = \left(Q, Q_{k-1,M}\right)$, with his own message M' so that verification ratio is performed. Hash-function is public, so he can compute the value $h' = H\left(M'\right)$. While signing, the image of the point $Q_M = hQ$ is computed, so attacker is to select such a value $Q_{M'} = h'Q$, hat verification ratio $e_{l_S^{e_S}}\left(\psi(P_M), Q_{k-1,M}\right) \stackrel{?}{=} e_{l_S^{e_S}}\left(P_M, h'Q\right)^{deg\psi}$ is performed. The value Q is a part of the signature and cannot be replaced, the attacker is to find M' for which $h = h'$, i.e., to solve the hard problem of finding hash-function collisions.

5 Conclusions

The analysis of approaches to group authentication shows that it is the threshold signature that ensures full anonymity of signers and guarantees participation of a certain number of users in signing, what can be of practical use in distributed systems of digital industry.

The security of the most modern threshold signature schemes is guaranteed by the problems that seem to be not so hard in future. Elliptic curve isogenies not only relate to post-quantum problems, but also provide new functionality for building distributed authentication protocols.

The scheme proposed is interactive, full anonymous, correct, unforgeable and dynamic. It can be modified to reduce the user's private key and group public key lengths, and, as a result, to reduce the time of key and signature generation. Another interesting area is adaptation of other post-quantum group authentication mechanisms to the tasks of distributed systems security.

Acknowledgements The reported study was funded by RFBR, project number 20-37-90106.

References

1. Poltavtseva, M.A.: Evolution of data management systems and their security. In: International Conference on Engineering Technologies and Computer Science (EnT), pp. 25–29 (2019)
2. Vasil'ev, Y.S., Zegzhda, D.P., Poltavtseva, M.A.: Problems of security in digital production and its resistance to cyber threats. Aut. Control Comp. Sci. **52**, 1090–1100 (2018)
3. Stepanova, T., Pechenkin, A., Lavrova, D.: Ontology-based big data approach to automated penetration testing of large-scale heterogeneous systems. In: ACM International Conference Proceeding Series, pp. 142–149 (2015)
4. Zegzhda, D., et al.: Cyber attack prevention based on evolutionary cybernetics approach. Symmetry **12**(11), 1931 (2020)
5. Lavrova, D., Zaitceva, E., Zegzhda, P.: Bio-inspired approach to self-regulation for industrial dynamic network infrastructure. In: CEUR Workshop Proceedings, vol. 2603, pp. 34–39 (2019)
6. Pavlenko, E., Zegzhda, D., Shtyrkina, A.: Criterion of cyber-physical systems sustainability. In: CEUR Workshop Proceedings, vol. 2603, pp. 60–64 (2019)
7. Kalinin, M., Krundyshev, V.: Sequence alignment algorithms for intrusion detection in the internet of things. Nonlinear Phenom. Complex Syst. **23**(4), 397–404 (2020)
8. Aleksandrova, E.B., Shtyrkina, A.A., Yarmak, A.V.: Post-quantum group-oriented autentication in IoT. Nonlinear Phenom. Complex Syst. **23**(4), 405–413 (2020)
9. Aleksandrova, E., Shtyrkina, A., Iarmak, A.: Post-quantum primitives in information security. Nonlinear Phenom. Complex Syst. **22**(3), 269–276 (2019)
10. De Feo, L., Jao, D., Plut, J.: Towards quantum-resistant cryptosystems from supersingular elliptic curve isogenies. https://eprint.iacr.org/2011/506. Accessed 2021/04/29
11. Aleksandrova, E., Shkorkina, E., Kalinin, M.: Organization of the quantum cryptographic keys distribution system for transportation infrastructure users. Autom. Control. Comput. Sci. **53**(8), 969–971 (2019)
12. Rostovtsev, A., Stolbunov, A.: Public-Key Cryptosystem Based on Isogenies. https://eprint.iacr.org/2006/145. Accessed 2021/04/29
13. Srinath, M.S.: Isogeny-based quantum-resistant undeniable blind signature scheme. Int. J. Netw. Secur. **20**(1), 9–18 (2018)
14. Jao, D., Soukharev, V.: Isogeny-based quantum-resistant undeniable signatures. In: International Workshop on Post-Quantum Cryptography, pp. 160–179, Springer, Cham (2014)
15. Aleksandrova, E.B., Shtyrkina, A.A.: Directed Digital Signature on Isogenies of Elliptic Curves. Aut. Control Comp. Sci. **52**, 1059–1064 (2018)
16. Aleksandrova, E.B., Shkorkina, E.N.: Using undeniable signature on elliptic curves to verify servers in outsourced computations. Aut. Control Comp. Sci. **52**, 1160–1163 (2018)
17. Mahalle, P., Prasad, N., Prasad, R.: Threshold cryptography-based group authentication (TCGA) scheme for the Internet of Things (IoT). In: 2014 4th International Conference on Wireless Communications, Vehicular Technology, Information Theory and Aerospace & Electronic Systems (VITAE), pp. 1–5. IEEE (2014)
18. Itakura, K.: A public-key cryptosystem suitable for digital multisignatures. NEC J. Res. Dev. 71(1983)
19. Chaum, D., Van Heyst, E.: Group signatures. Workshop on the Theory and Application of of Cryptographic Techniques, pp. 257–265. Springer, Berlin, Heidelberg (1991)
20. Shoup, V.: Practical threshold signatures. In: International Conference on the Theory and Applications of Cryptographic Techniques, pp. 207–220, Springer, Berlin, Heidelberg (2000)
21. Ito, M., Saito, A., Nishizeki, T.: Secret sharing scheme realizing general access structure. Electr. Commun. Japn. (Part III: Fundamental Electronic Science) **72**(9), 56–64 (1989)

An Approach for the Robust Machine Learning Explanation Based on Imprecise Statistical Models

Lev Utkin, Vladimir Zaborovsky, Vladimir Muliukha, and Andrei Konstantinov

Abstract An approach for a robust modification of some explanation methods by using imprecise statistical models is proposed. The imprecise models produce sets of probability distributions such that every distribution from the sets can be a candidate for constructing a robust explanation model. Two types of explanation models will be considered and modified. The first type is based on modification of the factual local explanation method like LIME where the linear approximation is implemented to explain the black-box model. This type of models aims to search for important features which impact on the prediction of an example of interest. The second type uses counterfactual explanation which is extended to be robust. We assume that outcomes of the black-box model compose a set of values, for example, the class probabilities as a prediction vector of an input example. An example of the imprecise contaminated model is given. It is shown how the maximin optimization problem implementing the explanation method is reduced to the standard linear or quadratic optimization problems.

Keywords Machine learning · Explainable intelligence · XAI · Counterfactuals · Imprecise statistical models · Optimization

1 Introduction

We observe a rapid development of new methods and models of artificial intelligence, in particular, of deep machine learning models nowadays. A success in applying machine learning models to various applied tasks, especially to medicine [1], meets an important problem of interpreting or explaining the corresponding predictions provided by deep learning models [2]. In other words, the problem is to explain why we get a certain model prediction for a certain patient or for a group of patients, what features or characteristics of the patients, for example, temperature, blood pressure, etc. significantly impact on the obtained prediction, what features are responsible for

L. Utkin (✉) · V. Zaborovsky · V. Muliukha · A. Konstantinov
Peter the Great St.Petersburg Polytechnic University, Saint-Petersburg, Russia
e-mail: utkin_lv@spbstu.ru

C. Jahn et al. (eds.), *Algorithms and Solutions Based on Computer Technology*,
Lecture Notes in Networks and Systems 387,
https://doi.org/10.1007/978-3-030-93872-7_11

the prediction. This problem stems from the fact that a doctor, stating a diagnosis and making decisions related to a disease treatment, has to have an explanation of the stated diagnosis in order to choose a corresponding treatment. However, the most efficient models, for example, deep neural networks, are black-box that is only their input data and predictions are known for users, but it is difficult or just impossible to be aware what features and their values in input data make the models actually arrive at their decisions [3].

The problem concerns not only with medical applications. A lot of applied task solved by means of machine learning models require to have some interpretation or explanation of their predictions or decisions. This implies that explanations of predictions provided by a machine learning model can help users to better understand the obtained results. Explanations should be an important option of many machine learning models especially in such applied areas as medicine, construction engineering, control, etc. Importance of explanations led to development of various methods providing tools for getting explanations of predictions [4–7].

All explanation methods can be divided into two large groups defined by a number of examples for explanation: global and local. Methods from the first group explain the black-box model on the whole dataset or its part whereas local methods derive explanation locally around a test example, for example, in medicine, they explain an inferred disease of a certain patient, but not all patients. It should be noted that the local methods are of the most interest in many applications. Therefore, we focus on local explanations below. In fact, most local models can be represented as simple meta-models, including linear models, decision trees, which approximate the black-box model. The linear model for explanation is used in one of the most popular method called the Local Interpretable Model-agnostic Explanations (LIME) [8], which uses understandable linear models to locally approximate the predictions of black-box models. The linear assumption of the explanation model implies that coefficients of the corresponding variables in the linear function approximating the black-box model at the point of interest (the patient) can be viewed as quantitative impacts on the prediction.

The explanation methods can be also divided into two groups defined by the simplicity of explanations for users. They are defined by the human-friendly character of explanations. The first group consists of methods which try to explain by answering the direct (factual) question: "why did the model do that" or "what is a reason of the disease". Answers on these questions explain why a certain prediction was obtained and what features of an example led to a decision. However, users of the machine learning models often think in the contrastive or counterfactual manner by getting answers on the questions: "why this prediction was made instead of another prediction" or "what change of the patient characteristics would lead to a different disease". Often, we are not directly interested in all important features that led to a prediction, but we would like to know features that need to change so that the prediction would also change. The corresponding explanations are called counterfactual explanations or counterfactuals [9], and they form the second group. These explanations are often more intuitive and human-friendly than factual explanations from the first group.

It is important to point out that a lot of explanation methods have been developed in accordance with the above simplified and incomplete classification of methods. However, one of the crucial difficulties of many methods is a lack of robustness to cases of a small amount of training data or outlier data. The main problem here is that prediction of the black-box model may by unreliable and inaccurate. As a result, the explanation may be also incorrect. One of the ways to overcome this difficulty is to develop robust models.

A lot of robust machine learning models assume that each data point or example in a training set can move around within the Euclidean ball. These models stem from perturbations of the training data due to corrupted data, small amount of data, etc., which are usually described by an additive noise. The main assumption underlying these models is that all examples in the training set have equal probabilities or weights. This assumption may be violated when the training set is small. In order to relax it, it is natural to assume that probabilities of examples may be contaminated or changed. One of the statistical models taking into account these changes is the imprecise ε-contaminated model [8], which produces a set of probability distributions over examples of the training set. According to the model, a precise "true" probability distribution over the training set is unknown, but it is known that it belongs to the set of distributions. So, in contrast to the approach with perturbations of the training data, we assume that each probability assigned to every data point can move around within a set of probabilities under certain restrictions defined by the imprecise statistical model.

The above approaches cannot be applied to explanation models because they are trained on the perturbed examples or generated dataset, but not on the original dataset. This peculiarity differs the explanation models from the black-box models. Therefore, we propose to consider a general robust modification of explanation methods by using imprecise statistical models and the corresponding sets of probability distributions. Moreover, two types of models will be considered. The first type is based on modification of the factual local explanation method like LIME where the linear approximation is implemented to explain the black-box model. The second type uses counterfactual explanation which is extended to be robust. We assume that outcomes of the black-box model is a set of values, for example, the class probabilities as a prediction vector of an input example. The main idea behind the proposed modification is to assume that the obtained class probabilities is imprecise and they are viewed as one of the probability distributions which are produced by the imprecise statistical model, for example, by the imprecise ε-contaminated model.

2 Materials and Methods

First, we briefly consider the explanation method LIME [8], which is one of the popular methods. Suppose that a black-box model implements a function $f(\mathbf{x})$, where $\mathbf{x}$ is the input feature vector consisting of d features. The set of values of the function f depends on the problem solved, for example, if the binary classification is considered, then the function takes values from the set {1,2}. The basic idea behind LIME is to approximate the black-box model (or the function f) with a simple function $g(\mathbf{x})$ in the vicinity of the point of interest $\mathbf{x}_0$, whose prediction by means of f has to be explained. The important condition for the function $g(\boldsymbol{x})$ is that it has to belong to a set of explanation models G, for example, it is the linear function, i.e., $g(\mathbf{x}) = \mathbf{bx}$, where $\mathbf{b}$ is a vector of coefficients which has to be found in order to indicate the important features; φ is some function which is defined by our assumptions about function g and its output. According to LIME, a new dataset consisting of N perturbed samples $\mathbf{x}_k$ is generated to construct the function g, $k = 1,\ldots, N$. Predictions corresponding to the perturbed samples are nothing else but outcomes of the explained black-box model, that is $f(\mathbf{x}_k)$. Moreover, every generated sample is assigned by weight w_k which is defined for point $\mathbf{x}_k$ in accordance with its proximity to the point of interest $\mathbf{x}_0$ by using a distance metric, for example, the Euclidean distance.

The explanation model in accordance with LIME is trained on new generated samples by solving the following optimization problem:

$$\underset{g \in G}{\operatorname{argmin}} L(f, g, w_{\boldsymbol{x}}) + R(g).$$

Here L is a loss function, for example, mean squared error, which measures how the explanation function g is close to the prediction of the black-box model f; $R(g)$ is the model complexity.

The resulting local linear model $g(\mathbf{x}_0)$ can be obtained by solving the above optimization problem. This model explains by analyzing its coefficients, i.e., large coefficients correspond to important features responsible for the prediction $f(\mathbf{x}_0)$.

3 Results

Let us briefly consider a definition of the counterfactual explanation proposed by Wachter et al. [7]. According to [4], a counterfactual explanation of a prediction can be defined as the smallest change to the feature values of an input original example that changes the prediction to a predefined outcome.

We again have the prediction function $f(\mathbf{x})$. A counterfactual denoted as $\mathbf{z}$ for a given input feature vector $\boldsymbol{x}$ is computed by solving the following optimization problem:

$$\min_{z \in \mathbf{R}^m} L(f(\mathbf{x}), f(\mathbf{z})) + C\theta(\mathbf{z}, \mathbf{x}),$$

where $L(f(\mathbf{x}), f(\mathbf{z}))$ denotes a loss function which establishes a relationship between the black-box model predictions $f(\mathbf{x})$ and $f(\mathbf{z})$; $\theta(\mathbf{z},\mathbf{x})$ is a penalty term for deviations of $\mathbf{z}$ from the original input $\mathbf{x}$, which is defined by using a distance between $\mathbf{z}$ and $\mathbf{x}$, for example, the Euclidean distance; $C > 0$ denotes the regularization strength.

The function $L(f(\mathbf{x}), f(\mathbf{z}))$ encourages the prediction of $\mathbf{z}$ to be different in accordance with a certain rule than the prediction of the original point $\mathbf{x}$, i.e., the prediction of $\mathbf{z}$ has to be different from the prediction of $\mathbf{x}$. The penalty term $\theta(\mathbf{z},\mathbf{x})$ encourages to minimize the distance between $\mathbf{z}$ and $\mathbf{x}$ with the aim to find nearest counterfactuals to $\mathbf{x}$.

The above optimization problem can be extended by including additional terms which restrict the distance between $\mathbf{z}$ and $\mathbf{x}$. In particular, there are the counterfactual explanation algorithms that use a term which makes counterfactuals close to the observed data. It can be done, for example, by minimizing the distance between the counterfactual $\mathbf{z}$ and the k nearest observed data points [11] or by minimizing the distance between the counterfactual $\mathbf{z}$ and the class prototypes [12].

4 Discussion

Robust models have been widely used for solving machine learning problems in order to relax some strong assumptions underlying the standard classification models [13]. However, they are used to robustify the machine learning models themselves. The same can be done for the explanation meta-model. Therefore, our aim is to develop a general algorithm for explaining the machine learning survival models, which could be robust to the small amount of training data and to outliers in the training or testing sets.

Robust models based on imprecise probabilities apply different strategies defined by a way of selecting a probability distribution of the class probabilities from a set of distributions produced by an imprecise model. Therefore, in order to construct a robust model, we have to consider main ways for selecting the probability distribution. One of the ways is to select the "worst" probability distribution providing the largest value of the expected loss. It corresponds to the maxim in (pessimistic) strategy in decision making, which can be interpreted as an insurance against the worst case because it aims at minimizing the expected loss in the least favorable case [14]. This is the most popular strategy in the robust modelling. Another way is to select a distribution which minimizes the expected loss and corresponds to the minim in (optimistic) strategy which cannot be called robust. However, it is interesting as another extreme strategy.

It should be noted that the loss function L is the weighted distance between functions f and g defined by the metric L_p metric. Suppose that the function $f(\mathbf{x}_k)$ has the outcome in the form of class probability vector $\mathbf{p}_k$. Denote the outcome of the function $g(\mathbf{x}_k)$ as $\mathbf{q}_k$. In fact, $\mathbf{q}_k$ is the vector of the approximated class probabilities obtained by means of the explanation linear model. Moreover, we assume that the vector $\mathbf{p}_k$ is unknown the set of probability distributions P has to be considered instead of this vector. For example, if we use the imprecise ε-contaminated model, then the set P is of the form:

$$P(\varepsilon, \mathbf{p}_k) = \{\mathbf{p} : (1 - \varepsilon)\mathbf{p}_k + \varepsilon \boldsymbol{r}\}.$$

Here ε is the tuning contamination parameter; $\boldsymbol{r} = (r_1, \ldots, r_m)$ is an arbitrary probability distribution and $r_1 + \cdots + r_m = 1$. The rate ε reflects how "close" we feel that $\mathbf{p}$ must be to $\mathbf{p}_k$.

For generality, we will consider functions h_i, $i = 1, \ldots, d$, of features instead of the features themselves, i.e., we study the linear model

$$\mathbf{b} \cdot \mathbf{h}, \text{ where } \mathbf{h} = (h_1(x_1), \ldots, h_d(x_d))^{\mathrm{T}}.$$

So, every generated point $\boldsymbol{x}_k$ produces the set of distributions $P(\varepsilon, \boldsymbol{p}_k)$. By returning to the optimization problem written for LIME, by using the maximin strategy and the distance between functions f and g, we write the following optimization problem:

$$\max_{\mathbf{p}_k \in P(\varepsilon, \mathbf{p}_k)} \min_{\mathbf{b}} \sum_{k=1}^{N} w_k \left\| \mathbf{p}_k - \mathbf{B}\mathbf{h}_k \right\|_p + R(g).$$

Here $\mathbf{B}$ is the matrix of coefficients. We use the $m \times d$ matrix $\mathbf{B}$ instead of the vector $\mathbf{b}$ as it is made in LIME because every row of the matrix corresponds to a certain class. As a result, the above maximin optimization problem can be regarded as a way for robust computing the coefficients $\mathbf{B}$ which define the important features.

In order to simplify the above optimization problem, we suppose that $R(g) = 0$. Moreover, we take the L_1 distance, i.e., $p = 1$. In this case, we get

$$\max_{\boldsymbol{p}_k \in P(\varepsilon, \boldsymbol{p}_k)} \min_{\mathbf{b}} \sum_{k=1}^{N} w_k \sum_{j=1}^{m} \left| p_{k,j} - q_{k,j} \right|.$$

Let us denote $z_{k,j} = \left| p_{k,j} - q_{k,j} \right|$ and $\mathbf{z}_k = (z_{k,1}, \ldots, z_{k,m})$. Then we can rewrite the problem as follows:

$$\max_{\boldsymbol{p}_k \in P(\varepsilon, \boldsymbol{p}_k)} \min_{\mathbf{b}, \mathbf{z}_k} \sum_{k=1}^{N} w_k \sum_{j=1}^{m} z_{k,j}.$$

subject to

$$z_{k,j} \geq p_{k,j} - q_{k,j},\ j = 1, \ldots, m,\ k = 1, \ldots, N,$$

$$z_{k,j} \geq -p_{k,j} + q_{k,j},\ j = 1, \ldots, m,\ k = 1, \ldots, N.$$

Let us fix the matrix of coefficients $\mathbf{B}$. It can be seen from the above problem that maximization with respect to the distribution $\mathbf{p}_k$ does not depend on maximization with respect to $\mathbf{p}_l$, where $l \neq k$. This implies that the maximization problem can be separately solved for every k. Hence the objective function can be rewritten as:

$$\min_{\mathbf{b},z_k} \sum_{k=1}^{N} w_k \max_{\mathbf{p}_k \in P(\varepsilon,\mathbf{p}_k)} \sum_{j=1}^{m} z_{k,j}.$$

One can see that optimization problems are linear with $\boldsymbol{p}_k$, but the objective function depends on $\mathbf{B}$ through $q_{k,j}$. Therefore, the whole problem cannot be directly solved by well-known methods. To overcome this difficulty, note, however, that all vectors $\mathbf{p}_k$ belong to the set of distributions $P(\varepsilon, \mathbf{p}_k)$. This set has T extreme points denoted as $\boldsymbol{\pi}_k^{(i)} = \left(\pi_{k,1}^{(i)}, \ldots, \pi_{k,m}^{(i)}\right)$, $i = 1, \ldots, T$. According to some general results from linear programming theory, an optimal solution to the above problem is achieved at extreme points of the simplex, and the number of its extreme points is T.

In order to take into account the restrictions of the probability distributions, we can simply extend constraints for $z_{k,j}$ by adding the following constraints:

$$z_{k,j} \geq \pi_{k,1}^{(i)} - q_{k,j},\ j = 1, \ldots, m,\ k = 1, \ldots, N,\ i = 1, \ldots, T,$$

$$z_{k,j} \geq -\pi_{k,1}^{(i)} + q_{k,j},\ j = 1, \ldots, m,\ k = 1, \ldots, N,\ i = 1, \ldots, T.$$

Indeed, the variable $z_{k,j}$ cannot be smaller than all expressions on the right hand of inequalities. With this subtle technique, we realize the maximization problem of the sum of $z_{k,j}$ over $\mathbf{p}_k \in P(\varepsilon, \mathbf{p}_k)$. Finally, we get the following optimization problem:

$$\min_{\mathbf{b},z_k} \sum_{k=1}^{N} w_k \sum_{j=1}^{m} z_{k,j},$$

subject to

$$z_{k,j} \geq \pi_{k,1}^{(i)} - \mathbf{b}_j \mathbf{h}_k,\ j = 1, \ldots, m,\ k = 1, \ldots, N,\ i = 1, \ldots, T,$$

$$z_{k,j} \geq -\pi_{k,1}^{(i)} + \mathbf{b}_j \mathbf{h}_k,\ j = 1, \ldots, m,\ k = 1, \ldots, N,\ i = 1, \ldots, T.$$

Here $\mathbf{b}_j$ is the j-th row of the matrix $\mathbf{B}$; $\mathbf{h}_k$ is the vector of functions $h_1(x_{k,1}), \ldots, h_d(x_{k,d})$ of $\mathbf{x}_k$.

In sum, we have obtained the standard linear optimization problem whose solution does not meet any difficulties. It can be solved by one of the well-known methods.

If to use the regularization term $R(g)$ of the form $\|\mathbf{B}\|_2^2$, then we get the standard quadratic optimization problem which can be also solved by well-known methods.

In the case of the counterfactual explanation, the problem is similarly solved. The optimization problem for computing the counterfactual $\mathbf{z}$ becomes

$$\max_{\substack{p \in P(\varepsilon, \mathbf{p}) \\ \mathbf{q}_k \in P(\varepsilon, \mathbf{q}_k)}} \min_{\mathbf{z}_k \in \mathbf{R}^m} \sum_{j=1}^{m} \left| p_j - q_{k,j} \right| + C\theta(\mathbf{z}, \mathbf{x}).$$

Here $\mathbf{z}_k$ are candidates for the counterfactual explanation generated around the point $\mathbf{x}$. It can be seen from the above optimization problem that we use now many imprecise ε-contaminated models. The model $P(\varepsilon, \mathbf{p})$ is used to take into account the imprecision of $f(\mathbf{x}) = \mathbf{p}$. The models $P(\varepsilon, \mathbf{q}_k)$ are used to implement robustness of the class probabilities $\mathbf{q}_k$ produced by generated points $\mathbf{z}$. It should be noted that the term $\theta(\mathbf{z}, \mathbf{x})$ does not depend on the class probability distributions. The above maximin optimization problem can be solved in two ways. The first way is to again use extreme points of $P(\varepsilon, \mathbf{q}_k)$ and $P(\varepsilon, \mathbf{p})$ in order to remove the maximization problem. This way has been considered above. The second way is to replace the maximization problem by the dual one which is a minimization problem. As a result, we get the overall problem.

It is important to note that the above optimization problem is rather complex due to a huge number of constraints produced by combinations of extreme points $P(\varepsilon, \mathbf{q}_k)$ and $P(\varepsilon, \mathbf{p})$. This difficulty can be partially overcome by means of considering the dual problem (the second way). Another way is to use an approximate solution when it is assumed that predictions of all candidates $\mathbf{z}_k$ for the counterfactual are precise. In this case, the maximin problem becomes similar to the same optimization problem derived for the case of the robust modification of LIME.

5 Conclusions

A general approach for robust modification of the explanation methods has been presented in the paper. Its basic idea is to introduce imprecise statistical models for class probabilities which are regarded as outcomes of the explained model. We have shown the incorporation of one imprecise model, namely, the imprecise ε-contaminated models. However, there exists several models with interesting properties, for example, the imprecise pari-mutuel model [10], the constant odds-ratio model [10], the Kolmogorov–Smirnov bounds [15], which can be also used for getting robust explanation models. This is a direction for further research. Moreover, the proposed approach has been considered for the class probability distributions as one of the possible outcomes. Actually, the approach in the same way with small changes can be applied to different outcomes. It is very interesting to note that the approach can be also applied to explanation of the neural network layers. Indeed, normalized activations of units belonging to a layer can be regarded as the class probabilities. This trick allows us to determine important features at layers of the neural network and to observe the important feature change with layers.

We did not provide numerical results of the approach because the paper aimed to give a general scheme which can be afterwards detailed with specific applications and examples.

Acknowledgements The reported study was funded by RFBR according to the research project № 19-29-01004.

References

1. Holzinger, A., Langs, G., Denk, H., Zatloukal, K., Muller, H.: Causability and explainability of artificial intelligence in medicine. WIREs Data Min. Knowl Discov. **9**, e1312 (2019)
2. Ilin, I., Levina, A., Lepekhin, A., Kalyazina, S.: Business requirements to the IT architecture: a case of a healthcare organization. Adv. Intell. Syst. Comput. **983**, 287–294 (2019)
3. Moreira, M.W.L., Rodrigues, J.J.P.C., Kumar, N., Saleem, K., Illin, I.V.: Postpartum depression prediction through pregnancy data analysis for emotion-aware smart systems. Inf. Fusion **47**, 23–31 (2019)
4. Adadi, A., Berrada, M.: Peeking inside the black-box: a survey on explainable artificial intelligence (XAI). IEEE Access **6**, 52138–52160 (2018)
5. Guidotti, R., Monreale, A., Ruggieri, S., Turini, F., Giannotti, F., Pedreschi, D.: A survey of methods for explaining black box models. ACM Comput. Surv. **51**, 1–42 (2019). Article 93
6. Molnar, C.: Interpretable Machine Learning: A Guide for Making Black Box Models Explainable (2019). https://christophm.github.io/interpretable-ml-book/
7. Murdoch, W.J., Singh, C., Kumbier, K., Abbasi-Asl, R., Yua B.: Interpretable machine learning: definitions, methods, and applications (2019). arXiv:1901.04592
8. Ribeiro, M., Singh, S., Guestrin, C.: Why should I trust you? Explaining the predictions of any classifier (2016). arXiv:1602.04938v3
9. Wachter, S., Mittelstadt, B., Russell, C.: Counterfactual explanations without opening the black box: automated decisions and the GPDR. Harv. J. Law Technol. **31**, 841–887 (2017)
10. Walley, P.: Statistical Reasoning with Imprecise Probabilities. Chapman and Hall, London (1991)
11. Dandl, S., Molnar, C., Binder, M., Bischl, B.: Multi-objective counter- factual explanations (2020). arXiv:2004.11165
12. Looveren, A.V., Klaise, J.: Interpretable counterfactual explanations guided by prototypes (2019). arXiv:1907.02584
13. Xu, H., Caramanis, C., Mannor, S.: Robustness and regularization of support vector machines. J. Mach. Learn. Res. **10**, 1485–1510 (2009)
14. Robert, C.P.: The Bayesian Choice. Springer, New York (1994)
15. Johnson, N.L., Leone, F.: Statistics and Experimental Design in Engineering and the Physical Sciences. Wiley, New York (1964)

Big Data Simulation for Demand Forecasting in Retail Logistics

Sergey Svetunkov, Mokhinabonu Agzamova, and Adiba Nuruddinova

Abstract The article discusses intricate interdependencies between improving the accuracy of final demand forecasting in retail companies and efficiency of supply chains. The target of research is "Semya" Trading House LLC, one of the main retailers in Kaliningrad and the Kaliningrad Region. The importance of the task of improving the accuracy of final demand forecasting for this retailer is complicated by the fact that it operates under difficult conditions of historically formed enclave position of the region and under the effect of "counter-sanctions" on behalf of the Russian Federation. The big data on final demand for diversified products, that the retailer has, is the subject of simulating the short-term demand forecasting purposes. Methods such as exponential smoothing, vector autoregression and complex autoregression are considered in order to address this problem. Key problems and challenges for the development of an automated system for short-term final demand forecasting are identified and for their possible solutions using modeling based on big data analysis are outlined herein.

Keywords Final demand · Demand forecasssting · Exponential smoothing · Autoregressions · Logistics chains · Supply · Big data analysis

S. Svetunkov (✉)
Peter the Great St.Petersburg Polytechnic University, St.Petersburg, Russia
e-mail: sergey@svetunkov.ru

M. Agzamova
Department of the Ministry of Employment and Labor Relations of the Republic of Uzbekistan, Tashkent, Uzbekistan

A. Nuruddinova
Department "Information Security Provision", Tashkent University of Information Technologies Named After Muhammad Al-Kharizmi, Tashkent, Uzbekistan

C. Jahn et al. (eds.), *Algorithms and Solutions Based on Computer Technology*, Lecture Notes in Networks and Systems 387,
https://doi.org/10.1007/978-3-030-93872-7_12

1 Introduction

Companies operating in the retail sector perform special functions in the economic system: acquisition of goods on wholesale markets, their delivery, storage and sale o on retail markets. At the same time, the dominating aspect of a retail business is the final demand as it determines the range, quality, and quantity of goods that a retailer acquires from manufacturers. Therefore, a retailer must forecast demand; otherwise, there may be either shortage or overstocking goods. Lack of proper demand forecasting results in shortage of goods, which forces a retailer to either operate in an emergency mode causing losses or to form additional warehouse stocks in order to smooth out the unpredicted growth in demand. Yet, the latter option can also generate losses: if there is no growth in demand for the product, or, even worse, the demand for the product decreases contrary to expectations, the retailer will face the problem of overstocking and damage to goods. It also leads to obvious damage to the retailer.

The complexity of final demand forecasting is that it is influenced by many different factors ranging from weather conditions to consumer income levels. Furthermore, the number of these factors and conditions is so significant that it is impossible to take into account all of them. Therefore, there is a need to restrict oneself to account only the major factors [2]. This inevitably reduces the accuracy of economic forecasts. The problems becomes even more complex when the demand for any product is discrete. Therefore, ideally, a retailer should use discrete forecasting methods, but they are still at the research stage and are not ready for widespread use in practice [2].

The object of our research is "Semya" Trading House (Kaliningrad, Russia), which is one of the largest retailers in Kaliningrad and the Kaliningrad region. The features of the Kaliningrad region, as an enclave territory of Russia, affect many business processes of Kaliningrad companies including supply chain management. Let us dwell on the main distinctive features of the supply of goods to the retail chains of the Kaliningrad Region:

1. Higher proportion of direct imports in supplies. This feature emerges because delivery of goods from other regions in Russia is associated with the passage of customs control and rather complex logistic processes thus, making direct import a more profitable solution.
2. Limited choice of suppliers from other regions in Russia. Since the delivery of goods to the Kaliningrad Region is associated with the passage of customs controls, not all manufacturers of goods from other regions in Russia are ready to work with retail chains in Kaliningrad under direct contracts. This leads to a high dependence of chains on suppliers since replacing one supplier with another implies rather high transaction costs. Moreover, transactions cannot be realized in a short time.
3. Since suppliers from other regions in Russia, due to these reasons, cannot actively work in the region, there is a high dependence of distribution networks in Kaliningrad and the region on local suppliers.
4. Limited supply of goods from local suppliers. The small size of the region and certain difficulties in business operation, which involve close interaction with

other organizations, lead to the fact that local suppliers cannot meet the existing demand of retail chains for goods.

5. Increased risks of late delivery of goods due to customs controls. Because goods from Russia and foreign countries pass through customs, there are cases of untimely arrival of goods due to delays at customs (which may be associated with both queues at customs control points and errors in paperwork or cargo consolidation). This factor becomes critical in the delivery of perishable goods since the delay leads to a reduction of shelf life of the goods and an increase in the losses of the retail network.
6. Consumption features of the population of the region. The residents of the Kaliningrad Region are historically accustomed to an assortment that differs from the typical assortment of most regions in Russia since they have an opportunity to travel to neighboring countries and buy goods that not available in the retail chains of the region (including goods banned for import into the territory of Russia due to counter-sanctions). The need for such goods is partially met by small trade enterprises that sell imported goods under preferential taxation regimes, and through illegal import of goods by individuals for unauthorized trade (the so-called "Polish tents").

To neutralize the factors that can lead to shortage of goods on store shelves, retailers have to insure themselves against the risk of the shortage of goods on the shelves. This insurance is associated with increased costs for the company and can be implemented in two ways [3, 4]:

1. building a buffer stock of goods in the company's warehouses (directly in stores or at a distribution center, if the company has one)
2. a shift in focus from direct contracts with suppliers to deliveries through distributors with a representative office in the Kaliningrad Region

The first way is a classic solution that is common among large retailers, where the consolidation of supplies to stores through a distribution center can reach 100%. This approach allows, on the one hand, to have a buffer stock in case of stockouts of goods (in case of high turnover of goods and in terms of delivery with a deferred payment, this does not create an additional financial burden for the company), on the other hand, it is associated with additional costs for maintaining the commodity stocks.

The maximum effect from the consolidation of goods at a distribution center is achieved if the distribution center can select and ship goods to stores according to the demand of the store without filling a store's warehouse with goods. The limitation, in this case, is the capacity of the distribution center since with a relatively small size of retail chains and a wide range of goods, the distribution center is unable to provide daily selection and shipment of all necessary goods to stores in accordance with their needs. Therefore, under these conditions, the task of selecting the optimal discrete delivery of goods to the store is relevant, which directly depends on the quality of sales forecasting of each particular store.

The second way is associated with an increase in the purchasing price of goods by the company since the distributor assumes the costs of organizing the buffer stock of the goods. At the same time, by consolidating orders for several retail players, a distributor, on the one hand, can compensate for errors in forecasting the demand by individual trading companies by adjusting their orders for the supply of goods to the region in order to reduce their costs for storing stock. On the other hand, when the distributor makes errors in these adjustments, the trading network may be left without goods in the event of a sudden increase in demand for them from the population.

Despite the fact that the distributor has a liability under the contract to provide the trading network with goods in accordance with requests, this condition is not always fulfilled. Moreover, in the context of shortage of goods, the distributor faces a rapidly increasing an uncertainty about the possibility of a sales network to receive the goods i due to competition between different sales networks for commodity stocks available at the distributor's warehouse.

Besides, operations with distributors and local suppliers are often conducted on terms of delivery of goods directly to retail outlets. At the same time, a supplier is not interested in frequent deliveries of goods to stores since they increase its transport costs. As a result, goods arrive in stores in large quantities and add up to stocks of goods already available in the store and reduce their overall turnover. Errors in forecasting the sales at a particular store can form the significant stock on hand of low-turnover goods, which reduces the efficiency of a store and a chain as a whole [5, 6].

Obviously, quality of forecasting sales both for the chain as a whole and for each store separately is key for effective supply chain management regardless of operational arrangements with the supplier. The task of increasing the accuracy of forecasting demand in retail is relevant from the standpoints of economic practice and economic science.

In this paper, we will focus on a possibility to increase accuracy of final product demand forecasting using modern technologies of digital economy, primarily, the big data processing techniques.

2 Materials and Methods

In the view of the above, the most urgent of the many tasks of final demand forecasting in retail is short-term forecasting. This greatly facilitates the solution of the problem pertaining to economic forecasting of final demand since a medium-term and long-term forecasting methodology differs considerably from a short-term forecasting methodology and it is rather difficult to create a multilevel forecasting system [7–11].

Theoretically, the best solution to the problem of short-term demand forecasting in retail would be to build a system of multivariate predictive models, but according to our research, drivers of final demand are diverse. Their influence on demand is prolonged over time and is so volatile in relation to each individual product that multivariate predictive models of final demand can be successfully applied to forecasting sales volumes in general, but not in particular, and only for the medium- and long-term forecasting.

The diversified nature of retail trade turnover is so significant that it is not economically profitable to use individual forecasting models for each individual commodity or even enlarged product groups. Costs of developing and debugging a system of individual predictive models by far significantly exceed the results from its use.

The impossibility of using multivariate models in final demand forecasting in retail stores necessitates the use trend models assuming that the positive and negative influence of factors can be leveled out and reflected in the trends in demand changes. At the same time, the task of applying some unified universal model of short-term forecasting becomes urgent, the weights of this model could be adjusted for each commodity nomenclature reflecting the features of its demand.

This assumption makes it easier to forecast the final demand. In fact, it narrows down to the need to choose a universal forecasting model that would be suitable for modeling trends that are most varied in form and dynamics in the demand for a product ranging from monotonous trends to seasonal fluctuations. Two groups of models are the most adequate for tackle this problem:

1. autoregressive models and
2. exponential smoothing models.

The fundamental difference between them is discussed below.

Autoregressive models represent the dependence of the predicted index y_{t+1} on one or more previous values of the same index, observed earlier—y_t, y_{t-1}, y_{t-2}, etc.

In general form, the autoregressive model can be written as follows:

$$\hat{y}_{t+1} = f(y_t, y_{t-1,\dots,} y_{t-\tau}) \tag{1}$$

where $\hat{y}_{t+1}$ are the values of the analyzed index predicted for the next step of observation, $y_{t-\tau}$ is the actual value of the same index at the previous time point $(t-\tau)$.

Nonlinear autoregressions are practically not used in economic forecasting, while linear autoregressions denoted as $AR(\tau)$ are used.

The weights of autoregressive models are often derived using the least squares method (*LSM*). $AR(\tau)$ models are mainly difficult to use because of the need to choose indices with the greatest impact on forecast results from among a variety of previous indices. To this end, autocorrelation functions are calculated and researched and the lags are detected according to the properties of these functions. This is a labour-intestive task. Therefore, in recent years, experts in forecasting have been increasingly turning to the construction of autoregressive models according to the principle of "information criterion"- sequentially complicating the model, increasing the number of variables until the forecasting accuracy calculated by the model ceases to increase [12]. This has become possible with the significant development of digital hardware used by economists.

The exponential smoothing model is much simpler than the autoregressive model [13]:

$$\hat{y}_{t+1} = \alpha y_t + (1 - \alpha)\hat{y}_t \quad (2)$$

Here, the task comes down to calculating the optimal value of the smoothing parameter α for each predicted product (or product group). At some point, it was proved that the variation range of the smoothing constant lies in the interval from zero to two [14], but so far, in the overwhelming majority of this model's use cases feature erroneous practical use variation range of the parameter α in the interval from zero to one. Interestingly, for example, that these reduced ranges referred to as "classical" ranges are "embedded" into widely used MS Excel software product and a transition to the transcendental set of exponential smoothing (with constant smoothing varying from one to two) is impossible.

Exponential smoothing models are simple and reliable making them fit for wide application in practice when working with large databases. Development of powerful computing systems for retail, which some time ago were considered economically unreasonable and technically impossible, now is now considered as a possibility for large and even medium-sized retailer companies in the context of rapid growth of digital technologies and concurrent reduction in the cost of software and hardware.

Simplicity and reliability of mathematical forecasting models are an important, but not a decisive factor, in processes retail supply chains optimization underpinned by the need to improve accuracy of final demand forecasting.

Modern digital technologies are capable of powerful operations with large databases. Therefore, realizing that exponential smoothing models acceptably cope with the task, more complex models can be implemented. Such alternatives, as shown above, are autoregressive models. They are more complex, it is more difficult to calculate their weights than to find a smoothing constant for exponential smoothing models. However, it is worth it since exponential smoothing models perform worse than autoregressive models when the simulated demand has lags distributed over time. In exponential smoothing models, all observations are important, and their importance decreases with observations waning over time. In autoregressive models, observations that are apart by several observations from the predicted index the closest observations are more important for forecasting.

Our research of the statistical data from the "Semya" Trading House LLC, Kaliningrad, shows that the demand for many commodity items has lags distributed over time, not only to 1 and 2 days but also to 7 and 365 days. This is easily explained by the cyclical nature of demand, for example, every Friday night consumers shop for the weekend. This is the reason for the cyclical demand of 7 days. Models of exponential smoothing cannot take this circumstance into account—the further from the current value the statistical data are, the less they are considered in short-term forecasting of demand. This is the first aspect.

The second aspect is the analysis of big data on demand for the entire range of goods sold at Trading House LLC that has shown that there is a cross-correlation between them and a change in the volume of sales of certain goods leading to a change in the volume of sales of other goods. This applies to related goods and interchangeable goods. Therefore, when predicting the final demand in retail, it makes sense to use vector autoregression *VAR* [15] or complex autoregression *CAR* [16].

In the case under consideration, the first-order vector autoregression *VAR(1)* has the following form:

$$\begin{vmatrix} \hat{y}_{1(t+1)} \\ \dots \\ \hat{y}_{n(t+1)} \end{vmatrix} = \begin{vmatrix} a_{11} \dots a_{1n} \\ \dots \dots \dots \\ a_{n1} \dots a_{nn} \end{vmatrix} \begin{vmatrix} y_{1t} \\ \dots \\ y_{nt} \end{vmatrix} \tag{3}$$

In this model, in order to successfully predict the final demand for a product, it is necessary to estimate the values of unknown weights *nxn* from the available big data.

Since autoregressions with distributed lags, the sizes of which were mentioned above, are more accurate in predicting the final demand, then ideally it is necessary to use *VAR* (1,2,7,365):

$$\begin{vmatrix} \hat{y}_{1(t+1)} \\ \dots \\ \hat{y}_{n(t+1)} \end{vmatrix} = \begin{vmatrix} a_{11} \dots a_{1n} \\ \dots \dots \dots \\ a_{n1} \dots a_{nn} \end{vmatrix} \begin{vmatrix} y_{1t} \\ \dots \\ y_{nt} \end{vmatrix} + \begin{vmatrix} b_{11} \dots b_{1n} \\ \dots \dots \dots \\ b_{n1} \dots b_{nn} \end{vmatrix} \begin{vmatrix} y_{1(t-2)} \\ \dots \\ y_{n(t-2)} \end{vmatrix} + \begin{vmatrix} c_{11} \dots c_{1n} \\ \dots \dots \dots \\ c_{n1} \dots c_{nn} \end{vmatrix} \begin{vmatrix} y_{1(t-7)} \\ \dots \\ y_{n(t-7)} \end{vmatrix}$$
$$+ \begin{vmatrix} d_{11} \dots d_{1n} \\ \dots \dots \dots \\ d_{n1} \dots d_{nn} \end{vmatrix} \begin{vmatrix} y_{1(t-365)} \\ \dots \\ y_{n(t-365)} \end{vmatrix} \tag{4}$$

The use of such models in retail, even with modern information technologies, turns out to be impossible. Working with big data involves processing large amounts of data, and in this case, we are talking about processing large datasets and parallel computations of huge arrays of intermediate data—matrices of *VAR* weights with thousands of optimized weights.

The use of complex autoregressive *CAR* can reduce the dimension of this complex problem by half. In the two-dimensional case, the *CAR*(1) model in vector form has the following form:

$$\begin{vmatrix} \hat{y}_{1(t+1)} \\ \hat{y}_{2(t+1)} \end{vmatrix} = \begin{vmatrix} a_0 & -a_1 \\ a_0 & a_1 \end{vmatrix} \begin{vmatrix} y_{1t} \\ y_{2t} \end{vmatrix} \tag{5}$$

It can be written in a more compact form due to the complex form of notation:

$$\hat{y}_{1t+1} + i\hat{y}_{2t+1} = (a_0 + ia_1)(y_{1t} + iy_{2t}) \tag{6}$$

With complex variables, only one complex factor is subject to estimation. In the case with a similar *VAR*, four weights are subject to estimation.

In the case of forecasting three variables, the *CAR*(1) model in complex form is written as follows:

$$\begin{vmatrix} \hat{y}_{1(t+1)} \\ \hat{y}_{2(t+1)} \\ \hat{y}_{3(t+1)} \end{vmatrix} = \begin{vmatrix} a_{01} & -a_{11} & a_{13} \\ a_{11} & a_{01} & a_{23} \\ a_{31} & a_{32} & a_{33} \end{vmatrix} \begin{vmatrix} \hat{y}_{1t} \\ \hat{y}_{2t} \\ \hat{y}_{3t} \end{vmatrix} \quad (7)$$

the *VAR(1)* model for three variables has the following form:

$$\begin{vmatrix} \hat{y}_{1(t+1)} \\ \hat{y}_{2(t+1)} \\ \hat{y}_{3(t+1)} \end{vmatrix} = \begin{vmatrix} a_{11} & a_{12} & a_{13} \\ a_{21} & a_{22} & a_{23} \\ a_{31} & a_{32} & a_{33} \end{vmatrix} \begin{vmatrix} y_{1t} \\ y_{2t} \\ y_{3t} \end{vmatrix} \quad (8)$$

Comparing the *VAR*(1) and *CAR*(1) models for three variables, it is easy to see that for the *CAR*(1) model, only seven real weights need to be estimated for statistical data, and for the *VAR*(1) model—nine weights. That is, by switching to a complex form of data presentation, it is possible to significantly reduce the number of estimated weights, and this is essential when there is a need to optimize several thousand weights. Therefore, with the modern development of digital technologies and available software tools, *CAR* models should be a preferred option.

Unfortunately, the theory of mathematical statistics of a complex random variable is not yet complete. Research in this area is ongoing [16–20]. There are some methodological problems in estimating the coefficients of such models.

Bayesian information criterion (*BIC*) is used to select the best model. This criterion is very sensitive to the increase in the number of coefficients of the models. As our research has shown, the *CAR* model is always the best if this criterion is used.

3 Results

To solve the problem of final demand forecasting in full for subsequent optimization of the entire supply chain, a short-term forecasting system based on an exponential smoothing model was developed at "Semya" Trade House LLC operating in Kaliningrad and the Kaliningrad region. The work was performed in two stages: at the first stage, optimal values of the smoothing constant were calculated for each product or product group using the algorithms of the numerical method, and, at the second stage, the obtained optimal values of the smoothing constant were used in the demand forecasting system.

The features of these demand models applied to final demand forecasting can be demonstrated with some typical examples.

Thus, a study of the demand for bananas at "Semya" Trade House LLC, Kaliningrad, from 01/01/18 to 03/08/2020 showed that the optimal smoothing constant is 1.02 in this case.

The optimal exponential smoothing model for forecasting the demand for "Green Mark" vodka also has a smoothing constant, which lies is in the transcendental set and equals to 1.001.

If, in these two cases, we apply the "classical" (in the range from zero to one), rather than the actual boundaries of variation in the smoothing constants (from zero to two), then we would be faced with the fact that the best value of the smoothing constant is one. This means that it is necessary to use an elementary forecasting method such as NAÏVE. Obviously, this degrades the accuracy of the forecast.

The optimal smoothing constant for "Krasnodarskiy" rice was 0.999956, that is, it is within the classical limits.

Since the range of goods sold through the logistics network of "Semya" Trade House LLC, Kaliningrad contains several thousand items for several hundred observations of the demand for a product, appropriate software products for working with big data should be used to process such data. We used "PYTHON" software for these purposes. It turned out that for 27% of these series, the optimal value of the smoothing constant is in the transcendental set, and 73% in the classical set (at that, very close to one).

The introduction of this system in a test mode has shown its high efficiency.

4 Discussion

The second stage of research work on short-term forecasting of the final demand for the products of the "Semya" Trade House LLC is to use CAR models. The need to process large amounts of statistical data for the simultaneous estimation of a large number of unknown weights explained the poor accuracy of the first constructed models. As our research shown, this was due to the fact that almost all data series have similar dynamics. This means that it is necessary to evaluate numerous arrays of weights of predictive models under multicollinearity conditions, which leads to serious computational difficulties due to the weak conditionality of the matrices of the LSM systems.

Apparently, the problem should be considered as a decomposition problem—to evaluate individual elements of the CAR model, and then to synthesize a general model of short-term forecasting.

5 Conclusions

More accurate short-term forecasts were obtained for the entire product range at Semya" Trade House LLC to introduce a system of short-term final demand forecasting for goods at this company using exponential smoothing models. As a result, the operation of the entire logistics chain of this company was stabilized.

A further increase in the accuracy of final demand forecasting for goods sold by "Semya" Trade House LLC is possible with the full implementation of a forecasting system based on the system of CAR models with distributed lags. According to our estimates, this will increase the accuracy of economic short-term forecasts by approximately 20%.

With rapid development of digital technologies and hardware in the coming years, it will become possible to build a system for final demand forecasting based on more accurate VAR models, which, according to our preliminary calculations of the comparative accuracy of these two models, will further increase the accuracy of short-term forecasting of final demand by approximately 12%. As a result, this will ensure measured operation of the logistics network of "Semya" Trade House LLC—one of the largest retailers in Kaliningrad and the Kaliningrad Region—and increase the efficiency of the supply system as a whole.

Acknowledgements The article was prepared under the financial support from RFBR. Grant No. 19-010-00610/19 "Theory, Methods and Techniques for Forecasting Economic Development through Autoregressive Models of Complex Variables".

References

1. Syntetos, A.A., Babai, Z., Boylanc, J.E., Kolassa, S., Nikolopoulos, K.: Supply chain forecasting: theory, practice, their gap and the future. Eur. J. Oper. Res. **252**, 1–26 (2016)
2. Ali, M.M. Centralized demand information sharing in supply chains, Unpublished PhD thesis. Buckinghamshire New University, Brunel University, UK (2008)
3. Ilin, I.V., Koposov, V.I., Levina, A.I.: Model of asset portfolio improvement in structured investment products. Life Sci. J. **11**(11), 265–269 (2014)
4. Izotov, A., Rostova, O., Dubgorn, A.: The application of the real options method for the evaluation of high-rise construction projects. In: E3S Web of Conferences, vol. 33, pp. 03008 (2018)
5. Sims, C.: Macroeconomics and reality. Econometrica **48**(1), 1–48 (1980)
6. Nuriddinova, A.G.: Development of educational complex as a factor of perfection of labour marketing. In: The Conditions of Knowledge. Economy Proceedings of the 5th Uzbekistan-Indonesia International Joint Conference on Globalization, Economic Development, and Nation Character Building, pp. 29–30 (2015)
7. Athanasopoulos, G., Ahmed, R.A., Hyndman, R.J.: Hierarchical forecasts for Australian domestic tourism. Int. J. Forecast. **25**(1), 146–166 (2009)
8. Neusser, K.: Time Series Econometrics. Springer International Publishing (2018)
9. Toroptsev, E.L., Marahovski, A.S., Duginsky, R.R.: Interdistribution modeling of transition processes. Econ. Anal. Theory Pract. **19**(3), 564–585 (2020)
10. Gelper, S.E.C., Wilms, I., Croux, C.: Identifying demand effects in a large network of product categories. J. Retail. **92**(1), 25–39 (2016)
11. Ilin, I.V., Iliashenko, O.Y., Klimin, A.I., Makov, K.M.: Big data processing in Russian transport industry. In: Proceedings of the 31st International Business Information Management Association Conference, IBIMA 2018: Innovation Management and Education Excellence through Vision 2020, pp. 1967–1971 (2018)
12. Wilms, I., Barbaglia, L., Croux, C.: Multi-Class Vector Autoregressive Models for Multi-Store Sales Data. KU Leuven, Faculty of Economics and Business (2016)
13. Vu, K.M.: The ARIMA and VARIMA Time Series: Their Modelings, Analyses and Applications. AuLac Technologies Inc. (2007)
14. Svetunkov, S.G.: Extreme Cases of Brown's Method. Economic Sciences: Scientific Notes of UlGU, issue 2. Part 1. Publishing house of SVNTs, Ulyanovsk (1997)

15. Lütkepohl, H.: New Introduction to Multiple Time Series Analysis. Springer, Berlin Heidelberg (2005)
16. Svetunkov, S.: Complex-Valued Modeling in Economy and Finance. Springer Science+Business Media, New York (2012)
17. Tavares, G.N., Tavares, L.M.: On the statistics of the sum of squared complex Gaussian random variables. IEEE Trans. Commun. **55**(10), 1857–1862 (2007)
18. Adili, T., Schreier, P.J., Scharf, L.L.: Complex-valued signal processing: the proper way to deal with impropriety. IEEE Trans. Signal Process. **59**(11), 5101–5125 (2011)
19. Jia, Y., Yang, X.: Adaptive complex-valued independent component analysis based on second-order statistics. J. Electr. Comput. Eng. **2016**, 7 (2016). Article ID 2467198. https://doi.org/10.1155/2016/2467198
20. Trampitsch, S.: Complex-Valued Data Estimation Second-Order Statistics and Widely Linear Estimators. Alpen-Adria-Universität Klagenfurt, Masterarbeit (2013)

Developing and Investigation of Heterogeneous Distributed Reconfigurable Platforms for Supercomputing

Alexander Antonov, Vladimir Zaborovskij, Ivan Kiselev, and Denis Besedin

Abstract The paper discusses our approach in developing and investigation of heterogeneous distributed reconfigurable platforms for supercomputing focusing on improves performance of heterogeneous distributed computing system sees as a solver of specific fundamental and applied tasks. Proposed decision associated with "constructive reconfiguration" of available digital processing and communication infrastructure due to adapt computational resources to the specific features of chosen algorithmic strategies. Solutions under consideration are based on developing hyper-convergent hierarchical platform, reconfigurable edge accelerator and high bandwidth access pipes from and between each processing units and distributed array of task memory. The discussed problem is analyzed from the standpoint of designing features of heterogeneous edge node as a modular building component of multilevel distributed hyper-converged platform. Pro and contra analysis of proposed approach as well as efficiency of high level Synthesis Tools in creating specialized hardware components are based on sorting algorithms as a typical example of complex computing procedures. The features, which define the properties of hyper-convergent platform and the developed Reconfigurable Accelerator of the heterogeneous node, are horizontal-vertical scaling architecture and the presence of four FPGAs on each edge board, which allows implementing four independent algorithmic specific accelerators. The term efficiency that used in the paper is defined integral function of platform performance, which estimates time of task solving as well as hardware "cost" in utilized FPGA resources. Carried out simulation modeling and a comparative analysis for the wide range of sorting algorithms that implemented on a universal processor and on the Reconfigurable Accelerator of the heterogeneous node clearly show that the accelerators created by Xilinx's HLS tool can in several time speed up sorting algorithm.

Keywords Hardware Acceleration · Sorting algorithms · High-level synthesis · Reconfigurable Hardware Accelerator · FPGA · Machine learning · DC cloud

---

A. Antonov (✉) · V. Zaborovskij · I. Kiselev · D. Besedin
Peter the Great St. Petersburg Polytechnic University, Saint-Petersburg, Russia
e-mail: antonov@eda-lab.ftk.spbstu.ru

C. Jahn et al. (eds.), *Algorithms and Solutions Based on Computer Technology*, Lecture Notes in Networks and Systems 387,
https://doi.org/10.1007/978-3-030-93872-7_13

1 Introduction

There are rapidly growing demands on high-performance computing for Big Data analysis, Data Mining and Management tasks. A modern trend in the development of high performance computing systems is an attempt to accelerate them through the introduction of Distributed Reconfigurable Heterogeneous High-Performance Computing systems (DRH HPC). Such systems, that form a distributed infrastructure, can be classified as hyper-converged platforms, which consist of several computing platforms that are combined into a horizontally scalable cluster [1, 2].

It is well known that scaling the resources in such platform (processors, memory, channels) is associated with certain difficulties, which are caused by various requirements for the use of computing resources from system level and application algorithms. As a result, in accordance with specific application task which needs often arise in "on fly" mode and required only one type resource, for example a RAM, requiring the connection of an additional industry hardware node, which is the carrier of different resources that are not required "right now", which makes the such standard solution not rather effective [2].

For data storage application, in such platforms widely used software-defined storage (SDS) solutions that combine all available resources into a common pool and provide them to cluster computing nodes or virtualized applications as their needs arise. Virtualization as a technology for cooperative use and software reconfiguration of shared resources can take into account the features of implemented applications and data structures but only within the specific hardware restrictions that apply to all standard industrial platforms. The abilities to adapt to the features of the algorithms used for solving problems and to increase the integrated computing performance of a hyper-converged platform can be significantly expanded by using the reconfigurable hardware components. Eventually quick to deploy, fast to scale, and hyper efficient for each running algorithms are basic high level requirements that reflects trends of transformation of HPC technologies of all apps, streamlining architecture taking advantage of artificial intelligence technologies and self-managing decision to distribute workloads, to reconfigure hardware of the nodes and to protect distributed data. The potential of existing hyper-converged solutions could be improved by "on a fly" reconfiguration computing nodes delivers the simplicity, efficiency, and economy to the corporate HPC platforms that used for different advent fundamental and wide shared applied research, that can overcome and prevents performance problems via balancing memory resources and workload placement [3].

2 Materials and Methods

The possibilities of modern Hyper-converged high-performance computing (HHPC) systems every year enhanced by the increased energy-performance efficiency of hardware (TFlops/W), wide availability of heterogeneous platforms and algorithms for distributed processing of high speed data flows, development of neuromorphic computing platforms, and computation in memory technologies [4, 5]. At the same time, not only the capabilities, but also the complexity of HHPC systems constantly increase. This is why the idea of using artificial intelligence technologies to manage computing processes, as well as applied "machine learning" methods to optimize or even self-reconfigure computing nodes in order to increase their real performance by taking into account the features of implemented algorithms and processed data structures had started to be discussed. Modern HHPC systems are used to solve various classes of problems that differ in both "algorithmic complexity" and requirements for the architecture and hardware resources of processors, memory, and network interfaces. Thus, the "fundamental" class traditionally includes NP—complete tasks related to conducting large amounts of calculations in modeling, searching for conditions for the feasibility of Boolean functions, including recognition or classification of objects based on machine learning methods. The theoretical difficulties [6] associated with the requirement of formal solvability of such problems are overcome by constructively taking into account the restrictions on the accuracy of calculated solutions, which must be comparable to the accuracy of the source data. From the view point of solvability problem, the above tasks can be reduced to: (1) "direct" programming digital " output "of a system using mathematical description and initial conditions of modelling process with known accuracy, and (2) "inverse" problems or building algorithms that calculate with a given accuracy the output data that corresponding to known input data or initial conditions.

3 Results

3.1 The Proposed Architecture of Reconfigurable Heterogeneous Distributed HPC System

The creation of a distributed heterogeneous hyperconvergent computing platform that allows us to effectively solve both "forward" and" reverse " problems opens up new opportunities for applying computing technologies in various fields of science and technology (Fig. 1) [1, 2].

The infrastructure of the investigated hyper-convergent platform consists of three levels:

- predictive analytics and "explanation" of the resulting solutions, the energy-computational efficiency of the level corresponds to 2–4 GFlops/W.
- data aggregation of modeling results and creating datasets for machine learning systems, energy-computational efficiency 5–10 GFlops/W.
- reconfigurable accelerators of algorithms and computational pipe between heterogeneous processors and data arrays of application tasks, energy-computational efficiency 10–20 GFlops/W.

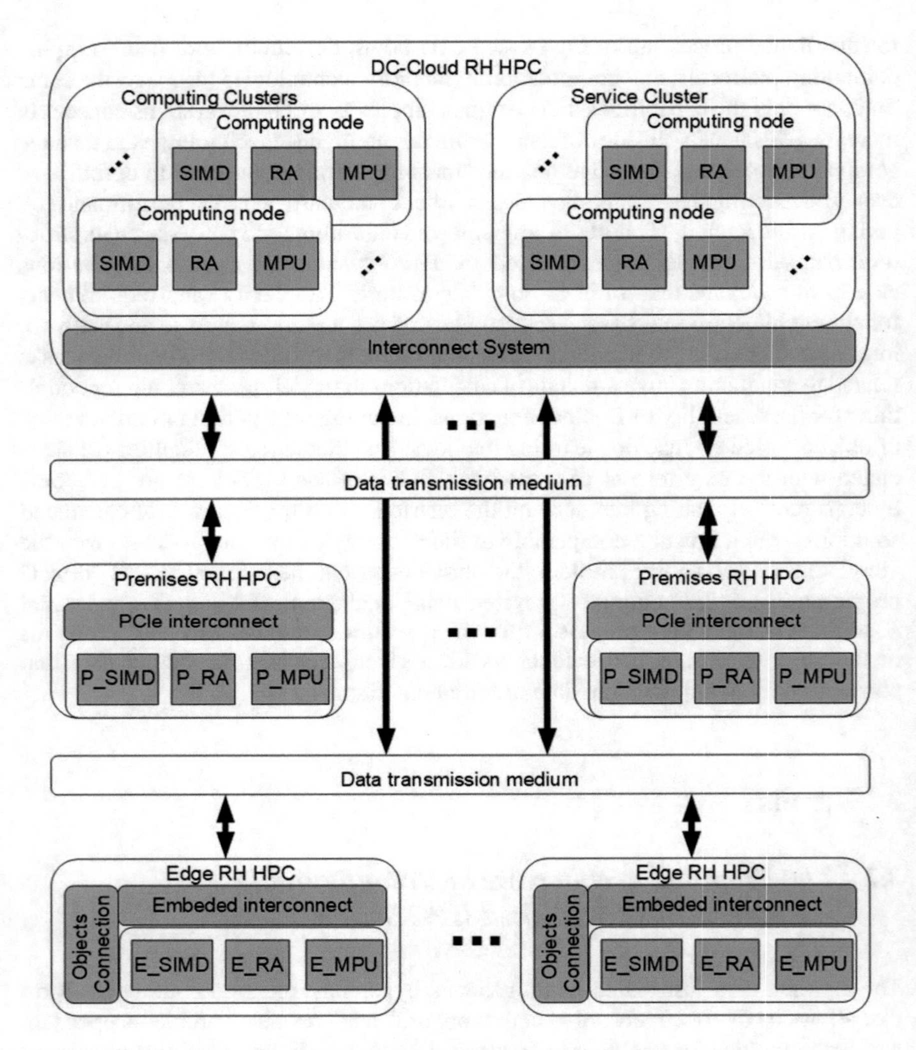

Fig. 1 The proposed architecture of reconfigurable heterogeneous distributed HPC system

Proposed organization of the hyper-convergent platform allows to maintain different software computing models such as procedure «call-on mention /pass-by-reference» and the procedure «call by value/transfer by data context» when the arguments of a procedure are evaluated before they are passed to the procedure. The such functional and algorithmic flexibility, which also complemented by the capabilities of the hardware reconfiguration of edge node (Fig. 2) [1, 2], allows to apply this hyper-convergent platform to solves all listed above "direct" and "inverse" tasks.

Modern DRH HPC consists of: multiprocessor units (MPUs) [7, 8], Single Instruction Multiple Data (SIMD) accelerators [9], commonly known as General Purpose

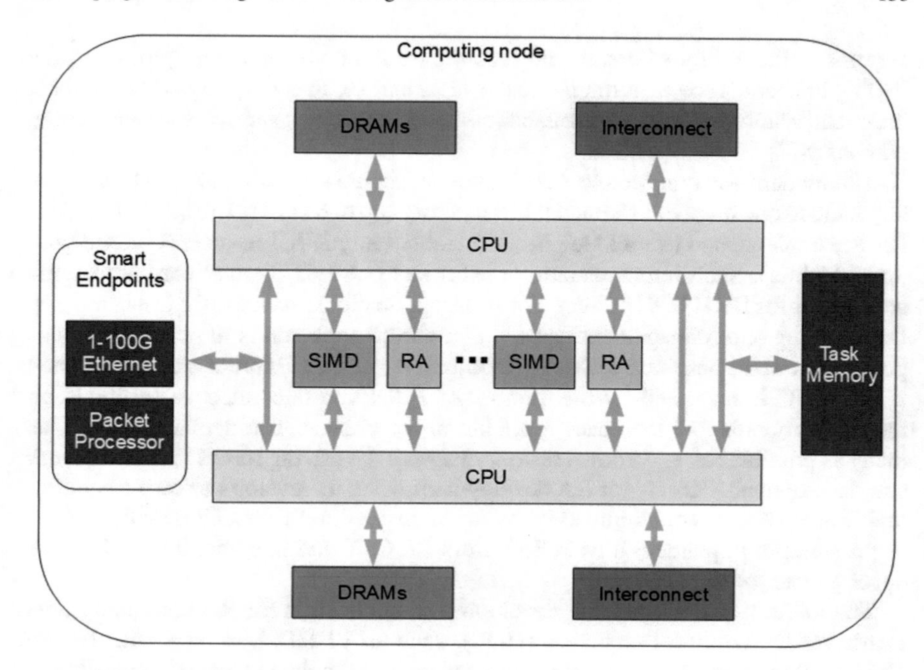

Fig. 2 The proposed architecture of the computing node

Graphics Processing Units (GPGPU), and Reconfigurable Accelerators (RA), based on Field Programmable Gate Arrays (FPGAs) [10, 11]. FPGA is an Integrated Circuit (IC) that can change its internal structure in accordance with the particular task. A modern FPGA consists of programmable logic cells (LCELL), which can perform any functions of logic or memory, and a hierarchy of programmable matrices, which can connect nearly all LCELL in FPGA together to implement complex logic and memory functions. The modern FPGAs contain not only LCELL and the programmable matrices, but also digital signal processing units (DSP), integrated memory units (BRAM), high-throughput memory units (HBM), hardware-implemented controllers for external DDR4 memory and multigigabits transceivers for external PCIe interfaces and 10-100G Ethernet ports. State-of-art FPGAs can be configured "on the fly"; it means that ones can be configured during the normal operation of the device. Some modern FPGAs support partial configuration and reconfiguration via PCIe and Ethernet. Partial reconfiguration of the FPGA means that a part of the FPGA can be reconfigured, while the rest of the FPGA continues to solve the current task [10].

DRH HPC system allows you to create highly specialized computational "pipes" for solving the particular tasks. The computational "pipe" can consist of just Reconfigurable Accelerator (RA), a reconfigurable FPGA based accelerator, or, for solving a complex task, can include MPUs, SIMD accelerators and Reconfigurable Accelerators working together. A huge advantage for the performance of such DRH HPC systems is the ability to create and reconfigure the computational "pipes" on the fly [1], in accordance with the particular task, and so, to satisfy to one of the most important criteria for high-performance computing systems: performance and energy efficiency [2].

The modern Reconfigurable Accelerators available on market [10, 11] traditionally include one huge FPGA and DDR4 memory (up to 64 GB). The most of modern RA are implemented as Full Height (FH) and ¾ Length PCI Express 3.0 Card with up to 16 lines. Such implementation means that one PCI Express card is just one accelerator for DRH HPC system. For some applications, such as Big Data analysis, Data Mining and Management tasks, the preferred approach is to have as many as possible of RA connected to MPU. The number of PCI Express cards connected to the MPU is restricted by the number of PCI Express connectors available on the motherboards and the space available in the chassis. The demand to have, as many as possible RA and known restrictions were the driving forces for our research and development. Targets for RA development were: to develop our own Reconfigurable Accelerator: containing as many the largest Xilinx Kintex UltraScale FPGAs as possible; to implement it as PCI Express 3.0 Card and integrate it with Xilinx's development tools.

The traditional procedure for developing an application for RA commonly uses Hardware Description Languages (HDL), such as VHDL, Verilog HDL, System Verilog. This approach is very time-consuming and requires painstaking work both at the development stage and at the debugging stage, since it is necessary to operate with logical functions and assemble complex systems from small pieces, between which it is also necessary to establish the correct connections and check them during debugging [12].

The modern approach is a development flow, which uses High-Level Synthesis (HLS) tools provided by leading manufacturers of FPGA programmable logic, such as Xilinx [12] and Intel PSG [13], and some other companies engaged in the development of HLS tools, for example, Mentor Graphics [14].

Modern HLS tools allow not only to synthesize a hardware solution, often called an synthesized implementation, for an algorithm described by high-level programming languages such as C or C++, but also to simulate the source code and the implementation for verifying the correctness with respect to the expected results. Modeling simulation of the source code and the synthesized implementation use a common test, often referred as self-checking testbench, described in C or C++.

The application part of our research was connected with searching the sorting algorithm, for which the hardware implementation has superior performance comparing with a software implementation of any sorting algorithm and hardware implementations of any other sorting algorithms. The target was to find the sorting algorithm and implement it on one FPGA used in the developed RA.

3.2 The Developed Reconfigurable Accelerator

The developed Reconfigurable Accelerator has a structure [2] highlighted on Fig. 3.

The developed Reconfigurable Accelerator consists of:

- KU115—Xilinx's Kintex UltraScale KU115 FPGA [15]. It is the largest Kintex UltraScale device available.
- DDR4 16 GB—SO-DIMM DDR4 memory module. Each module has 16 GB capacity and is connected to FPGA by 72 bits data bus.
- PCIe Switch—switch of PCI Express interface. It has the host port with 16 lines and four slave ports with 8 lines each. The switch provides non-blocking connection from the host with any FPGA and between FPGAs. The architecture of the Reconfigurable Accelerator provides the ability to connect all KU115 FPGAs among themselves.

The developed RA has a balanced throughput between components:

- The peak throughput between each KU115 FPGA and its own DDR4 memory is about 120 Gb/s.
- The peak throughput between each KU115 and the host is up to 64 Gb/s.

Due to the fact that the peak performance of the channel to memory is approximately two times greater than the peak performance of the channel to the host, it is possible to write data to DDR4 memory from the host and simultaneously process data in FPGA.

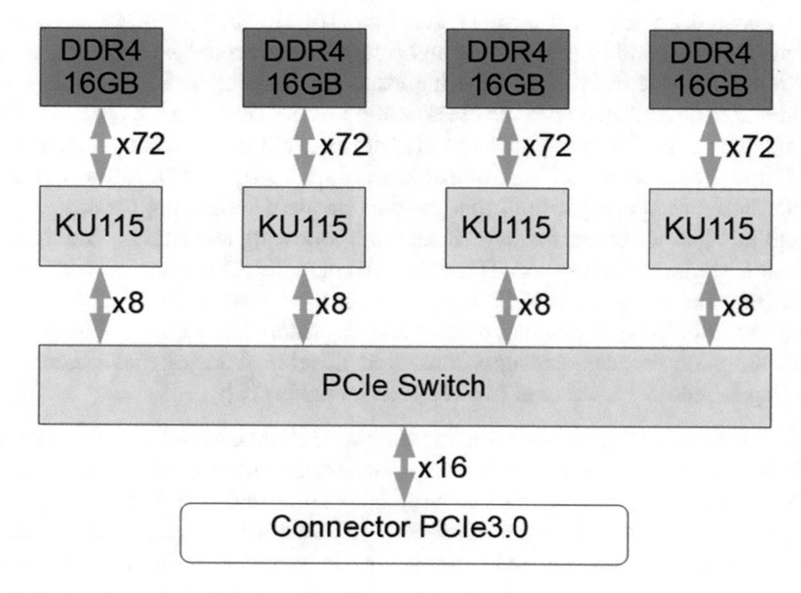

Fig. 3 Internal structure of the developed Reconfigurable Accelerator

For an application developer, and for an operating system (OS) of the MPU as well, the developed Reconfigurable Accelerators has four independent accelerators, each has its own DDR4 memory, connected to PCI Express interface. The Reconfigurable Accelerator was implemented as Full Height (FH) and ¾ Length PCI Express 3.0 Card. The expected peak power consumption of the RA is 450 W.

To integrate the Reconfigurable Accelerator into Xilinx development tools such as Vivado, Vivado HLS, Vitis (with support of OpenCL standard) we developed:

- The "board description", which is a set of design constraints and descriptions for facilitating the integration procedure of the board with Vivado and Vivado HLS tools.
- The "hardware platform", which is the hardware design for each FPGA, enabling to connect of the RA with the host through the PCI Express, and to reconfigure each FPGA on the fly.
- A set of drivers, enabling integration of the Reconfigurable Accelerator with Linux OS $\times$ 64, deployed on the host computer.

3.3 The Objects of the Application Part of the Research

When choosing sorting algorithms [16] for research, libraries of already implemented algorithms such as MatLab and IP library from Xilinx were considered. MatLab library contains many commonly used sorting algorithms described in .m language. Among them are comb sorting, merge, heap, quick sort, insertion, bubble, bucket, cocktail, choice, shell sorting algorithms [16]. IP library from Xilinx contains merging, inserting and bitonic sorting algorithms described in C++ [17].

Summarizing the both lists of the algorithms, we distinguished those types of algorithms, in accordance with the classification in Fig. 1, that were objects for the research. Thus, the following types of algorithms were distinguished: exchanging algorithms, selection algorithm, insertion sorting, merging algorithm, counting algorithms and unique algorithms that use non-standard sorting approaches.

From this list, we chose for our research the following algorithms: Gnome sort, Comb sort, Merge sort, Heap sort, Binary tree, Bucket sort, Counting sort and Bitonic sort. During the research, it is necessary to consider, that Xilinx's HLS tool has limited capabilities to implement a recursive algorithms. Therefore, non-recursive modifications for the recursive algorithms were developed during the research.

During the research, we used two research methods [18]:

- Simulation modeling. It means a simulation of each selected algorithm on the different computing architectures: on a universal processor and on synthesized FPGA based hardware, i.e. on Reconfigurable Hardware Accelerator.
- Comparative analysis. It means comparing efficiency of software implementation and a synthesized FPGA based hardware implementation of the same algorithm.

Since the architectures of a universal processor and a reconfigurable FPGA are different, the term efficiency, used in our research, was defined as a function of performance and Hardware "cost", estimated in utilized FPGA resources for a particular implementation. The term performance is defined as:

- Software implementation performance, estimated as a time, which is necessary to sort a randomly generated array by using a universal processor.
- Hardware implementation performance, estimated as a time, which is necessary to sort a randomly generated array by using a synthesized hardware implementation.

The research results are highlighted on Fig. 4. It visualizes performance estimations for both: the software implementations, for all algorithms investigated, and hardware implementations, for the fastest algorithm, Merge sort algorithm [19].

On Fig. 4 we used the following terminology:

- Name of the algorithm (CPU)—the software implementation of the pointed algorithm.
- Name of the algorithm (FPGA)—the hardware implementation of the pointed algorithm by using Vivado HLS.
- Array size—a number of elements in the input and output arrays.
- Time (for CPU implementation)—software implementation performance, a time in seconds, which is necessary to sort a randomly generated array by using a universal processor.
- Time (for FPGA implementation)—synthesized hardware performance, a time in seconds, which is necessary to sort a randomly generated array on hardware

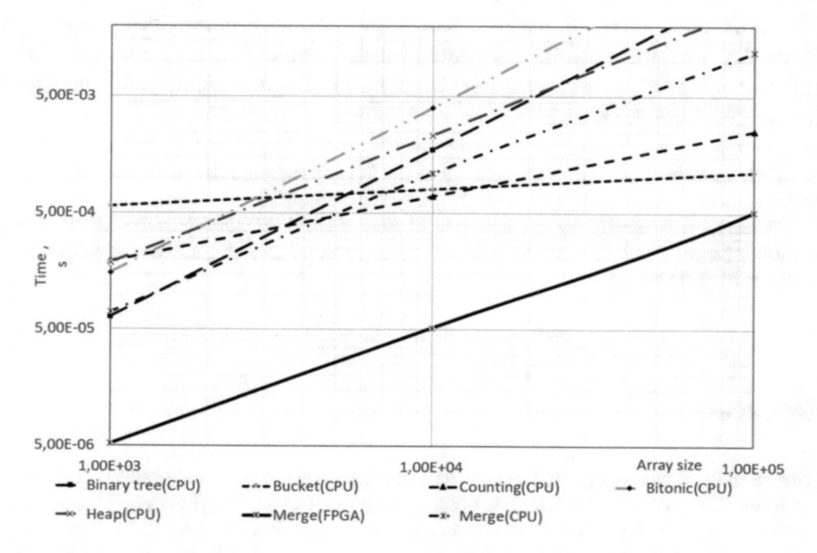

Fig. 4 Performance estimations

synthesized by Vivado HLS. It is calculated by multiplying the estimated Clock Period on II (Initiation Interval).

Clock Period—it is a minimum possible clock period for the synthesized hardware solution, estimated by Xilinx's HLS tool. II—it is the Initiation Interval, the number of clock cycles required by the synthesized hardware before to accept a new array for processing, estimated by Xilinx's HLS tool.

4 Discussion

The nearest future researches will deal with performance evaluation of developed and deployed reconfigurable accelerator. We expect to get significant leap in performance for the tasks related with Deep Neural Networks inferenced in Supercomputer Center.

The second direction for future researches is implementation and performance investigation of HPC EDGE unit developed in accordance with proposed architecture.

5 Conclusions

The Merge sort is the algorithm that can be recommended to speed up the process of sorting arrays with the number of elements up to 1.00E + 05 and for implementation on the developed Reconfigurable Accelerator.

The developed Reconfigurable Accelerator allows implementing up to four concurrently working sorting algorithms on the single PCI Express card, instead of implementing only one sorting algorithm while using a reconfigurable accelerator with one FPGA on the board.

Acknowledgements The authors are grateful to the Supercomputer Center 'Polytechnic' [20] for the help in gaining access to the resources of the supercomputer.

The research was funded by the Ministry of Science and Higher Education of the Russian Federation as part of Worldclass Research Center program: Advanced Digital Technologies (contract No. 075-15-2020-934 dated 17.11.2020)

References

1. Antonov, A., Zaborovskij, V., Kalyaev, I.: The architecture of a reconfigurable heterogeneous distributed supercomputer system for solving the problems of intelligent data processing in the era of digital transformation of the economy. Cybersecur. Issues **33**(5), 2–11 (2019). https://doi.org/10.21681/2311-3456-2019-5-02-11

2. Ilin, I.V., Iliashenko, O.Y., Klimin, A.I., Makov, K.M.: Big data processing in Russian transport industry. In: Proceedings of the 31st International Business Information Management Association Conference, IBIMA 2018: Innovation Management and Education Excellence through Vision 2020, pp. 1967–1971 (2018)
3. Antonov, A., Zaborovskij, V., Kisilev, I.: Specialized reconfigurable computers in network-centric supercomputer systems. High Availab. Syst. **14**(3), 57–62 (2018). https://doi.org/10.18127/j20729472-201803-09
4. Dongarra, J., Gottlieb, S., Kramer, W.: Race to Exascale. Comput. Sci. Eng. **21**(1), 4–5 (2019). https://doi.org/10.1109/MCSE.2018.2882574. (February)
5. Usman Ashraf, M., Alburaei Eassa, F., Ahmad Albeshri, A., Algarni, A.: Performance and power efficient massive parallel computational model for HPC heterogeneous exascale systems. In: IEEE Access, vol. 6, pp. 23095–23107 (2018). https://doi.org/10.1109/ACCESS.2018.2823299
6. Haidar, A., Jagode, H., Vaccaro, P., YarKhan, A., Tomov, S., Dongarra, J.: Investigating power capping toward energy-efficient scientific applications, In: Concurrency and Computation Practice and Experience, pp. 1–14 (2018). https://doi.org/10.1002/cpe.4485
7. Le Fèvre, V., Herault, T., Robert, Y., Bouteiller, A., Hori, A., Bosilca, J.G., Dongarra, J.: Comparing the performance of rigid, moldable and grid-shaped applications on failure-prone HPC platforms. Parallel Comput. **85**, 1–12 (2019)
8. Intel Xeon. [Online]. https://www.intel.com/content/www/us/en/products/docs/processors/xeon/2nd-gen-xeon-scalable-processors-brief.html. Accessed 19 Apr 2020
9. IBM PowerPC9. https://www.ibm.com/it-infrastructure/power/power9. Accessed 19 Apr 2020
10. NVIDIA Tesla V100. https://www.nvidia.com/en-us/data-center/tesla-v100/. Accessed 19 Apr 2020
11. Xilinx FPGA. https://www.xilinx.com/. Accessed 1 Feb 2020
12. Intel FPGA. https://www.intel.com/content/www/us/en/products/programmable.html. Accessed 19 Apr 2020
13. IDE Vivado HLS. https://www.xilinx.com/video/hardware/vivado-hls-tool-overview.html. Accessed 19 Apr 2020
14. Intel HLS compiler. https://www.intel.com/content/www/us/en/software/programmable/quartus-prime/hls-compiler.html?wapkw=HLS. Accessed 19 Apr 2020
15. Catapult HLS. https://www.mentor.com/hls-lp/catapult-high-level-synthesis/. Accessed 19 Apr 2020
16. UltraScale and UltraScale+ FPGA product Table (2019). https://www.xilinx.com/products/silicon-devices/fpga/virtex-ultrascale.html#productTable. Accessed 19 Apr 2020
17. Sorting Methods. https://www.mathworks.com/matlabcentral/fileexchange/45125-sorting-methods?focused=3805900&tab=function. Accessed 19 Apr 2020
18. Vitis_Libraries. https://github.com/Xilinx/Vitis_Libraries. Accessed 19 Apr 2020
19. Antonov A., Besedin D., and Filippov A.: Research of the efficiency of high-level synthesis tool for FPGA based hardware implementation of some basic algorithms for the big data analysis and management tasks. In: Proceedings of the FRUCT'26, pp. 23–29, April 2020

Algorithms for Labour Income Share Forecasting: Detecting of Intersectoral Nonlinearity

Stanislav Rogachev and Bakytbek Akaev

Abstract We examine different algorithms to forecast labour share for 18 KLEMS-classified economic sectors, 12 European countries. The choice is driven by data availability. For each sector 11 specifications of time component in CES production function with factor-augmenting technical change are tested. This includes comparing models with linear, nonlinear time and the same with structural breaks. Then, three degrees of models 'power' are proposed to characterize whether a model is consistent and valid for prediction. Here, residuals stationarity and autocorrelation as well as regressors and structural breaks statistical significance are investigated. To sum up main results, models with structural break in nonlinear time component show better predictive power according to the derived criteria. Next, overall labour share decline cannot be stated as only 7 sectors out of 18 have decreasing trend in more than one third of cases (countries). Additionally, each country sectors are grouped by LS forecast average value into four interval categories.

Keywords Labour share · Factor-augmenting technical change · CES production function · Labour share intersectoral decomposition · Labour share forecasting

1 Introduction

Within digital era and persistent automation processes humans indeed face challenge of becoming redundant—especially those who refuse to consider constant learning as a lifestyle. Therefore, it is tempting to count falling labour share as a commonly accepted fact of modern economy, which indeed is not true. An outline

S. Rogachev (✉)
National Research University Higher School of Economics, Kantemirovskaya 3, St. Petersburg, Russia
e-mail: srogachev@hse.ru

B. Akaev
International University of Kyrgyzstan, Bishkek, Kyrgyzstan

C. Jahn et al. (eds.), *Algorithms and Solutions Based on Computer Technology*, Lecture Notes in Networks and Systems 387,
https://doi.org/10.1007/978-3-030-93872-7_14

of steady empirical research "has provided economists—though not the lay public—with grounds for optimism that, despite seemingly limitless possibilities for labor-saving technological progress, automation need not displace labor as a factor of production" [1, p. 3; 2].

Still, vast majority of recent research on the topic of labour share[1] constantly reconcile this optimism with the reality and investigate what LS is impacted by and how. For instance, McKinsey discussion paper [3] review five factors[2] that impact U.S. labour share, report general characteristics of capital share (the opposite to LS) for each economic sector. Accordingly, globalization aspect, accounting for self-employed, "superstar" firms effect, or equivalently the concentration of sales in a relatively limited number of large enterprises is regarded as possible explanations of LS decline [4–7]. Similarly, the decreasing price of investment goods, regarded as a constituent part of technical progress, is said to yield global labour share decline [8].

To prove the contrary, no decline in the corporate sector LS of the chosen European countries[3] cast doubts on the global LS decline itself [9]. Moreover, deeper theoretical investigation into how technologies replace human labour states that 'old' tasks are inclined to automation but the emergence of new tasks that cannot be operated without human revive the situation [10]. However, later the effect of displacement for 'old' tasks in manufacturing was empirically found to be more significant than 'reinstatement effect' of new tasks [11].

In current paper the issue of technical progress impact on LS is concerned but from a rather different angle. Broadly, the paper indeed tries to answer the question how technical progress impacts labour share, which is close to above-mentioned discussion on how technologies replace human labour. The innovation is that the inter-industry (intersectoral) decomposition of labour share is considered, accompanied by nonlinearity inclusion in the model. In this sense, our study is conducted in style of [12–14] who were empirically investigating factor augmentation, elasticity of substitution and direction of technical change in framework of CES-production function. The relevance of such empirical study is confirmed by the facts of no common opinion on the value of elasticity of substitution and no significant difference between its inter-industry and country estimates depicted in [15] meta-analysis.

The practical research significance may be easily outlined by referring to LS reduction as to economic feudalism of the twenty-first century [16]. Yet, probably in

[1] Labour Share or Labour Income Share is referred to as LS hereafter.

[2] The contribution of the factor in total LS decline (since 1998) is indicated in parentheses [2, p. 2].
"Supercycle and boom-bust effects" (33%).
"Rising and faster depreciation due to higher capital stocks and a shift to intangible assets with shorter life cycles" (26%).
"Superstar effects—which see a small proportion of large firms capturing a disproportionately larger share of economic profit than their peers" (18%).
"Capital substitution of labor and automation" (12%).
"Globalization and decreased labor bargaining power" (11%).

[3] The authors exclude self-employed and residential activities from their analysis, which is based on historical data only. France, Italy, Germany and the UK are referred to as "four major European economies" [7, p. 9].

style of Box and Jenkins philosophy[4] the data on LS dynamics is provided in graphs only with no empirical quantitative modeling. Nevertheless, it is reasonably argued that upon the LS decrease the most negative aspect is not that people become inferior to technology, but that benefits from technological progress are distributed in favor of owners of capital. This eventually deepen the motivation of current research by leading to the idea of LS as a proxy of social equality.

2 Materials and Methods

Technical change direction is a key concept in models that incorporate innovations as investments of labor and capital. Usually, in the long-term perspective, the economy resembles a standard economic growth model with technological progress conditioned only by labor intensity which means that the share of labor income in GDP is constant in this case [17]. Nevertheless, the factors of production income shares may vary in case technological change is caused by capital intensification (p.e. technological innovations). In case the process of technological innovation (automation and robotics) creates new and relatively more complex tasks for humans, then a dynamic equilibrium exists, and consequently, the economy grows at a constant rate [10].

Reviewing economic growth theory and factor-augmenting technical change, we find a comprehensive toolbox of CES production function modifications as well as a profound literature review [18]—particular articles are analysed further [18, pp. 785–786]. First, estimation results based on CES functions with Hicks-neutral technical progress are compared to models with biased technical change[5] which report significantly lower elasticity of substitution values [12]. Second, empirical testing of normalized CES production function was pioneering in [20]; the authors argue that this, accompanied by the good data consistency performed in their research, avoids potential bias of parameter estimates though significantly complicates the mathematical apparatus. Similarly, co-evaluation of the production function and its first-order conditions proved to be more effective than evaluating single equations [13]. The authors use Monte-Carlo simulation of the CES production function and state that normalization of the production function improves parameter estimates. Next, the issue of intersectoral decomposition is connected to the described above methodology of factor-augmenting technical change incorporated by CES production function [14].

Furhermore, these results are extended to France and six more European countries (though for later time span) and it is reported that LS remains constant over time upon elimination or clarifying of three following features that potentially impose bias unless regarded in the model—first, the initial period for the empirical analysis, second, self-employed people, and third, the residential income [21]. Finally,

---

[4] "Let the data speak for themselves".

[5] Biased Technical Change is referred to as technical progress bias to highly qualified labor. Education here is complementary to new technologies [19]

structural breaks in the relative labor intensity parameter are designed to account for nonlinearity in explaining the decline in the labor share of income [22]. However, the need for cross-sectoral modeling is pointed out for better understanding of the situation.

To summarize, all mentioned research agree that elasticity of substitution does not exceed unity. This justifies that the process of replacing labor with capital is not explosive. Thus, the proposition of LS being constant over time has all chances to be true.

The current paper may be categorized as empirical since it tries to estimate the relationship between production factor relative price and their relative supply. We exploit Constant Elasticity of Substitution (CES) function with technical change parameter to estimate the effect of substitution of labor by capital and make forecasts. According to recent research [10, 22] the decomposition of elasticity and forecasts by economic sectors is a valuable extension in the analysis of capital-labor substitution. Such inter-sectoral decomposition was conducted for the U.S. using CES production function with factor-augmenting technical change [14] but neither residuals stationarity nor functional nonlinearity was concerned. At the same point, in each economic sector there should be an intrinsic non-linearity which may be captured either by seeking for the structural break or by implementing the best non-linear form of the model equation. The former method includes applying SupWald test and the latter method implies the empirical search for the best model specification thereby conducting the comparison of these models.

Formula (1) considers the direction of technological progress. In order to allow for the aforementioned non-linearity in the model, the following transformation from the initial production function (1) to the regression models (5) and (6) has to be considered.

$$Y(L,K) = C\left(\alpha(A_tL)^{\frac{\sigma-1}{\sigma}} + (1-\alpha)(B_tK)^{\frac{\sigma-1}{\sigma}}\right)^{\frac{\sigma}{\sigma-1}} \tag{1}$$

Here α is technical change shift parameter, σ—capital-labor elasticity of substitution,

A_t, B_t—labor and capital intensity parameters respectively, L and K—labor and capital stocks respectively.

The ratio of marginal products of (1) may be written as follows (2).

$$\frac{\frac{\partial Y}{\partial L}}{\frac{\partial Y}{\partial K}} = \frac{w}{r} \tag{2}$$

By its definition labor income share θ can be written as follows (3).

$$\theta = \frac{wL}{wL + rK} \tag{3}$$

Differentiating (1) in order to implement in (2) let obtain the following (4).

$$\frac{w}{r} = \frac{\alpha}{1-\alpha}\left(\frac{K}{L}\right)^{\frac{1}{\sigma}}\left(\frac{A_t}{B_t}\right)^{\frac{\sigma-1}{\sigma}} \tag{4}$$

By taking logarithms (4) can easily been transformed to linear in parameters model (5 and 6)

$$ln\left(\frac{w}{r}\right) = ln\left(\frac{\alpha}{1-\alpha}\right) + \frac{1}{\sigma}ln\left(\frac{K}{L}\right) + \frac{\sigma-1}{\sigma}ln\left(\frac{A_t}{B_t}\right) + \varepsilon \tag{5}$$

$$ln\left(\frac{K}{L}\right) = \sigma\, ln\left(\frac{1}{\alpha}-1\right) + \sigma\, ln\left(\frac{w}{r}\right) + (\sigma-1)ln\left(\frac{A_t}{B_t}\right) + \varepsilon \tag{6}$$

Allowing for nonlinearity of the time component (5) and (6) may be written as (7) and (8) respectively.

$$ln\left(\frac{w}{r}\right) = ln\left(\frac{\alpha}{1-\alpha}\right) + \frac{1}{\sigma}ln\left(\frac{K}{L}\right) + \frac{\sigma-1}{\sigma}\boldsymbol{\varphi}(t) + \varepsilon \tag{7}$$

$$ln\left(\frac{K}{L}\right) = \sigma\, ln\left(\frac{1}{\alpha}-1\right) + \sigma\, ln\left(\frac{w}{r}\right) + (\sigma-1)\boldsymbol{\varphi}(t) + \varepsilon \tag{8}$$

Equations (7) and (8) can be represented in terms of following estimators—see Eqs. (9) and (10) respectively.

$$ln\left(\frac{w}{r}\right) = \beta_0 + \beta_{KL}ln\left(\frac{K}{L}\right) + \beta_t\boldsymbol{\varphi}(t) + \varepsilon \tag{9}$$

$$ln\left(\frac{K}{L}\right) = \delta_0 + \beta_{wr}ln\left(\frac{w}{r}\right) + \delta_t\boldsymbol{\varphi}(t) + \varepsilon \tag{10}$$

$\boldsymbol{\varphi}(t)$ here is nonlinearity operator which may count either for structural break or for model specification. In further analysis we concentrate on the following time component $\boldsymbol{\varphi}(t)$ modifications.

$\boldsymbol{\varphi}(t) =$ 0—Hicks-Neutral models with no time.
$\boldsymbol{\varphi}(t) =$ t—linear time component.
$\boldsymbol{\varphi}(t) =$ $\frac{1}{\beta_t}(\beta_{t\_b}t_b + \beta_{t\_a}t_a)$—linear model with structural break point in time.
$\boldsymbol{\varphi}(t) =$ $ln(t+\lambda)$—logarithmic time component.
$\boldsymbol{\varphi}(t) =$ t^{λ}—time component incorporating nonlinearity as power function.
$\boldsymbol{\varphi}(t) =$ $e^{\lambda t}$—exponential time component.
$\boldsymbol{\varphi}(t) =$ $\frac{1}{1+e^{\lambda t}}$—logistic function time component.
$\boldsymbol{\varphi}(t) =$ $\ln(1+\lambda e^{-t})$—logarithmic logistic time component.
$\boldsymbol{\varphi}(t) =$ $\frac{1}{\beta_t}\left(\beta_{t\_b}t_b^{\lambda_b} + \beta_{t\_a}t_a^{\lambda_a}\right)$—time component incorporating nonlinearity as power function with structural break.

$\varphi(t) = \frac{1}{\beta_t}\left(\beta_{t\_b}e^{\lambda_b t_b} + \beta_{t\_a}e^{\lambda_a t_a}\right)$—exponential time component with structural break.

$\varphi(t) = \frac{1}{\beta_t}\left(\frac{\beta_{t\_b}}{1+e^{\lambda_b t_b}} + \frac{\beta_{t\_a}}{1+e^{\lambda_a t_a}}\right)$—logistic function time component with structural break.

The modelling and forecasting procedure includes estimating a separate equation from formulae (7) and (8) with each instance $\varphi(t)$. Here, it is important to mention that parameter λ value is determined iteratively by estimating a set of models within a certain interval of λ value and taking the best one by the criterion of the least RSS.[6] For each model specification the interval has its own length with boudaries determined empirically as the broadest possible.[7] Structural breaks are sought for with Sup Wald test [23]. In summary, nonlinear time components accompanied by structural break analysis is designed to explain nonlinearity inherent to the data.

Several criteria were applied to investigate if a model is consistent and valid for prediction. First, three types of ADF tests[8] were conducted to control for stationarity in residuals [24]. Next, autocorrelation in residuals was also considered via four tests—Durbin-Watson and Breusch-Godfrey for three lags. Remarkably, autocorrelation was mostly present in cases of no stationarity in residuals, which is an expected outcome and stands for no estimation bias for models with stationary residuals. Then, the main regressor significance (either $ln\left(\frac{K}{L}\right)$ in class of models (5) or $ln\left(\frac{w}{r}\right)$ in class of models (6)) was stated as a compulsory condition for a model to be valid for analysis and prediction. The last but not the least, in models with structural break point in time Sup-Wald statistic must have been significant at 5% level.

Thus, the following three degrees of model power were proposed (see Table 1 for summary). $$$ (large dark-blue points on Figs. 1, 2, 3 and 4)—the most powerful: all the regressors are statistically significant, the model has high R-squared adjusted value (>0.8), the main regressor estimate β_{KL} or β_{wr} is positive, all three modifications of ADF-test show residuals stationarity and Durbin-Watson test or one of the three modifications of Breusch-Godfrey[9] tests show no autocorrelation; $$ (medium-size purple points on Figs. 1, 2, 3 and 4)—powerful: all the conditions from $$$ are fullfilled but one regressor is not statistically significant OR all the regressors are statistically significant, the model has relatively high R-squared adjusted value (>0.5), the main regressor estimate β_{KL} or β_{wr} is positive, but ADF with trend does not reject the null hypothesis of no stationarity and Durbin-Watson test or one of the three modifications of Breusch-Godfrey tests show no autocorrelation; $ (small orange points on Figs. 1, 2, 3 and 4)—satisfactory: all the conditions from $$$ are fullfilled but two regressors are not statistically significant OR one regressor is not statistically significant, the model has relatively high R-squared adjusted value

[6] RSS—Residual Sum of Squares.

[7] This was done by searching for the local minimum in RSS vector of the models estimated on λ interval.

[8] ADF with trend, AFD with drift, ADF without trend and drift (moving from relatively more strictly stationarity conditions weaker ones respectively)—see [22].

[9] Durbin-Watson and Breusch-Godfrey tests may further be referred to as DW and BG tests respectively.

Table 1 Three degree of model selection criteria

Model power degree	Residuals stationarity	Residuals autocorrelation	β_{KL} or β_{wr} significance and "+" sign	Other regressors significance	R-squared adjusted
The most powerful $$$	All modifications of ADF test show stationarity	DW test **or** BG tests show no autocorrelation	Compulsory	Compulsory	>0.8
Powerful $$	ADF with trend shows no stationarity	DW test **and** BG tests show no autocorrelation	Compulsory	Constant term **or** one trend component insignificant	>0.5
Satisfactory $	ADF with trend, lagged ADF show no stationarity	DW test **and** BG tests show no autocorrelation	Compulsory	Constant term **and** one trend component insignificant	>0.5

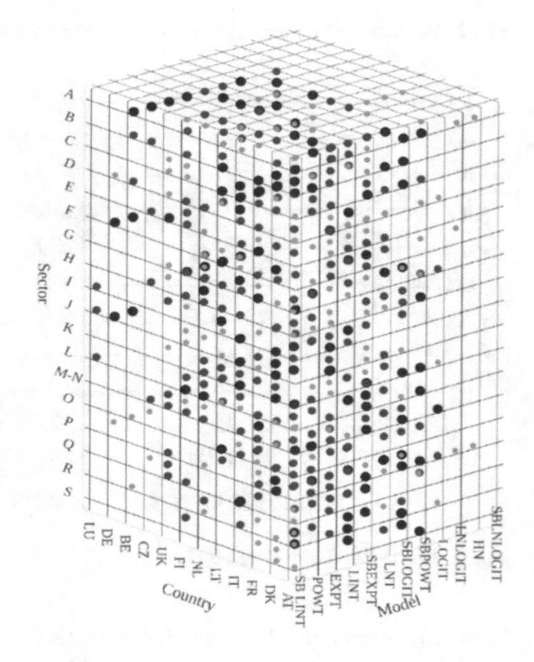

Fig. 1 Cube of selected models by model power criteria

(>0.5), the main regressor estimate β_{KL} or β_{wr} is positive, ADF with trend does not reject the null hypothesis of no stationarity and Durbin-Watson test or one of the three modifications of Breusch-Godfrey tests show no autocorrelation.

In current paper KLEMS statistical databases methodology [25] were exploited for eleven selected EU economies and the UK. Data collection process resembles technique, which used KLEM database [26] on the U.S intersectoral data [14]. However,

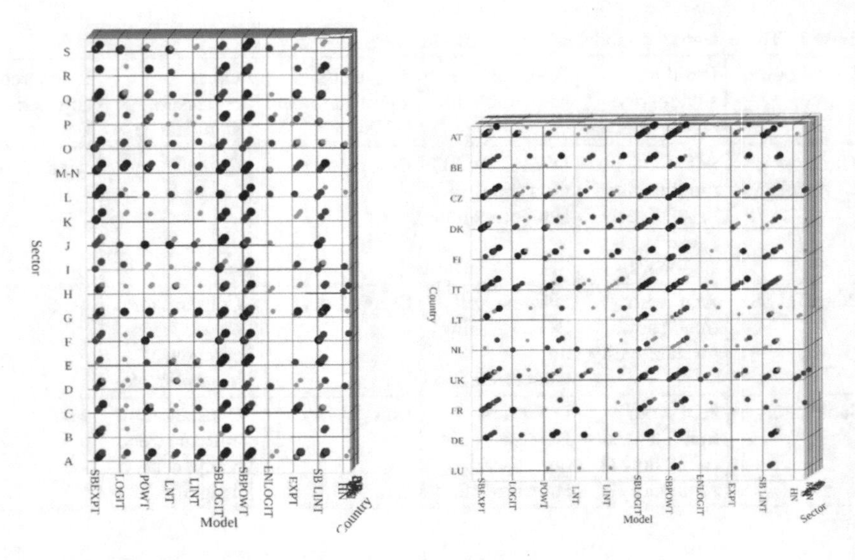

Fig. 2 Projection of the model selection per industry and per country

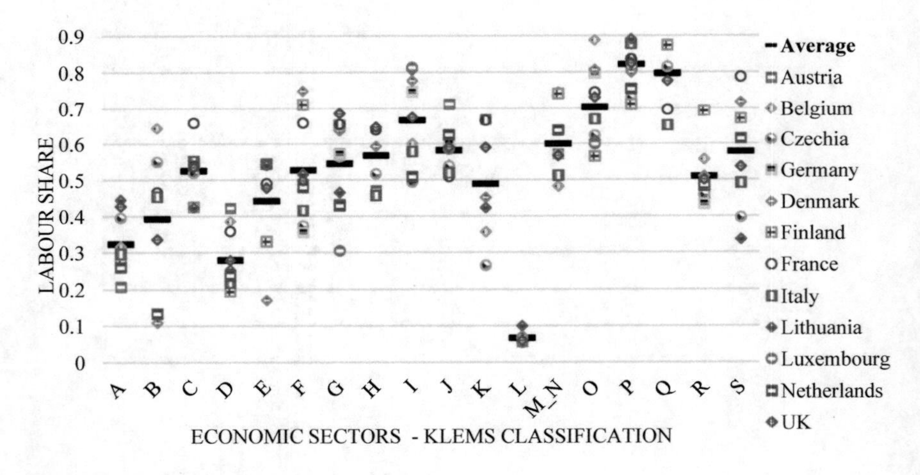

Fig. 3 Intersectoral averaged LS Forecasts

instead of two-digit (35-industries) intersectoral decomposition one-digit KLEMS level (18 industries) is presented—see Table 2—which allows to obtain relatively more aggregate information.

Econometric estimation required data collection for the following variables. For clarity denotions are indicated in parentheses. First, Y (GVA_2010) is Gross Value Added in constant 2010-prices taken in millions of national currency. Second, L (EMP) is a number of persons employed in thousands. Third, K (Kq_GFCF) is net Capital stock in 2010 prices, taken in millions of national currency. Next variables

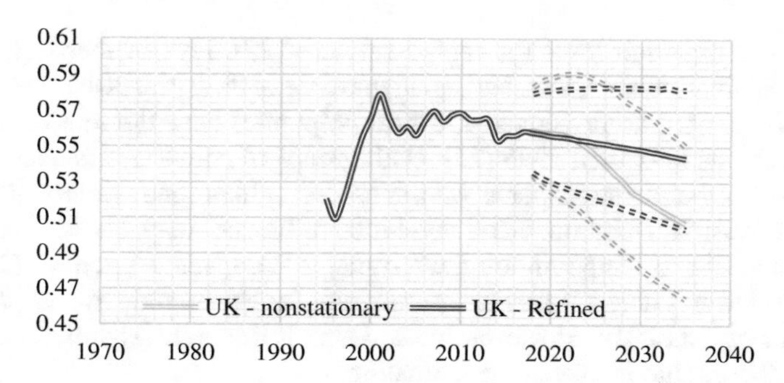

Fig. 4 UK: Industry-weighted LS forecasts: refined models

Table 2 One-digit KLEMS industry classification

A	Agriculture, forestry, and fishing
B	Mining and quarrying
C	Total manufacturing
D	Electricity, gas, steam, and air conditioning supply
E	Water supply; sewerage; waste management and remediation activities
F	Construction
G	Wholesale and retail trade; repair of motor vehicles and motorcycles
H	Transportation and storage
I	Accommodation and food service activities
J	Information and communication
K	Financial and insurance activities
L	Real estate activities
M–N	Professional, scientific, technical, administrative and support service activities
O	Public administration and defence; compulsory social security
P	Education
Q	Health and social work
R	Arts, entertainment, and recreation
S	Other service activities

Source (Stehrer, 2019, pp. 18, 41)

were calculated—wL (COMP_2010) is total compensation of employees in constant prices,[10] w are average wages, a ratio of COMP_2010 to EMP, and r are capital rental prices.[11]

The choice of countries was driven by data availability. The latest data is provided for 29 European countries but only 21 of them have intersectoral decomposition at

[10] $wL = COMP \times \frac{GVA\_2010}{GVA}$, where GVA is in current prices.

[11] $r = \frac{GVA\_2010 - COMP\_2010}{Kq\_GFCF}$.

NACE Rev. 2 one-digit level for capital services [27]. Five of these 21 countries (Hungary, Spain, Romania, Sweden, Japan) have the ending year 2016, and four more countries (Estonia, Latvia, Slovenia, Slovak Republic) have the first reported year for capital inputs later than 1995 [25]. Both groups of countries are excluded from the analysis for the current paper as we impose the unified time span for all analysed economies. Austria, Belgium, Germany, Finland, France, Italy, Netherlands, Spain, Sweden, and United Kingdom are the ten major European economies [26]. From this list we are not able to provide estimations for Spain and Sweden due to the ending year of data 2016. However, we add Czech Republic, Denmark, Lithuania, and Luxembourg into intersectoral estimation.

3 Results

Using OLS we estimated 2 376 pairs of equations which are 12 countries multiplied by 18 economic sectors and by 11 modifications of the time component, i.e. each sector included testing 11 above-mentioned time component modifications. Hicks-neutral models were also estimated for each sector. Appendix 1 provides intersectoral estimates of CES production function for the UK as a single estimation example out of 264 similar estimated sets of models—see Tables 7 and 8.

Applying models selection criteria, mentioned in methodology section, the 3D cube (Fig. 1) is derived. It depicts the density of 'powerful' models that are valid for forecasting. For instance, the projection on the plane (Sector, Model)—see Fig. 2—shows that models with structural breaks in overall produce more so called 'powerful' models in terms of stationarity and in terms of the issues mentioned above as criteria for model power. Moreover, Hicks-neutral models as well as those with logarithmic and loglogistic time shown the most obvious weakness. At the same time, though linear models with no structural breaks are not perfect, they are more 'powerful' than model with logarithmic time. In addition, it can be clearly seen at the plane (Country, Model) that models with structural breaks in time 'SBEXPT', 'SBLOGIT', 'SBPOWT', and 'SBLINT' have more points (specifically the dark-blue and the purple ones) than other models. This, models with structural break have stronger predictive power compared to their analogues without breaking point in time whereas in models which do not consider structural breaks the models incorporating $\varphi(t) = t^{\lambda}$ are the most efficient.

Figure 3 reports between-country variance of intersectoral labour share forecasts, averaged on the forecasting period. Large black rectangles illustrate sectoral average forecast (compounded as between-country average of time-average LS forecast of each country). Table 3 may be helpful to briefly outline from country perspective the interval in which its sectoral time-average LS forecasts lie.

Table 5 contain information on forecast LS trend which is determined as follows: increasing or decreasing LS must have 3 percentage points growth or fall respectively over 18 forecast years, whereas constant LS has less than 3 percentage points

Table 3 Sectoral labour share interval forecast average values by country

LS value	<0,3	0,3–0,5	0,5–0,7	>0,7
Austria	A,L	B,D,H,R	C,G,K,M–N,S	J,P
Belgium		A,D,K,M–N	B,F,G,H,I,J,R	O,P,Q
Denmark	A,B,L	K,R	G,H,J	F,I,M–N,O,P,Q,S
Germany	D	F,R	G,J	I,O
Italy	A,D	B,F,G.H,R	C,E,I,J,M–N,O,P,Q,S	P
Lithuania	D,L	A,C,G,K,S	E,F,M–N,O	I,P
Luxembourg	L	G	J,O	I,P
Netherlands	A,B,D	F,G,R	C,I,J,M–N,S	P
UK	D,L	A,B,E,F	C,G,H,I,J,K,M–N,R,S	O,P,Q
Finland	A,D	C,E	G,O,R,S	F,M–N,P,Q
France	L	B,D,E	C,F,G,H,J,K,Q	O,P,S
Czech Republic	D,K,L	F,I,R,S	B,C,E,H,J,M–N,O	P,Q

variance over the forecast period. Table 4 also shows the overall coverage of countries considered in this research with 'powerful' models. Here it can be seen that Germany and Luxembourg are bad leaders with maximum numbers of economic sectors for which the no forecasts can be produced due to insufficient predictive power of models. Simultaneously, sectors F, H, K, L, Q may be outlined as the least covered by countries empirics.

Tables 4 and 5 allow to conclude that the majority of economic sectors are not expected to fall in future. Nevertheless, 7 sectors out of 18 forecast decreasing LS in

Table 4 Labor share trends by economic sector

Sector Country	A	B	C	D	E	F	G	H	I	J	K	L	M–N	O	P	Q	R	S
Austria	↗	↗	→	↗			→	↘			↗	↘	→		↗		↗	→
Belgium	↗	↗		↗		→	→		↗	↗	↘		→	↘		↗	→	
Denmark	↘	↘				↗	→	→	↗	↘	↘	↘	→	↗	↘	↗	↘	→
Germany				↘		↘	↘		↗	↗							→	
Italy	↗	↗	→		→	↘	↗	→	↗	↗			↗	→	→	→	↗	↗
Lithuania	↗		↘			↘	↗		↗		↘	↗		→	→			→
Luxembourg							↘		→	→		↗		↘	→			
Netherlands	↗	→	→	↘		↘	↘		↘	↗			→		→		↘	↘
UK	→	→	↘	↘	↗	↗	↗	↘	→	→	↗	↗	→	↘	↘	→	↗	→
Finland	↗		↘	→	→	↗	→						↗	↘	↘	↘	↗	
France		↗	↗	↗	→	→	↗				→			↗	↘	→		↗
Czechia	→	↗	↗	↗	↗	↘		↗	↗	↗	↘	↗	↗	→	↗	↗	↘	↗

Table 5 Shares of increasing, constant, and decreasing LS forecasts by sector

Sector	Increasing (%)	Constant (%)	Decreasing (%)	Total (%)
A	56	33	11	100
B	71	14	14	100
C	25	38	38	100
D	38	25	38	100
E	40	60	0	100
F	20	30	50	100
G	27	45	27	100
H	20	40	40	100
I	67	22	11	100
J	63	25	13	100
K	29	14	57	100
L	33	33	33	100
M–N	38	63	0	100
O	11	44	44	100
P	10	50	40	100
Q	29	57	14	100
R	44	22	33	100
S	38	50	13	100

more than one third of countries (which has the valid forecast according to introduced model power characteristics).

Figures 5 and 6 in the Appendix 2 provides profound information on intersectoral LS dynamics in graphs which helps to gain confidence on the figures of averaged foreacts. On 14 of 18 graphs convergence in forecasted LS values may be noted which supports the sectoral average forecast values from Fig. 4.

In addition, it was possible to aggregate LS forecast to the whole economy level from intersectoral decomposition only for the UK, Italy, France, Denmark, and Czech Republic. UK example (see Fig. 4) is helpful to illustrate downward bias imposed on LS forecast unless stationarity, autocorreation of residuals and other models' power criteria are considered. The confidence intervals (dashed lines) for the refined forecasts are on average narrower.

Other countries (see graphs on Fig. 7 in the Appendix 3) also show an upward shift of LS forecast for models refined with the criteria described in methodology of this paper.

4 Conclusions

The current paper is designed to apply intersectoral decomposition on labour share forecasting. This is an attempt to identify the labour share structure in dynamics and to investigate LS trend direction for each KLEMS one-digit economic sector. For the purpose of comparability we selected the countries with equal time span of the necessary data. However, generalization across countries seems to be inefecctive not only due to countries heterogeneity (this may be party solved by grouping up the countries with identical trends direction of forecasted LS), but due to intersectoral estimation of CES production function parameters. Particular cases (see blank spaces in Table 5) shown residuals nonstationarity and, therefore, were not valid for prediction.

Possible dataset extension faces the problem of data convergence before 1995. For example, previous data publications [28, F380] use different methods of capital stocks' calculation compared to the newest dataset [25], i.e. capital stocks are not listed in the former source but instead IRR and Capital compensation are reported. As soon as these nominal IRR are given for a different classification of industries than in before-mentioned newer data source, it became impossible to augment newer data with the older one. Thus, despite there are data for value added and number and compensation of employees for Austria, Germany, Italy, UK, which dates back from 1995, the above-described problem of different methods of capital stock calculation (can be easily seen on example of Italy—see [29, p. 3]) does not allow to conduct analysis on a longer time series than 1995–2017.

Nevertheless, there are possibilities to extend intersectoral data coverage for particular countries. For instance, good converging sample back from 1995 can be found for Denmark, Finland, France, and Sweden [30]. For Belgium data on capital stocks are not available. For Netherlands and Spain there is problematic convergence of labour compensation indicator LAB with the indicator COMP from newer databases.

Appendix 1. Estimation Example

See (Tables 6, 7 and 8).

Appendix 2. Intersectoral LS Forecasts

See (Figs. 5 and 6).

Table 6 Regression (8) estimation—UK, $\varphi(t) = \frac{1}{\beta_t}\left(\beta_{t_b} e^{\lambda_b t_b} + \beta_{t_a} e^{\lambda_a t_a}\right)$

	β_0	β_{KL}	β_{t_b}	β_{t_a}	R^2_{adj}	F stat	λ_b	λ_a	SB year	Sup F
A	−0.698	1.001***	−1.111***	0.407***	0.93	98.750***	−0.71	1.4e-15	2004	17.17***
B	−1.806***	1.089***	0.001***	1.100***	0.87	51.645***	0.33	−3.3e-16	2015	17.23***
C	4.777***	2.501***	−0.056***	−1.400***	0.95	127.468***	0.12	−0.03	2015	15.74***
D	−0.614***	1.103***	0.000***	−0.001**	0.85	43.551***	1.30	0.25	2011	11.39**
E	0.583***	2.639***	−1.083***	−3.017***	0.99	559.791***	−0.06	−0.11	2013	16.15***
F	1.257***	1.719***	0.000***	14.609***	0.95	134.985***	0.79	−0.25	2009	17.50***
G	2.904***	1.740***	−0.053***	−24.947***	0.99	869.939***	0.48	−1.25	1997	13.75***
H	0.684***	0.997***	0.129***	97.663***	0.95	129.960***	0.10	−0.35	2009	18.00***
I	9.019***	3.638***	−.514***	−28.333*	0.93	93.653***	−0.66	−0.36	2010	18.21***
J	−0.215	0.752***	0.000***	0.594***	0.86	46.530***	1.37	−0.18	2001	15.74***
K	3.263***	2.492***	0.400***	4.324***	0.84	38.066***	0.19	−0.20	2001	18.77***
L	−2.774***	0.938***	−0.019***	−1.29 + e142**	0.99	612.918***	0.17	−27.51	2005	15.89***
M–N	3.732***	2.251***	−0.037***	−17.811***	0.87	51.521***	0.64	−1.15	1997	16.56***
O	−0.846***	0.075	−1.089***	−0.572***	0.96	188.851***	−0.17	−0.18	1996	20.13***
P	8.402***	2.957***	−0.804***	−0.178***	0.82	33.826***	−0.16	0.08	1996	18.47***
Q	1.922**	1.292***	−0.029*	0.028***	0.72	19.706***	0.80	0.12	1996	15.23***
R	1.100***	1.476***	−0.009**	−1.1e + 7	0.95	140.837***	0.75	−3.69	1998	6.22
S	−1.054**	0.753***	0.0001***	0.377***	0.94	112.656***	0.46	−0.03	2011	14.01***

Table 7 Residuals stationarity and autocorrelation for regression (8)

ADF trend, lag = 0			ADF trend, lag = 1			ADF drift, lag = 0		ADF drift, lag = 1		ADF, lag = 0	ADF, lag = 1	D-W	Phillips-Perron		
tau3	phi2	phi3	tau3	phi2	phi3	tau2	phi1	tau2	phi1	tau1	tau1	p-value			
−4.65	7.21	10.82	−2.57	2.22	3.31	−4.77	11.4	−2.61	3.42	−4.88	−2.68	0.38	0.76	0.67	0.53
−3.49	4.06	6.09	−3.38	3.96	5.89	−3.54	6.28	−3.26	5.35	−3.63	−3.33	0.04	0.27	0.43	0.62
−3.12	3.48	5.14	−2.60	2.56	3.75	−3.19	5.17	−2.54	3.33	−3.25	−2.56	0.00	0.05	0.14	0.24
−7.39	18.19	27.28	−5.31	9.47	14.16	−7.57	28.7	−5.34	14.29	−7.75	−5.49	0.98	0.02	0.01	0.02
−4.08	5.58	8.37	−3.45	4.14	6.14	−4.18	8.76	−3.45	6.02	−4.29	−3.52	0.16	0.71	0.85	0.56
−3.67	4.53	6.77	−3.38	4.04	5.98	−3.71	6.93	−3.22	5.27	−3.81	−3.28	0.06	0.34	0.61	0.39
−3.03	3.08	4.60	−2.20	1.67	2.50	−3.11	4.86	−2.30	2.65	−3.19	−2.37	0.02	0.11	0.28	0.17
−3.73	4.72	7.08	−2.46	2.08	3.03	−3.78	7.16	−2.53	3.30	−3.87	−2.63	0.06	0.43	0.72	0.57
−3.25	3.54	5.27	−4.19	6.16	9.24	−3.18	5.10	−3.75	7.03	−3.27	−3.82	0.02	0.12	0.08	0.16
−5.51	10.31	15.44	−6.01	12.2	18.34	−5.58	15.6	−5.89	17.34	−5.70	−5.93	0.53	0.40	0.06	0.13
−5.01	8.38	12.56	−4.00	5.34	8.00	−5.12	13.1	−4.09	8.36	−5.25	−4.20	0.46	0.54	0.43	0.10
−5.35	9.68	14.48	−3.54	4.19	6.27	−5.49	15.1	−3.64	6.65	−5.61	−3.74	0.54	0.46	0.60	0.79
−4.32	6.35	9.52	−4.00	5.36	8.01	−4.46	9.97	−3.92	7.74	−4.57	−4.04	0.25	0.93	0.34	0.54
−1.63	0.95	1.40	−2.53	2.19	3.27	−1.70	1.47	−2.63	3.48	−1.75	−2.71	0.00	0.00	0.00	0.00
−3.40	3.89	5.82	−3.06	3.18	4.75	−3.50	6.14	−3.17	5.04	−3.59	−3.26	0.04	0.28	0.37	0.46
−2.09	1.53	2.28	−2.12	1.56	2.32	−2.16	2.35	−2.21	2.46	−2.22	−2.28	0.00	0.00	0.01	0.02
−3.65	4.76	7.10	−2.71	2.54	3.81	−3.76	7.13	−2.78	3.86	−3.83	−2.83	0.02	0.21	0.42	0.56
−5.28	9.40	14.08	−4.52	6.87	10.30	−5.40	14.6	−4.57	10.46	−5.52	−4.66	0.50	0.44	0.23	0.31

Table 8 Critical values for respective ADF-test

ADF trend, critical values			ADF drift, crit.values		ADF, crit.value
tau3	phi2	phi3	tau2	phi1	tau1
– 3,60	5,68	7,24	– 3,00	5,18	−1,95

Appendix 3. Aggregate LS forecasts

See (Fig. 7).

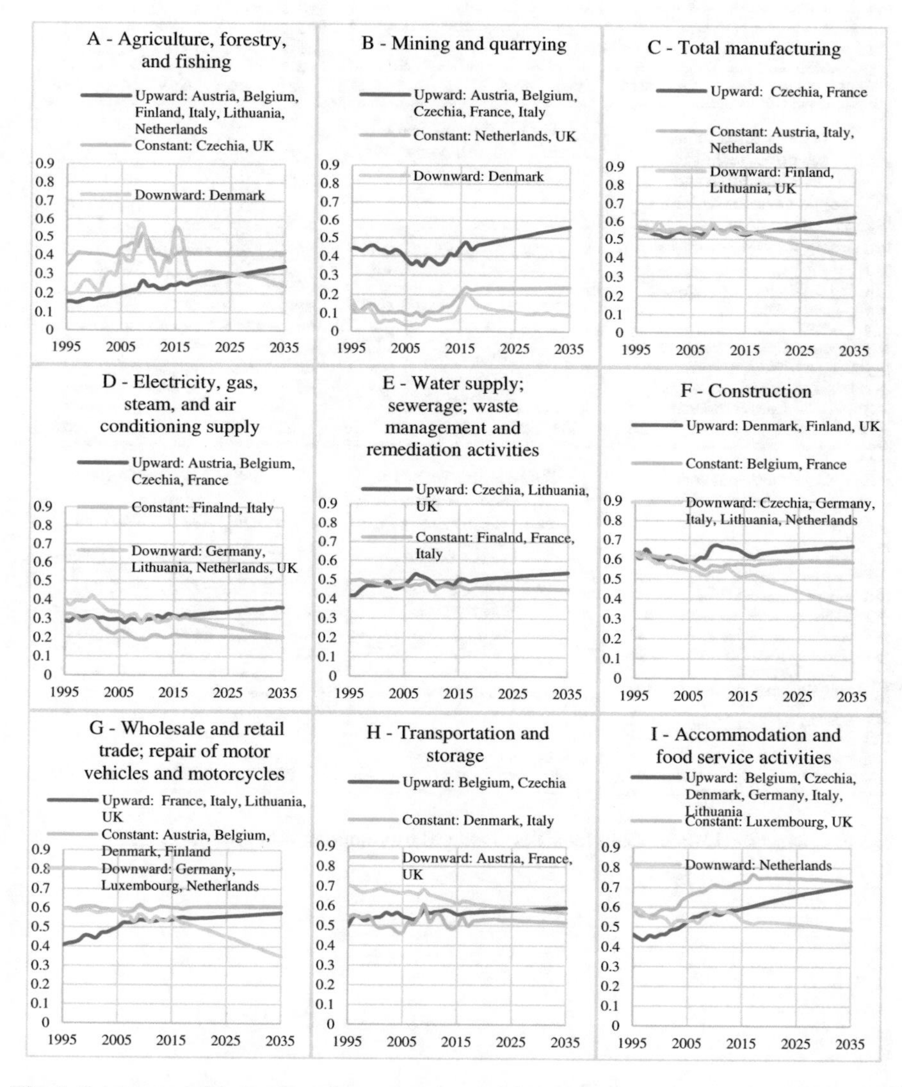

Fig. 5 Intersectoral labour share forecasts, averaged by countries with upward, constant, and downward labour share trend (sectors A-I)

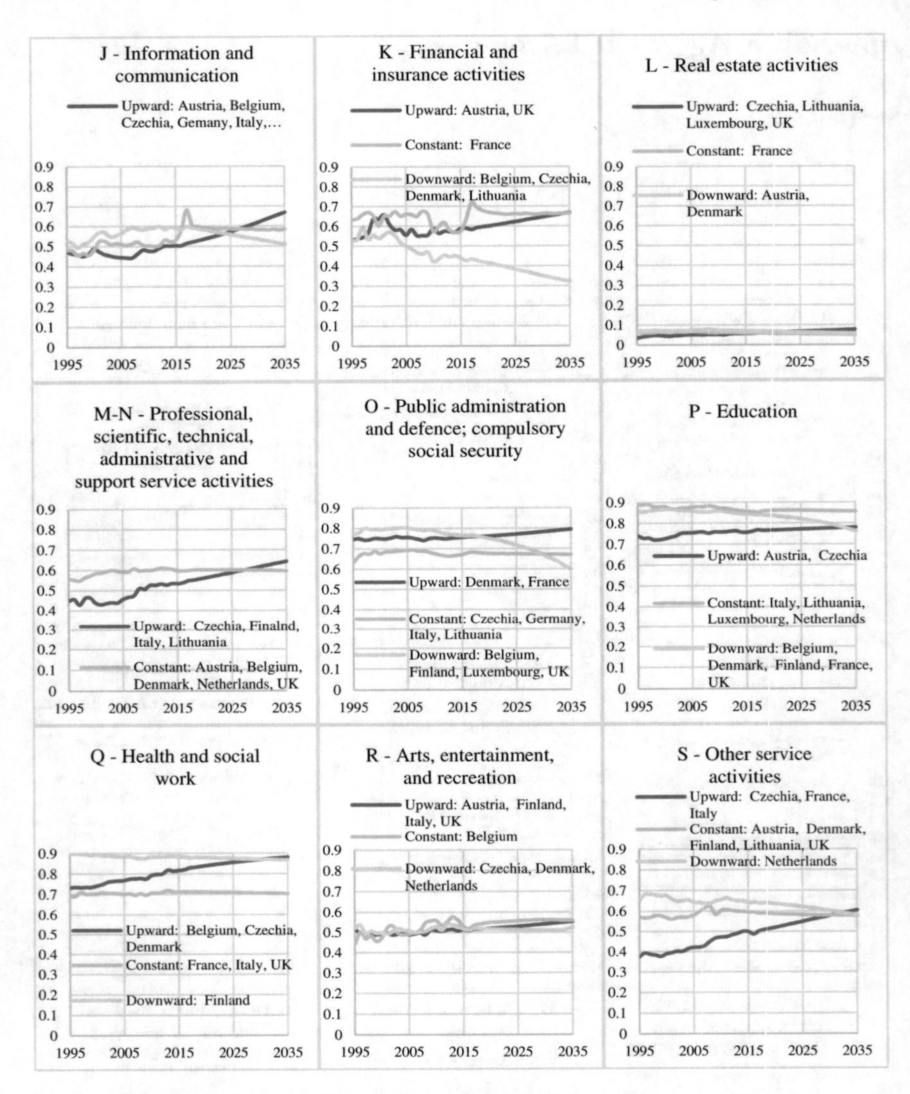

Fig. 6 Intersectoral labour share forecasts, averaged by countries with upward, constant,

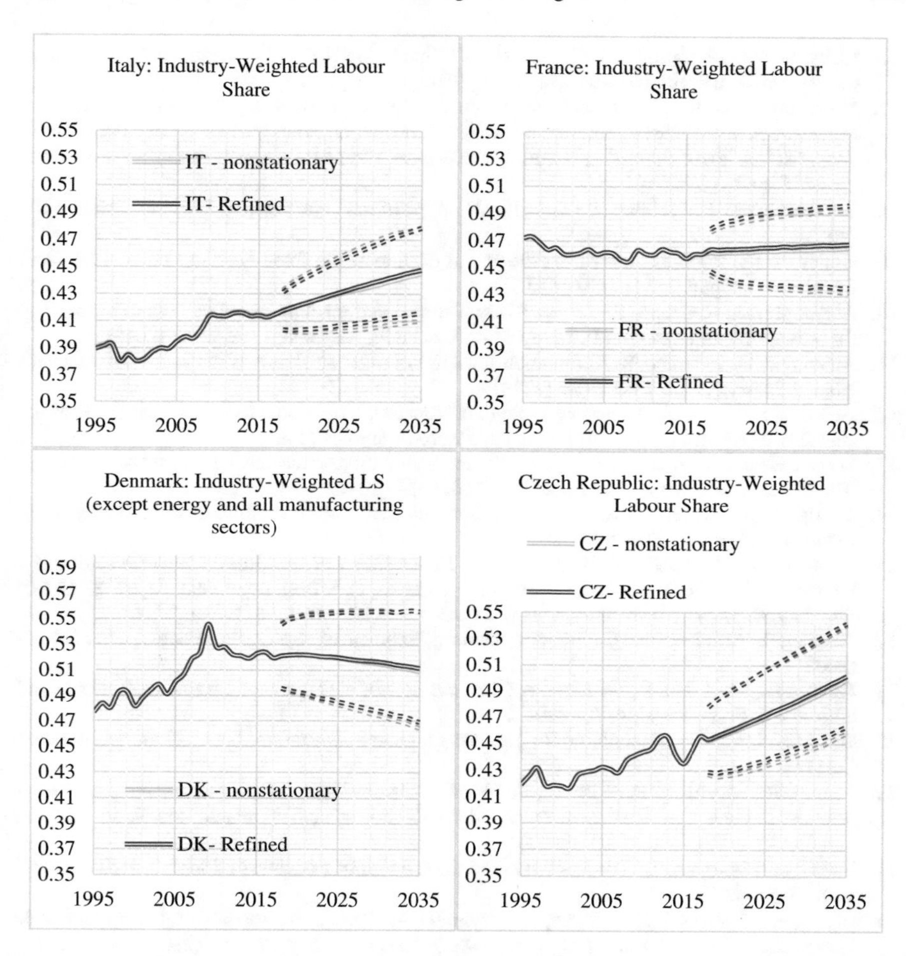

Fig. 7 Aggregate labour share forecasts convoluted from intersectoral forecasts for Czech Republic, Denmark, France, and Italy

References

1. Autor, D., Salomons, A.: Is automation labor share–displacing? Productivity growth, employment, and the labor share. Brookings Papers on Economic Activity, pp. 1–63 (2018)
2. Ilyashenko, O., Kovaleva, Y., Burnatcev, D., Svetunkov, S. Automation of business processes of the logistics company in the implementation of the IoT. In: IOP Conference Series: Materials Science and Engineering, vol. 940, no. 1, № 012006 (2020)
3. Manyika, J., Mischke, J., Bughin, J., Woetzel, J., Krishnan, M., Cudre, S.: A new look at the declining labor share of income in the United States (2019)
4. Autor, D., Dorn, D., Katz, L.F., Patterson, C., Van Reenen, J.: Concentrating on the fall of the labor share. Am. Econ. Rev. **107**, 180–185 (2017)
5. Autor, D., Dorn, D., Katz, L.F., Patterson, C., Van Reenen, J.: The fall of the labor share and the rise of superstar firms. Q. J. Econ. **135**, 645–709 (2020)

6. Elsby, M.W.L., Hobijn, B., Şahin, A.: The Decline of the U.S. Labor share. Brookings Papers on Economic Activity, vol. 2013, pp. 1–63 (2013)
7. Maydanova, S., Ilin, I. Strategic approach to global company digital transformation. In: Proceedings of the 33rd International Business Information Management Association Conference, IBIMA 2019: Education Excellence and Innovation Management through Vision 2020, pp. 8818–8833 (2019)
8. Karabarbounis, L., Neiman, B.: The global decline of the labor share. Q. J. Econ. **129**, 61–103 (2013)
9. Gutiérrez, G., Piton, S.: Revisiting the global decline of the (Non-housing) labor share. Am. Econ. Rev. Insights **2**, 321–338 (2020)
10. Acemoglu, D., Restrepo, P.: The race between man and machine: implications of technology for growth, factor shares, and employment. Am. Econ. Rev. **108**, 1488–1542 (2018)
11. Acemoglu, D., Restrepo, P.: Automation and new tasks: how technology displaces and reinstates labor. J. Econ. Perspect. **33**, 3–30 (2019)
12. Antras, P.: Is the U.S. Aggregate production function Cobb-Douglas? New estimates of the elasticity of substitution. Contributions in Macroeconomics (2004)
13. León-Ledesma, M.A., McAdam, P., Willman, A.: Identifying the elasticity of substitution with biased technical change. Am. Econ. Rev. **100**, 1330 (2010)
14. Young, A.T.: U.S. elasticities of substitution and factor augmentation at the industry level. Macroecon. Dyn. **17**, 861–897 (2013)
15. Knoblach, M., Roessler, M., Zwerschke, P.: The elasticity of substitution between capital and labour in the US Economy: a meta-regression analysis. Oxf. Bull. Econ. Stat. **82**, 62–83 (2020)
16. Freeman, R.B.: Who owns the robots rules the world. IZA World of Labor (2015)
17. Acemoglu, D.: Labor- and Capital-Augmenting Technical change. J. Eur. Econ. Assoc. **1**, 1–37 (2003)
18. Klump, R., McAdam, P., Willman, A.: The normalized CES production function: theory and empirics. J. Econ. Surv. **26**, 769 (2012)
19. Gimpelson, V., Kapeliushnikov, R.: Russian employee: education, occupation, qualification (2011)
20. Klump, R., McAdam, P., Willman, A.: Factor substitution and factor-augmenting technical progress in the United States: a normalized supply-side system approach. Rev. Econ. Stat. **89**, 183 (2007)
21. Koehl, L., Philippon, T.: The labor share in the long term: a decline? Economie et Statistique. **510**, 35–51 (2019)
22. Akaev, A., Devezas, T., Ichkitidze, Y., Sarygulov, A.: Forecasting the labor intensity and labor income share for G7 countries in the digital age. Technol. Forecast. Soc. Change **167**, 120675 (2021)
23. Andrews, D.W.K.: Tests for parameter instability and structural change with unknown change point. Econometrica **61**, 821–856 (1993)
24. Pfaff, B., Zivot, E., Stigler, M.: Package 'urca' unit root and cointegration tests for time series data. https://cran.r-project.org/web/packages/urca/urca.pdf. Accessed 22 Mar 2021
25. Stehrer, R., Bykova, A., Jäger, K., Reiter, O., Schwarzhappel, M.: Industry level growth and productivity data with special focus on intangible assets (2019)
26. Jorgenson, D.W.: 35 Sector KLEM (2007)
27. EUROSTAT: NACE rev. 2. (2008)
28. O'Mahony, M., Timmer, M.P.: Output, input and productivity measures at the industry level: the EU KLEMS database. Econ. J. **119**, F374–F403 (2009)
29. Gouma, R., Timmer, M.: Description of methodology and country notes for Italy. Groningen Growth and Development Centre (2012)
30. Jäger, K.: EU KLEMS growth and productivity accounts 2017 release, statistical module. In: The Conference Board (2018)

Rationale for Information and Technological Support for the Enterprise Investment Management

Anastasiia Grozdova, Svetlana Shirokova, Olga Rostova, Anastasiia Shirokova, and Anastasiia Shmeleva

Abstract The article justifies the feasibility of developing and implementing an automated system "Investment Planning and Control" to optimize the existing business processes of the oil refinery related to investment activities. The key business processes of the enterprise in the field of investment were analyzed, the main problems related to the development, approval and coordination of investment projects and programs, as well as control at the stage of their implementation were identified. The project stages are considered, as well as their peculiarities are identified taking into account the specifics of the enterprise activity. The process of development and implementation of an automated system allowing to manage investment activities at the research site is described, the effectiveness of this implementation is evaluated, recommendations for its further development and use are given.

Keywords Information system · Investment activities · Investment project management · Business processes · Project performance assessment · Project management

1 Introduction

Research into the design and development of information technology support for the processes of the refinery is important and necessary for improving the efficiency of the enterprise [1]. The role of automation increases every year and becomes more and more obvious for the management of companies.

The oil industry is a branch of the economy engaged in the extraction, processing, transportation, warehousing and sale of oil, as well as related petroleum products.

A. Grozdova · S. Shirokova (✉) · O. Rostova · A. Shirokova
Peter the Great St. Petersburg Polytechnic University, Saint-Petersburg, Russia

O. Rostova
e-mail: o.2908@mail.ru

A. Shmeleva
Marriott International Yerevan, Yerevan, Armenia

C. Jahn et al. (eds.), *Algorithms and Solutions Based on Computer Technology*,
Lecture Notes in Networks and Systems 387,
https://doi.org/10.1007/978-3-030-93872-7_15

In Russia, oil is an important export item. As a result, a large number of funds are invested in this industry.

Like any large enterprise, an oil refinery cannot exist without any information system. Complex purchasing, planning, production, sales processes that include sub-process groups cannot exist without any systematization and accounting [2]. Thus, the development of information systems for oil refining enterprises is one of the key points of competent management and implementation of the production process.

There are several types of information systems designed for the oil industry. The first is a system that is designed to create a single database, as well as to implement methods for collecting and storing information. This type of information system allows you to improve the efficiency of production management and access to data [3].

The second type of information systems is automated information systems for transporting raw materials in pipelines of the main type. Such information systems are developed for managers, as well as management personnel who are responsible for the entire range of work in this activity [4].

The third type of information systems is the system for maintaining a register of investment activities, which will ensure not only the formation of a database, but also the recording of financing and disbursement of funds. In addition, such systems have the ability to prepare and generate various analytical data, converting information into analytical reports [5].

It is worth noting that absolutely all companies that are engaged in oil and gas production, actively master modern technologies and introduce automated information systems. Many enterprises resort to the development of an information system by their IT departments, which makes the information product as unique and sharpened as possible for certain features of the management of both personnel and the production process.

Thus, the topic of research on the design features of information systems for oil refineries and their development is quite relevant for study and development.

2 Materials and Methods

The purpose of the study was to substantiate the need to develop and implement an automated system for managing the investment activities of an oil refinery.

The methodological basis of the study was the proceedings [6–8], as well as studies on the management of project activities of organizations [9, 10] and investment management [11, 12]. The development of these concepts in the direction of methodological approaches to assessing the effectiveness of information technology projects in business has been gained in research [13, 14].

The object of the study was an oil refinery, which is one of the five largest enterprises in the country in this area of activity and produces all types of fuel, as well as petrochemical and paint products for household chemicals and the construction industry.

The main tasks of creating an automated system "Investment Planning and Control" were to increase the efficiency of information exchange between enterprise divisions in the preparation of investment projects, ensure control over the progress of pre-competitive work of divisions, create lists of investment objects taking into account the readiness and criticality of the proposed works, as well as ensure the "continuity" of various stages of each investment project with the indication of a unique object code for it [15].

3 Results

The paper presents a detailed description of the research object, a description of existing IT solutions that are actively used in the enterprise, as well as a justification for the need to develop and implement NPP "Investment Planning and Control" to optimize the existing business processes of the enterprise related to investment activities.

The authors considered the issues of planning investment projects at the enterprise, namely: the principles of the investment policy of the enterprise, the process and stages of implementation of investment projects and investment programs.

The investment policy of the enterprise is reflected in the emerging strategic directions of development. These areas are detailed in the technical development plans, which contain investment projects in terms of technical re-equipment and reconstruction of existing or construction of new facilities of the enterprise [16].

The implementation of investment projects can include several stages [17, 18]:

- preparation of feasibility study, performance of design and survey works;
- execution of design and development of detailed documentation;
- procurement of process equipment requiring long-term manufacture;
- procurement of materials and equipment;
- execution of construction and installation, commissioning works;
- commissioning of facilities.

It should be noted that not all projects can be implemented within one calendar year, for example, due to the long period of equipment manufacture. Individual investment projects accepted for execution together constitute the investment program of the enterprise, which is drawn up for a period of 5 years. The project documentation and the project schedule are the initial information for the planning of the disbursement of funds.

As a rule, the funding plan differs from the implementation plan due to the availability of advance payments under contracts, advance payment for the manufacture of equipment, a temporary lag between the performance of work and their payment, fixed in the terms of payment under contracts.

To calculate the required funding for the coming year, the costs of the investment project are pre-detailed. Along with the implementation plan and the funding plan, an entry plan is drawn up, it is required to include funds allocated for depreciation of fixed assets in the budget of income and expenses of the enterprise [19].

Development, financing and commissioning plans are drawn up for the entire duration of the investment project. The planning horizon and detail are determined by the information available at that time. The list of investment projects, the financing or execution of which is required in the planned year, structured according to the characteristics of belonging to the type of construction is called the cover list of objects of technical re-equipment, reconstruction and new construction (hereinafter referred to as LOTRNC). Figure 1 shows the overall organization of investment activities.

The authors conducted an in-depth analysis of business processes to identify problems that require information support [20]. As an example, Fig. 2 presents the business process of including the initiative in the LOTRNC before optimization. The work provides a detailed description of the cover list, its structure and methodology for assigning a particular investment project to sections of this structure, as well as a

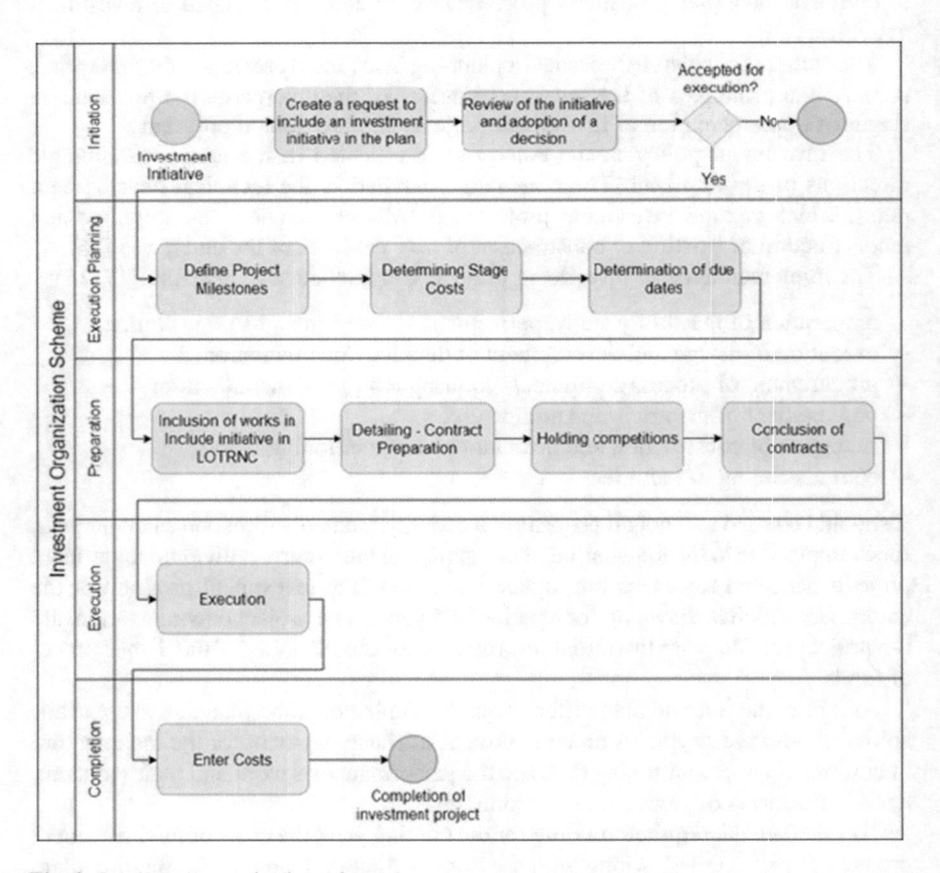

Fig. 1 Investment organization scheme

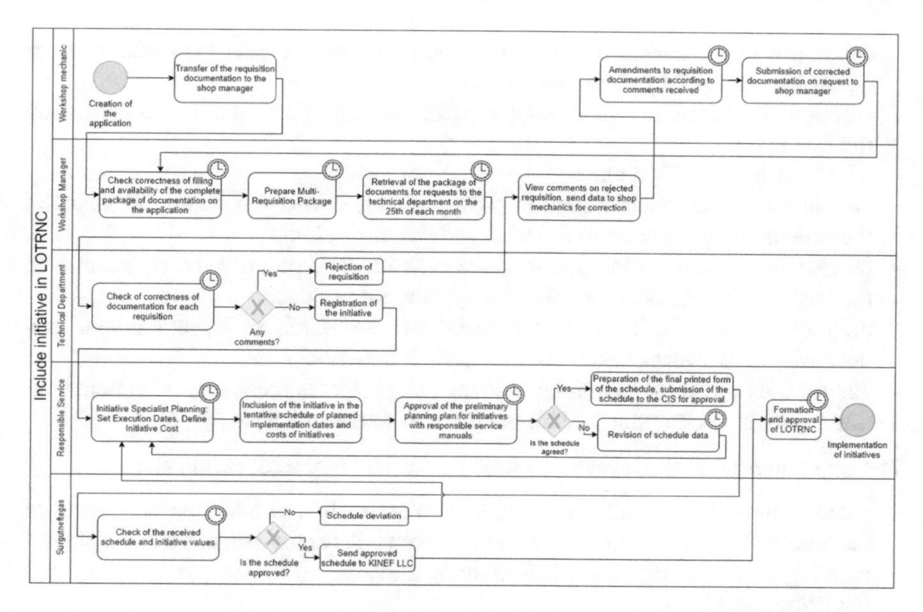

Fig. 2 Business process of inclusion of the initiative in the list of technical re-equipment, reconstruction and new construction facilities

method for compiling it, starting from the registration of the initiative, ending with the approval of the annual plan of the LOTRNC and the implementation of initiatives.

The diagram highlights the main stages of a rather complex process. Many stages take a long time. This time cost is due to the fact that the employees of the enterprise have to maintain all the documentation involved in the business process manually. The documentation includes: a package of documents for the initiative, acts of approval or rejection of initiatives, schedules of planned deadlines and costs, which have many iterations due to the complex process of coordination between departments of the enterprise. The human factor had a great influence on the implementation of the business process for the inclusion of the initiative in the LOTRNC.

The implementation of the automated system "Investment Planning and Control," as part of the described business process, will allow:

- minimize the impact of the human factor;
- to allow the employees of the enterprise more time to complete the stages of this business process;
- provide a single information space for employees to complete the steps of the business process.

The purpose of AS "Investment Planning and Control" is to automate the following processes of investment activity of the enterprise:

- maintenance of the list of planned investment projects;
- storage of justifications and related documentation for investment projects;

- development of execution plan for each investment project—indication of execution stages, their dates and cost;
- creation of a list of technical re-equipment, reconstruction and new construction facilities;
- development of design plans;
- enabling the automated system to provide detailed information on the stages of the investment project up to the level suitable for bidding;
- preparation of tenders for performance of works and purchase of equipment on the basis of information specified for investment projects;
- preparation of schedules of contracts for the purchase of equipment and performance of works based on the results of tenders;
- control over the implementation of investment projects accepted for implementation in terms of compliance with the planned dates and costs.

The cost planning of the investment project consists of the following steps:

- entering into the automated system, known at the time of planning, the stages of the investment project and the services responsible for the execution of the stages;
- entering details (if necessary information is available) for the planned stages of the investment project;
- specify positions for the planned stages of the investment project or their details;
- indication of the planned cost of the investment project stages known at the time of planning.

The study developed a plan for the development and implementation of the information system, the main stages of which are presented in Table 1. In addition, a detailed description of the process of development and implementation of AS "Investment Planning and Control" at the enterprise is given, as well as the appearance of software modules.

In order to make an investment decision, the commercial effectiveness of the information system implementation was assessed using the discounted cash flow methodology. The net present value and discounted payback period of the project were calculated.

Due to the fact that the company has developed and implemented an information system, the efficiency of work has increased significantly. A list of investment projects planned for execution was introduced, the storage of justifications and related documentation for investment projects was established. The formation of a plan for the execution of each investment project—the indication of the stages of execution, their timing and cost, as well as the formation of an in-house cover list of technical re-equipment, reconstruction and new construction objects are now carried out in an automated system, which significantly reduces the time for this work.

AS "Investment Planning and Control" made it possible for users to specify detailed information about the stages of the investment project up to the level suitable for the formation of tenders.

The main goals pursued in the development and implementation of the AS "Investment Planning and Control" were achieved, namely, to improve the efficiency

Table 1 Stages of information system development and implementation

Stages
1 Pre-design survey
1.1. Analyze business processes and existing enterprise IT architecture 1.2. Formulate clear implementation objectives, develop implementation plan
2. Project Team Generation and SOW Development
3. Building a new enterprise IT architecture
4. Development of the information system
5. Purchase of equipment
6. Setting Up the Information System
6.1. Setting Up IP According to Business Specifics 6.2. Integration of IE with existing systems 6.3. Data Synchronization
7. Installation and configuration of equipment
8. Writing User's Guide
9. Staff training
10. Commissioning
11. IS testing
12. Documentation Development
13. Pilot operation
14. IS debugging
15. Data Migration
16. Industrial operation

of information exchange between business units in the preparation of investment projects, as well as to provide information on the progress of investment projects for a plan of factual analysis of the timing and cost. The latter is a key aspect for making correct management decisions by the top management of the enterprise, since an accurate analysis of the information obtained from the system allows you to form clear plans for the implementation of investments, which is very significant for such a large enterprise.

After the introduction of the automated system, the company's employees began to quickly solve such tasks as ensuring control over the preparation and formation of the LOTRNC, ensuring the planning of financing of technical re-equipment, reconstruction, new construction facilities included in the VVTS, as well as providing this LOTRNC to the parent company.

The introduction of AS "Investment Planning and Control" allowed to reduce the share of manual labor of the company's employees in terms of planning investment projects, drawing up LOTRNC and forming.

Users working in the system showed active interest in the training process and, subsequently, working in it. The management of the enterprise has the possibility of long-term and accurate planning of the investment process in the organization.

4 Discussion

Such projects often face problems of not meeting deadlines, non-compliance of the project product with the required quality criteria, exceeding planned costs. In this regard, I would like to note the importance of organizing project activities as part of the development and implementation of an automated system. The project is the implementation of certain work, which requires preparation, careful analysis, elaboration of options, justification of allocated resources. Such a complex process in the organization undoubtedly requires sound management of information flow systems both internally and with external agents and the external environment. This means that in order to achieve positive results, each member of the project team and each interested person must have full up-to-date information about the project and its components, concentrated in one information resource. Successful project activities also involve the teamwork of a large number of employees [9].

The introduction of the information system largely determines the further activity and competitiveness of the enterprise. The information system allows the company to reduce transaction costs, receive additional income due to an increase in product turnover. In the case of the company in question, this means reducing the time and improving the accuracy of analytics and planning, correctly managing production processes, increasing its efficiency, monitoring the state of production capacities, as well as increasing the investment attractiveness of the company [17].

5 Conclusions

Thus, in accordance with the described business processes, the following are implemented in AS "Investment Planning and Control":

- possibility to create a list of the investment project works on the basis of templates with indication of planned deadlines, responsible services and divisions of the enterprise;
- the possibility of creating a list of tenders for the provision of services with indication of planned costs, recording the results of tenders and forming contract items on their basis;
- the possibility of obtaining a tool to monitor the implementation of the investment programs.

At the moment, the oil refinery successfully operates an information system, which was developed and implemented by the partner company. The system is constantly updated and improved. Employees willingly work in the system, cooperate with the information support service, offering the necessary innovations, show interest. The management of the enterprise plans to continue the process of automation of the company.

Information support for the investment process will allow the company to coordinate investment projects and programs with the strategic goals and objectives of the organization.

References

1. Anisiforov, A.B., et al.: Methods for Evaluating the Effectiveness of Information Technology Projects in Business. Publishing House of the Polytechnic University, St. Petersburg (2018)
2. Shirokova, S., Rostova, O., Chuprikova, A., Zharova, M.: Automation of warehouse accounting processes as an integral part of digital company. In: IOP Conference Series: Materials Science and Engineering 2020, vol. 940, no. 1, pp. 012016 (2020)
3. Lavrov, E.A., Zolkin, A.L., Aygumov, T.G., Chistyakov, M.S., Akhmetov, I.V.: Analysis of information security issues in corporate computer networks. In: IOP Conference Series: Materials Science and Engineering, vol. 1047, no. 1, pp. 012117 (2021)
4. Klimin, A.I., Pavlov, N.V., Efimov, A.M., Simakova, Z.L.: Forecasting the development of big data technologies in the Russian Federation on the basis of expert assessments. In: IBIMA 2018, pp. 1669–1679 (2018)
5. Grashchenko, N. et al.: Justification of financing of measures to prevent accidents in power systems. In: EMMFT 2018. Advances in Intelligent Systems and Computing, vol. 983. Springer, Cham (2019)
6. Jeston, D.: Business Process Management. A practical guide to successful project implementation. Alpina Digital, Moscow (2008)
7. Rudskoy, A., Borovkov, A., Romanov, P., Kolosova, O.: Reducing global risks in the process of transition to the digital economy. In: IOP Conference Series: Materials Science and Engineering (2019). https://doi.org/10.1088/1757-899X/497/1/012088
8. Rostova, O., Zabolotneva, A., Dubgorn, A., Shirokova, S., Shmeleva, A.: Features of using blockchain technology in healthcare. In: Proceedings of the 33rd International Business Information Management Association Conference, IBIMA 2019: Education Excellence and Innovation Management through Vision 2020. pp. 8525–8530 (2019)
9. Ilyin, I.V. et al.: Project Management. Fundamentals of theory, methods, project management in the field of information technology. Publishing house of the Polytechnic University, St. Petersburg (2012)
10. Mentsiev, A.U., Guzueva, E.R., Yunaeva, S.M., Engel, M.V., Abubakarov, M.V.: Blockchain as a technology for the transition to a new digital economy. J. Phys. Conf. Ser. **1399**(3), 1–5 (2019)
11. Zhuravlyov, V., Khudyakova, T., Varkova, N., Aliukov, S., Shmidt, S.: Improving the strategic management of investment activities of industrial enterprises as a factor for sustainable development in a Crisis. Sustainability 2019, vol. 11, pp. 6667 (2019)
12. Bulatova, E.I., Amirova, E.F.: Financial impact of digital technologies as a promising element of import substitution. Int. J. Financ. Res. **11**(5), 392–398 (2020)
13. Prisyazhnyuk, S.P., et al.: Indices of the effectiveness of information protection in an information interaction system for controlling complex distributed organizational objects. Autom. Control. Comput. Sci. **51**(8), 824–828 (2017)
14. Tselykh, A.B.: Evaluating the effectiveness of IT projects. "Balanced" approach. https://www.cfin.ru/management/practice/supremum2002/18.shtml. Accessed 17 Mar 2021

15. Zhahov, N.V., Krivoshlykov, V.S., Aleeva, E.A., Nesenyuk, E.S.: Inevi-tability of structural and economic reforms of regional economy. In: 33th Inter-National Business Information Management Association Conference, pp. 4392–4397. Granada, Spain (2019).
16. Khudyakova, T., Shmidt, A., Shmidt, S.: Implementation of controlling technologies as a method to increase sustainability of the enterprise activities. Entrep. Sustain. Issues **7**(2), 1185–1196 (2019)
17. Ilyin, I.V., et al.: Investment Management. Publishing house of the Polytechnic University, St. Petersburg (2017)
18. Amirova, E.F. et al.: Import substitution as an economic incentive mechanism for Russian commodity producers. Int. J. Civil Eng. Technol. T. **10**. № 2, 926–931 (2019)
19. Zharova, M., Shirokova, S., Rostova, O.: Management of pilot IT projects in the preparation of energy resources. In: E3S Web of Conferences, vol. 110, pp. 02033 (2019)
20. Repin, V.V.: Business processes. Modeling, implementation, and management. Moscow (2014)

Tools for Modeling Heat Flows from Buildings in the Context of Digital Transformation of the Urban Environment

Igor Evsikov, Timur Ablyazov, and Andrei Aleksandrov

Abstract Digital transformation of the human environment is one of the key tasks of the formation of the digital economy. In the context of urbanization and an increase of the load on infrastructure, the concept of a smart city is becoming more widespread; the increase in the energy efficiency of buildings and structures is highlighted among the ways of its implementation. In addition, digital transformation implies the creation of digital models of cities in order to visualize physical objects of the human environment. Within the framework of this paper, the authors studied various approaches to ensuring energy efficiency in smart cities, studied the issues of integrating 2D models from geo information systems into more visual 3D models of cities, and also assessed the importance of an interdisciplinary approach in the implementation of projects in the area of modeling individual elements of urban infrastructure. In the course of the study, the authors characterized the features of open data included in the OpenStreetMap geographic information system, and a script was written in the Grasshopper visual programming language that forms 3D models of buildings. As a result of the application of the developed script and formulas substantiating the process of transferring heat flows, a toolkit was developed for constructing a map-diagram of the distribution of heat flows from buildings, which made it possible to visualize the processes inherent in the course of energy consumption. The results of the work can be used to assess the energy efficiency of development and visualization of urban objects within the framework of building a digital model of the city.

Keywords Heat flows · 3D model · Open data · Energy efficiency · Interdisciplinary approach

I. Evsikov · T. Ablyazov (✉)
Saint Petersburg State University of Architecture and Civil Engineering, 4 Vtoraya Krasnoarmeiskaya, 190005 Saint Petersburg, Russian Federation

I. Evsikov
e-mail: ievsikov@lan.spbgasu.ru

A. Aleksandrov
South-Eastren Finland University of Applied Sciences, 3 Patteristonkatu, 50100 Mikkeli, Finland

C. Jahn et al. (eds.), *Algorithms and Solutions Based on Computer Technology*, Lecture Notes in Networks and Systems 387,
https://doi.org/10.1007/978-3-030-93872-7_16

1 Introduction

Currently, the processes of urbanization and digital transformation of the human environment are among the most significant trends in the development of modern cities. According to estimates by the World Bank, by 2050 the urban population may double its current level, which will lead to a concentration of 80% of GDP in cities [1]. Accordingly, it is necessary to solve the problem of increasing the load on the existing urban infrastructure based on the use of digital technologies within the framework of the smart city concept, covering the transformation of the human environment in many areas: transport, construction, economics and management, medicine and education, as well as energy consumption [2].

In our opinion, the issues of improving the energy efficiency of buildings and structures are a key element in increasing the comfort and safety of the human environment. At present, the distribution of heat flows from buildings has not been sufficiently studied while analyzing large data arrays in the framework of the study of the processes of transformation of urban areas, especially from the point of view of visualization of physical objects of the human environment, including the fact that tools for their study are not perfect today.

The development of the digital economy determines the spread of the practice of crowd sourcing in all spheres of human life, which was reflected in the creation of the OpenStreetMap project, which is an open platform with a database on the territory implemented by University College London in 2004 [3]. This project is designed to provide free access to geo data about the territory for individuals, organizations and the scientific community [4]. Moreover, geo data obtained from OpenStreetMap must be analyzed using digital technologies, since efficient processing of large amounts of information is impossible without automating this process. As part of the work, we will use such a research tool as the Grasshopper programming language, available for use by specialists in the area of architecture and construction, since it does not require deep knowledge of information technology [5].

The purpose of this work is to develop tools for constructing a 3D model of the territory and calculating the total heat flows, for which the following tasks will be solved: studying approaches to ensuring energy efficiency in smart cities, studying the possibilities of integrating 2D models from geo information systems (GIS) into more visual 3D models, as well as the development of tools for constructing a map-diagram of the distribution of heat flows from buildings.

2 Materials and Methods

Ensuring energy efficiency of buildings and structures in smart cities has been studied by many scientists. Ertugrul O.F. and Kaya Y. propose an extreme learning method to assess the cooling and thermal loads of buildings [6]. Issa Batarseh R.A.A. and

Rhoades S. consider digital platforms for energy efficiency of buildings and renewable energy production [7]. Galvão J.R. et al. offer to improve the energy consumption system through the use of LED street lighting [8]. Battista G. et al. study the issues of improving the energy consumption system of buildings and propose to use parametric analysis in order to find optimal solutions for changing data of these systems, taking into account the budget of the project [9].

Akcin M. et al. consider energy efficiency in terms of active and passive approaches [10]. The active approach includes the improvement of the sphere of energy consumption based on the introduction of new technological solutions [11], while the passive approach involves the constructive improvement of buildings and structures in order to bring them to modern energy efficiency requirements. The active approach is aimed at improving the performance of energy generating organizations, while the main goal of the passive approach is to reduce energy demand by reducing heat losses.

The active approach includes smart electricity grids, smart transportation of energy sources (gas, gasoline, coal, renewable sources, nuclear sources), as well as household energy production systems for their own needs (for example, from solar energy). The passive approach involves the study of the issues of natural ventilation of buildings and structures, the processes of their cooling, heating and lighting. We believe that it is precisely the passive approach to the problem of energy efficiency that is the most relevant in the conditions of the existing development of the territory, which is subject to transformation without deep technological innovations, the use of which is more accessible in the construction of new objects of the human environment.

Mathematically substantiated improvement of the sphere of energy consumption requires the availability of large arrays of data on the functioning of the territory, including information on various areas of the human environment. In order to build models for visualization of heat flows in a city, as a rule, open geographic information systems (GIS) are used, which consist of various subjects of the city's life. Distinctive characteristics of open systems are free distribution, the presence of an integral source code, non-discrimination and technological neutrality [12, 13].

Geo information data is studied by both social and technical sciences [14]. Open GIS systems provide ample opportunities for interdisciplinary research [15], since the data can be used not only in the study of energy efficiency issues, but also in improving the transport system, updating engineering communications, planning development, etc. It is now recognized that an interdisciplinary approach plays an important role in the development of cities, and the data generated in the process of energy consumption by organizations and the population is a source of information for establishing interdisciplinary interaction [16].

As part of this work, we used data from the non-commercial web mapping system OpenStreetMap (OSM) [17]. The OSM Project is an online resource containing cartographic information about roads, geometry and building materials. The database is stored on a server, constantly updated and replenished by a community of enthusiastic cartographers, GIS professionals and engineers. Information from OSM can be freely exported in file format with xml-structure. The OSM database stores the coordinates of the control points needed to draw the contours of buildings, roads, building boundaries, etc. All points that fall into the exported area of the map have their own unique identification numbers—id. The profiles used to describe buildings use references to the ids of the points that shape them, as well as a number of attributes that describe the characteristics of the object. For buildings, the key attributes are address, category, height, number of floors, wall and roof materials. The number of attributes for different buildings is different: not all buildings have a height attribute, more often floors are indicated and sometimes there is information about arches.

Thus, the use of data from open platforms makes it possible to find solutions in the sphere of improving the energy efficiency of buildings and structures. However, converting datasets into a 3D model requires the use of information processing technologies, which may differ depending on the purpose of the simulation.

3 Results

A script was written that forms the outlines of buildings in the visual programming language Grasshopper to assess heat flows and 3D-modeling of urban areas. The number of attributes for different buildings is different. Unfortunately, not all buildings have a height attribute—height, more often the number of floors is indicated—building: levels, sometimes there is information about arches using the min_height and building: min_level attributes (see Fig. 1).

If there is no height attribute, then the attribute responsible for the number of floors of the building is used, on the basis of which the height of the building is calculated (we assume a floor height of 3 m). In the absence of both attributes, the height of the building can be set, for example, according to the data of the GIS Housing and Communal Services [18].

To get 3D models of buildings from a *.osm file, a script written in the visual programming language Grasshopper was applied (see Fig. 2).

The obtained script draws the outlines of buildings, checks for the presence of the height attribute (height:), if it is absent, it looks for the building: levels attribute and, taking into account the floor height of 3 m, calculates the height of the entire

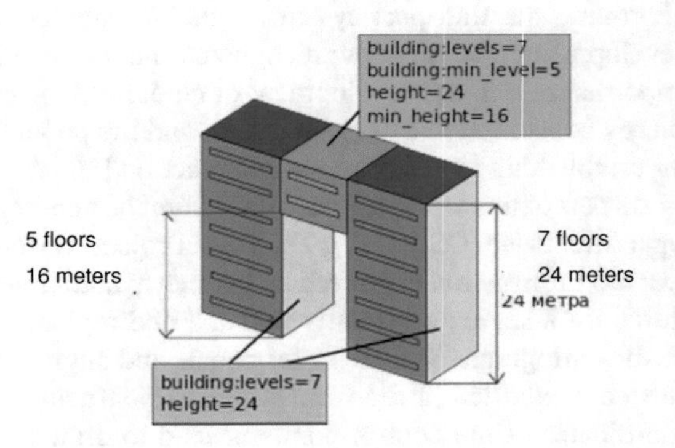

Fig. 1 Interpretation of building height attributes in xml file markup *.osm

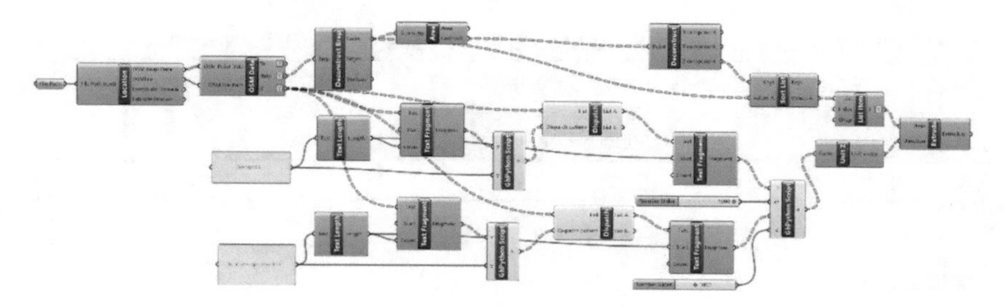

Fig. 2 Script in the visual programming language Grasshopper

building. If both attributes are missing, the building height is determined by the given default value. After all checks, the contour of the building is stretched to the obtained height and a perfect 3D object is formed.

After determining all the attributes, the outline of the building is formed and a 3D object is formed. When modeling a set of buildings adjacently located, it is not necessary to take into account the adjacent enclosing structures, since no heat transfer occurs through them. To get rid of such unnecessary geometry elements, it is enough to perform a Boolean Union of the geometry; Grasshopper has a ready-made solution for this.

As an example of visualization of the distribution of heat flows from buildings, let us discuss the area of St. Petersburg called "Vasilievsky Island". To build a 3D building model, the territory of Vasilyevsky Island is divided into square plots with an area of 30*30 m^2. Figure 3 shows the constructed 3D model of the structures of Vasilievsky Island, compiled using data from the OSM. There is the 1st line and Bolshoy prospect of Vasilievsky Island in the foreground.

To calculate the heat losses of buildings, taking into account the climate of the region, there are formulas described in [19, 20]. Thus, first, it is necessary to determine the indicator of the degree per day of the heating period (DDHP) according to the

Fig. 3. 3D-model of buildings on Vasilievsky Island area in St. Petersburg according to OSM data

Table 1 Standard values of the reduced heat transfer resistance at DDHP = 4000 °C· days/year

Walls	Coverings and ceilings over driveways	Attic coverings, over unheated undergrounds and basements	Translucent enclosing structures, except lanterns	Lanterns
2,8	4,2	3,7	0,63	0,35

formula (1):

$$DDHP = (t_{BH} - t_{OT})d_{OT}, \tag{1}$$

where t_{OT}, d_{OT}—is the average outside air temperature °C and the duration days/year, of the heating period, taken according to SP 131.13330.2012 standard for residential and public buildings for the period with an average daily outside air temperature of no more than 8 °C, and when designing treatment-and-prophylactic, children's institutions and boarding schools for the elderly it is no more than 10 °C [21].

t_{BH}—the design temperature of the internal air of the building, °C, taken when calculating the enclosing structures of the groups of buildings indicated in Table 1 (GOST 30,494–2011 [22]), in the interval 20–22 °C for residential buildings, hostels and hotels, in the interval 16–21 °C for treatment-and-prophylactic, preschool educational and general educational organizations, boarding schools. Let us take as the average value $t_{BH} = 20$ °C.

For St. Petersburg $t_{OT} = -1, 3$ °C, $d_{OT} = 213$ days for a period with an average daily air temperature of ≤ 8 °C and $t_{OT} = -0, 4$ °C, $d_{OT} = 232$ days for a period with an average daily air temperature of ≤ 10 °C [19]. Thus, the DDHP will be equal to 4536.9 and 4732.8 °C· days/year, respectively.

In SP 50.13330.2012 standard [19] there are values of the reduced resistance to heat transfer R_i^{TP}, for a different category of elements of the enclosing structure with DDHP = 4000 and 6000 °C· days/year. The DDHP = 4000 °C· days/year is the closest to the indicators of St. Petersburg, corresponding values of R_i^{TP} are given in Table 1.

To calculate the heat flux from different elements of enclosure structure Q_F^i the formula is required (2):

$$Q_F^i = K_i(t_{BH} - t_{Hap})S_i, \tag{2}$$

where $K_i = 1/R_i^{TP}$—heat transfer coefficient of the i-th element of the enclosure structure,

t_{Hap}—outdoor temperature, °C,

S_i—area of the i-th element of the enclosing structure, m^2.

All parameters in the program are changeable. You can set different coefficients of resistance to heat transfer for walls, windows, roofs and foundations, as well as vary the coefficient of glazing of facades. At the moment, all calculations are performed with one coefficient.

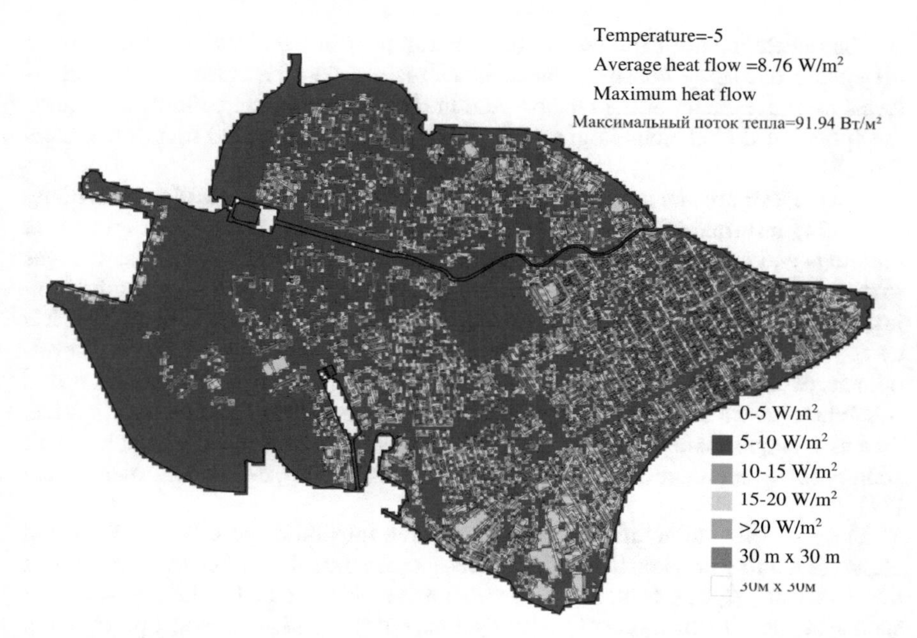

Fig. 4 Distribution of heat flux over the territory of Vasilievsky Island of St. Petersburg at an average air temperature −5 °C

A grid covering a given area divides buildings into separate parts, according to which the total heat flow Q_F inside each cell is calculated. The obtained data is divided into groups with a step of 5 W/m^2 and a separate color is assigned to each group.

Based on the data obtained, a map-diagram of the distribution of heat fluxes from buildings was compiled; where different ranges of heat emission are indicated by certain colors (see Fig. 4).

Thus, based on open data from OSM and the application of the above calculations, the city's territory was modeled taking into account heat flows from buildings, which can be used to further study the human environment in the area of energy efficiency of buildings.

4 Discussion

The concept of a smart city is based on a combination of various technological infrastructures that form the essence of smart cities [23]. A smart city involves the construction of an infrastructure that, based on the use of information and communication technologies, will ensure the interaction of many subsystems with each other using an integrated management mechanism [24, 25].

The development of the human life environment in the direction of improving the sphere of energy consumption includes maintaining the autonomy of various levels of vital activity of the territory, the introduction of green building standards, the spread of digital technologies in the energy, construction and transport sectors [26–28].

The most significant construction visualization technology in the digital economy is building information modeling (BIM) [29], which makes it possible to represent buildings and structures in a multidimensional format, as well as to make changes to the model to assess various options for construction, reconstruction and modernization. It is recognized that BIM can be applied at all stages of the project life cycle—from design to renovation of development [30]. The next stage in the development of BIM is city information modeling (CIM), with the help of which it is possible to visualize not only individual buildings, but to consider the entire urban area as a single modeled space. It is believed that city information modeling will soon become an integral part of the process of transforming the human environment [31].

The scientific community recognizes that the integration of GIS and BIM will allow conducting previously unavailable research in the sphere of urban environment development [32, 33]. Nevertheless, complex, universal tools for the interaction of such systems with each other have not yet been developed. The model proposed in this paper for the distribution of heat flows from buildings is one of the steps towards the transition from 2D data from GIS to 3D models of cities in the sphere of modeling energy efficiency of buildings.

Moreover, in the context of the digital transformation of the human environment, ensuring the energy efficiency of buildings is an interdisciplinary task, which is determined by the following factors [34]:

- globalization and accelerated flows of knowledge, technology, labor and financial capital;
- strengthening ties between previously segmental spheres of city life;
- the growing importance of modeling and forecasting in order to ensure the competitiveness of entrepreneurial entities;
- spreading of high technology decision-making;
- the need for continuous updating of competence to meet the requirements of the labor market in the digital economy;
- the emergence of the possibility of building networks of interaction of experts through the Internet without taking into account territorial affiliation.

Consequently, in the digital economy, the issues of modeling the physical processes inherent in the life of the territory are associated with interdisciplinary interaction in connection with the need to develop complex solutions. Modeling heat flows from buildings is no exception—the model proposed in this study not only allows integrating data from GIS to build one of the elements of the city's CIM model, but also opens up opportunities for improving the human life environment, taking into account the energy efficiency of existing buildings.

5 Conclusions

Thus, the development of the human environment based on the concept of a smart city is inextricably linked with the use of digital technologies in various aspects of the territory's life. In our opinion, one of the most important areas of urban transformation is the presentation of data arrays about the territory in the form of a 3D model in order to visualize physical objects of the human life environment. A visual representation of the territory and the processes inherent in its functioning will increase the level of interdisciplinary research.

The toolkit developed in this work for constructing a 3D model of the territory and calculating the total heat flux from buildings will make it possible to model the city from the point of view of energy efficiency of development. In addition, the proposed scheme for modeling heat flows is based on the use of open data, which meets modern trends in urban development within the framework of the concept of a smart city. As a result of the application of the proposed methods for assessing heat flows from buildings, it becomes possible to construct a map-diagram of territories, which can later be used to create a CIM-model of a Grasshopper modeling of the city.

Acknowledgements The article was written as part of the work on the Grant of the President of the Russian Federation MK-462.2020.6.

References

1. The World Bank. http://www.worldbank.org/en/topic/urbandevelopment/overview. Accessed 13 Apr 2021
2. Vishnivetskaya, A., Alexandrova, E.: «Smart city» concept. Implementation practice. In: IOP Conference Series: Materials Science and Engineering 497 (2019). https://doi.org/10.1088/1757-899X/497/1/012019. Accessed 10 Apr 2021
3. Jokar Arsanjani, J., Zipf, A., Mooney, P., Helbich, M.: An introduction to OpenStreetMap in geographic information science: experiences, research, and applications. In: Jokar Arsanjani, J., Zipf, A., Mooney, P., Helbich, M. (eds.) OpenStreetMap in GIScience. Lecture Notes in Geoinformation and Cartography. Springer, Cham. (2015)
4. Haklay, M., Weber, P.: OpenStreetMap: user-generated street maps. IEEE Pervasive Comput. **7**(4), 12–18 (2008)
5. Why use Grasshopper. https://softculture.cc/blog/entries/articles/zachem-ispolzovat-grasshopper-analiz-i-simulyatsia. Accessed 25 Apr 2021
6. Ertugrul, O.F., Kaya, Y.: Smart city planning by estimating energy efficiency of buildings by extreme learning machine. In: 4th International Istanbul Smart Grid Congress and Fair (ICSG), pp. 1–5. IEEE, Istanbul, Turkey (2016)
7. Amarin, R.A., Batarseh, I., Rhoades, S.: Efficient energy solutions enabling smart city deployment. In: 2016 Future Technologies Conference (FTC), pp. 791–795. IEEE, San Francisco, CA, USA (2016)
8. Galvão, J.R., Moreira, L.M., Ascenso, R.M.T., Leitão, S.A.: Energy systems models for efficiency towards smart cities. In: IEEE EUROCON 2015-International Conference on Computer as a Tool (EUROCON), pp. 1–6. IEEE, Salamanca, Spain (2015)

9. Battista, G., Evangelisti, L., Guattari, C., Basilicata, C., Vollaro, R.L.: Buildings energy efficiency: interventions analysis under a smart cities approach. Sustainability **6**(8), 4694–4705 (2014)
10. Akcin, M., Kaygusuz, A., Karabiber, A., Alagoz, S.: Opportunities for energy efficiency in smart cities. In: 4th International Istanbul Smart Grid Congress and Fair (ICSG), pp. 1–5. IEEE, Istanbul, Turkey (2016)
11. Hancke, G.P., Silva, B.C., Jr., Hancke, G.P.: The role of advanced sensing in smart cities. Sensors **13**(1), 393–425 (2012)
12. Free and Open Source Software for Geospatial Applications (FOSS4G) at the University of Colorado Denver. https://gis.ucar.edu/sites/default/files/uploads/Moreno_NCAR_FOSS4GMatureAlternative.pdf. Accessed 05 Apr 2021
13. Maurya, S. P., Ohri, A., Mishra, S.: Open Source GIS: A Review. In: National Conference on Open Source GIS: Opportunities and Challenges, pp. 150–155. Department of Civil Engineering, IIT (BHU), Varanasi (2015).
14. Bracken, L.J.: Interdisciplinarity and geography. In: Richardson, D., Castree, N., Goodchild, M.F., Kobayashi, A., Liu, W., Marston, R.A. (eds.) The international encyclopedia of geography, pp. 3746–3755. John Wiley & Sons, Chichester (2017)
15. Rickles, P., Ellul, C., Haklay, M.: A suggested framework and guidelines for learning GIS in interdisciplinary research. Geo Geography and Environment **4**(2), 1–18 (2017)
16. Ablyazov, T.: Application of an interdisciplinary approach to the implementation of projects to create a comfortable environment for human life. J. Environ. Treat. Tech. **8**(3), 1136–1139 (2020)
17. OpenStreetMap. https://www.openstreetmap.org. Accessed 27 Mar 2021
18. GIS Housing and Communal Services. https://dom.gosuslugi.ru. Accessed 27 Mar 2021
19. SP 50.13330.2012. http://docs.cntd.ru/document/1200095525. Accessed 27 Mar 2021
20. Drozdov, V.F.: Heating and ventilation. Heating. Textbook for building, universities. M., “Higher. School” (1976)
21. SP 131.13330.2012. http://docs.cntd.ru/document/1200095546. Accessed 27 Mar 2021
22. GOST 30494–2011. http://docs.cntd.ru/document/1200095053. Accessed 25 Mar 2021
23. Esmaeilian, B., Wang, B., Lewis, K., Duarte, F., Ratti, C., Behdad, S.: The future of waste management in smart and sustainable cities: a review and concept paper. Waste Manag. **81**, 177–195 (2018)
24. Minoli, D., Sohraby, K., Occhiogrosso, B.: IoT considerations, requirements, and architectures for smart buildings-Energy optimization and next-generation building management systems. IEEE Internet Things J. **4**(1), 269–283 (2017)
25. De la Hoz-Rosales, B., Camacho, J., Tamayo, I.: Effects of innovative entrepreneurship and the information society on social progress: an international analysis. Entrep. Sustain. Issues **7**(2), 782–813 (2019)
26. Sarma, U., Karnitis, G., Zuters, J., Karnitis, E.: District heating networks: Enhancement of the efficiency. Insights Reg. Dev. **1**(3), 200–213 (2019)
27. Taveres-Cachat, E., Grynning, S., Thomsen, J., Selkowitz, S.: Responsive building envelope concepts in zero emission neighborhoods and smart cities-a roadmap to implementation. Build. Environ. **149**, 446–457 (2019)
28. Tvaronaviciene, M.: Towards sustainable and secure development: energy efficiency peculiarities in transport sector. J. Secur. Sustain. Issues **7**(4), 719–725 (2018)
29. Ablyazov, T., Petrov, I.: Russian practice of providing a construction industry with information and communication infrastructure in conditions of a digital economy establishment. Adv. Econ. Bus. Manag. Res. **81**, 366–370 (2019)
30. Vishnivetskaya, A., Mikhailova, A.: Employment of BIM technologies for residential quarters renovation: global experience and prospects of implementation in Russia. IOP Conference Series: Materials Science and Engineering 497 (2019), https://doi.org/10.1088/1757-899X/497/1/012020. Accessed 10 Apr 2021
31. Shaikh, M.Z., Shah, D., Kawale, V., Anand, K.: Developing smart cities in a BIM environment. IOSR J. Mech. Civil Eng. **14**(5), 11–16 (2017)

32. Månsson, U.: BIM & GIS connectivity paves the way for really Smart Cities. Geo forum Perspektiv **14**(25), 19–24 (2015)
33. Ugurlu, D., Sertyesilisik, B.: Usage of BIM in smart cities. Int. J. Digital Innov. Built Environ. **8**(1), 17–27 (2019)
34. Gitelman, L., Magaril, E., Khodorovsky, M.: Interdisciplinary as heuristic resource for energy management. Int. J. Energy Prod. Manag. **1**(2), 163–171 (2016)

Generalization, Weighing and Coordination of Group Expert Assessments When Making Management Decisions

Evgeny Lutsenko, Tatyana Baranovskaya, and Alexander Trounev

Abstract One of the reasons for the acute system problems that exist in the conditions of digitalization of the economy (based on innovations generated by Russia's innovation complex) is the low consistency of group expert assessments in the process of making managerial decisions. Therefore, a new intellectual business service has become very popular and relevant: "Generalization, weighing and coordination of group expert assessments when making management decisions". When providing this intellectual service, we face a problem that consists in the absence of an algorithm and available software tools, i.e. a method for solving this problem. Therefore, the aim of the work is to develop a method that provides automated provision of this intellectual service. There is a requirement for the intellectual model on the basis of which this intellectual service will be executed: under the same conditions, to issue the same forecasts and recommendations as the group of experts. The core of the proposed method of achieving the goal is automated system-cognitive analysis. This method belongs to the methods of artificial intelligence and has its own personal-level software tools with a zero entry threshold. The main result of the work is the algorithm (methodology) and available software tools (open source software) that ensure the provision of this new intellectual business service. This study will be useful for experts who make management decisions in all subject areas. For example, to generalize, weigh and coordinate the group expert assessments of the medical council in the examination of difficult cases for diagnosis.

Keywords Algorithm · Weighting · Group expert assessments · Tools · Intelligent business service · Coordination

E. Lutsenko (✉) · T. Baranovskaya
Kuban State Agrarian University, St. Kalinina, Krasnodar Territory, 13, Krasnodar 350004, Russia

A. Trounev
A&E Trounev IT Consulting, Likalo LLC, Toronto, Canada

C. Jahn et al. (eds.), *Algorithms and Solutions Based on Computer Technology*, Lecture Notes in Networks and Systems 387,
https://doi.org/10.1007/978-3-030-93872-7_17

1 Introduction

In the conditions of digitalization of all spheres of modern society's activities based on innovations, expert methods of decision-making play a key role. Ultimately, the decision maker (DM) can reasonably be considered an expert in his field. As a rule, this DM has full-time and temporarily engaged advisers, experts and consultanxts, in fact, making up an expert group. However, unfortunately, it often happens in practice that both the experts themselves and their advisers are not unanimous in their assessments and decisions [1]. Thus, there is a problem of correct and scientifically based generalization, weighing and coordination of group expert assessments when making management decisions. We have considered the very concept of the "problem" quite traditionally: as a discrepancy, a contradiction between the actual and the desired (goal).

This is followed by a detailed classical definition of the problem to be solved in the study:

- Actual. Experts can give both agreed assessments and recommendations, and disagree with each other, as well as have different actual and formal degrees of competence, i.e. "different weight". Therefore, figuratively speaking, the result of the work of the expert group is "a whole bundle of multidirectional vectors of different lengths", in which each vector corresponds to a single solution. These solutions can be both feasible at the same time, and alternative (according to the system of determining factors) [2].
- Desired. However, the DM must somehow either choose from the whole system of proposed alternative solutions some one that is best according to certain criteria, or form some resultant solution from relatively compatible solutions [3].

It is clear that to do so, the DM must use a certain method that is adequate to solve this problem according to the current time and act corresponding to a certain algorithm with the help of a certain software toolkit.

Hence, "Generalization, weighing and coordination of group expert assessments when making management decisions" as a new intellectual business service is very popular and relevant.

To provide this intellectual service, a certain appropriate infrastructure is required, including scientific approaches and software tools that are understandable to the mass user, universal in terms of implemented functions and independent of the subject area [4].

2 Materials and Methods

In the following study [5], we can find the classification of organizations according to the criterion of the degree of centralization of management: "A unimodal organization is an organization that has a hierarchical structure. The pyramid of power in such an organization is crowned by an individual who has a decisive voice and is able to resolve all the differences that arise at lower levels. In a multimodal organization, there is no such ultimate authority, so we require an agreement between two or more autonomous responsible persons. "For multimodal organizations, the relevance of solving the problem examined in this work is quite obvious. But even for unimodal organizations, such a decision may be in demand if the DM makes it by himself, although taking into account the recommendations of experts.

Making control decisions in order to transfer the control object to a predetermined target state or states is the main function of the control system and one of the main elements of the control system and the control cycle. Thus, the DM should develop and make a decision based on the analysis of feedback information about the state of the control object and on the basis of its model, which quantitatively reflects the reaction of the control object to various control actions. This model should quantify the strength and direction of the influence of various values of factors on the control object.

In his fundamental textbook work [6], Frederick Winslow Taylor, the founder of the scientific organization of labor and enterprise management, wrote: "Happy are those companies that can enlist the services of experienced experts who have proper practical experience in the field of scientific management and have specifically studied its theoretical foundations."

However, these and other works on the organization of management do not offer a method, i.e., an algorithm and available software tools that provide a solution to the problem, so this work is relevant.

Therefore, it is proposed to solve this problem in an online environment and a developed infrastructure of automated system-cognitive analysis. This method of artificial intelligence meets all the necessary requirements [7].

3 Results

In E.V. Lutsenko's works [8] the author presents approaches that create real prerequisites for solving the problem. He proposes a method that includes an algorithm (methodology) and available software tools (open source software) that provides the provision of "Generalization, weighing and coordination of group expert assessments when making management decisions" intellectual service. Figure 1 has the proposed algorithm presented in a somewhat simplified form (a detailed up-to-date software implementation of this algorithm with notes is available online in the source test of the Eidos system at the following link: http://lc.kubagro.ru/__AidosALL.txt, search for: FUNCTION F3_7_9()).

Step 1: Follow the link: http://lc.kubagro.ru/aidos/_Aidos-X.htm to download and install the software tools, which currently is an intelligent system called "Eidos" on your computer, following the instructions on the site.

Step 2: Based on empirical observations and their expert assessments, we create a training sample. The numerical example considered in this article is based on

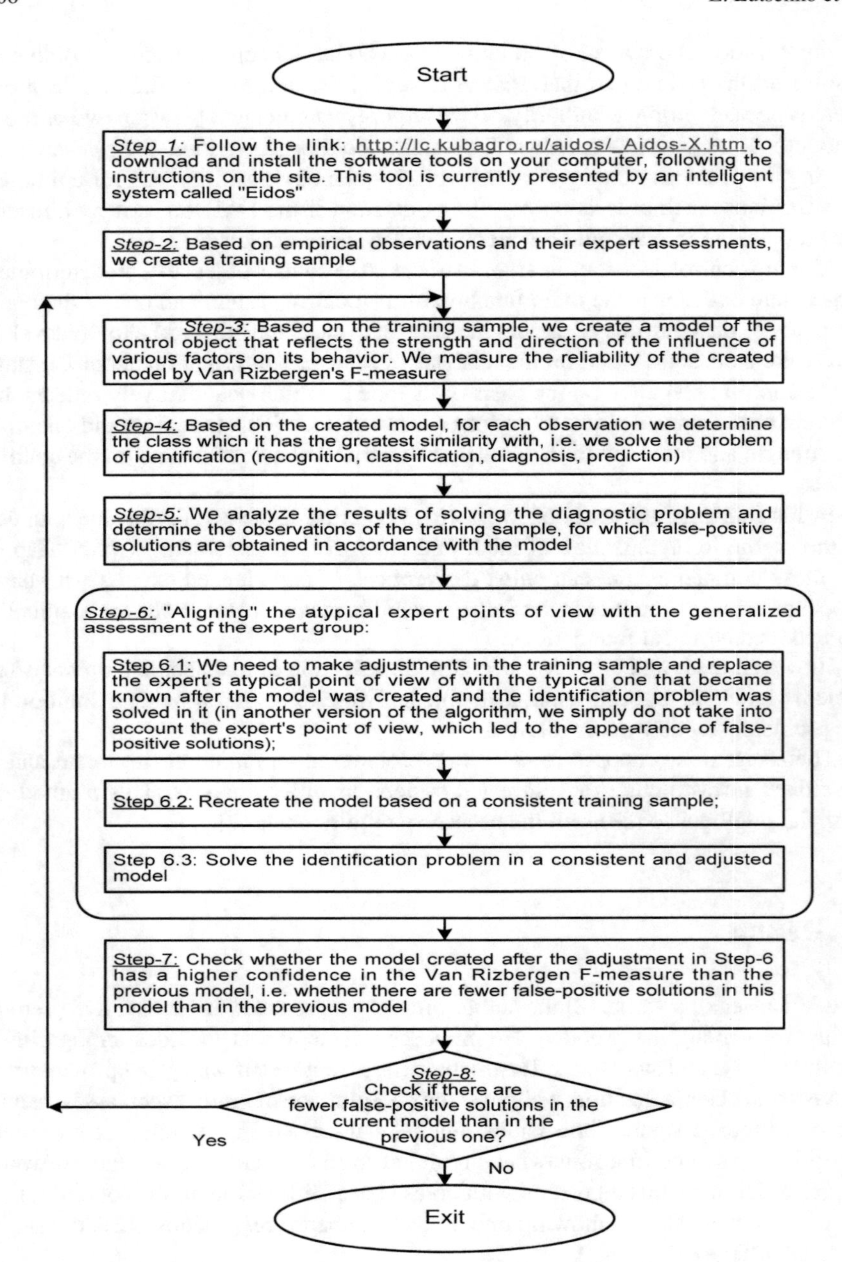

Fig. 1 Algorithm for providing intellectual services: "Generalization, weighting and coordination of group expert assessments when making management decisions". *Source* ("Compiled by the authors")

the source data of the Eidos intelligent cloud application No. 290, which can be downloaded from the Eidos cloud and installed for researches in the Eidos system application manager (mode 1.3).

Next, we have to consider the structure of the observation. On the one hand, the observation includes a description of the modeling object by indicating the degree of expression of its properties or the values of the factors (signs) affecting it. Both properties and factors can be measured in different types of scales (textual: (nominal, ordinal) and numeric) and in different units of measurement. On the other hand, observation is referred by experts in an "informal" way (based on experience, intuition and professional competence) to some generalizing categories (classes). Similar to property values or factor values, classes can also be gradations (values) of different types of scales (text: (nominal, ordinal) and numeric) and can be measured in different units of measurement.

A fact is the discovery of a certain property value in a modeling object of a certain class, or the transition of the control object to a certain future state corresponding to a certain class under the influence of a certain factor value. Thus, one observation, generally speaking, contains data on many facts.

From this definition of fact, we can come to conclusion that different experts, basically, will state different facts on the basis of the same observations, depending on what generalizing categories (classes) they will refer to the objects of observation. At the same time, we will assume that the measurement of the degree of expression of properties or values of factors is carried out by objective methods and is identical for different experts. At the same time, it is clear that this is also not always the case.

The "generalization" of expert assessments is implemented in the proposed algorithm due to the fact that the same observation in the training sample will be duplicated with the expert assessments of various experts (whether agreed with each other or not) with the same descriptive part (the degree of expression of properties and the values of factors) as many times as the amount experts is in the expert group.

Thus, when forming generalized images of classes, we can say that expert assessments are "weighed".

The different level of competence of experts and their weight can be taken into account at the stage of forming a training sample by posting the expert assessment results in it several times. When forming the model, the points of view of various experts will play a role corresponding to their weight.

Step-3: Based on the training sample, we create a model of the control object that reflects the strength and the direction of the influence of various factors on its behavior. We measure the reliability of the created model by Van Rizbergen's F-measure.

Step-4: Based on the created model, we determine the class it has the greatest similarity with, for each observation, i.e. we solve the problem of identification (recognition, classification, diagnosis, prediction). All these terms, in fact, are synonymous, but they are mainly used in their subject areas. For example, the use of the term "diagnosis" is typical for medicine and psychophysiology.

Step-5: The results of solving the diagnostic problem are now analyzed and we also determine the observations of the training sample, for which false-positive solutions are obtained in accordance with the model.

Some observations are correctly assigned by the system based on the model to the classes they belong to, according to the results of a weighted generalization of expert assessments, and some are assigned incorrectly (false-positive decisions). False attribution of observations by the model to classes (that some experts did not assign them to) means that the point of view of these specific experts, who made these decisions which led to false-positive decisions, does not match the point of view of the other participants in the expert group, is atypical or even erroneous when compared with the generalized agreed point of view of the expert group. Meanwhile, the intellectual model on the basis of which this intellectual servant will be provided is required: under the same conditions, to issue the same forecasts and recommendations as the group of experts. Therefore, in this case, it is necessary to adjust the model so that all of the experts' points of view would be consistent with each other.

Step-6: For "harmonizing" unusual experts' points of view and generalized assessment of the expert group we may perform the following:

Step-6.1: we make adjustments in the training set and we also change (adjust) the experts' unusual point of view to the typical one, which became known after the creation of the model and solving it problems of identification (in another type of the algorithm we just do not consider the point of view of such experts, that led to the emergence of false-positive decisions);

Step-6.2: to re-create the model based on the agreed training samples;

Step-6.3: solve the identification task in the consistent and adjusted model.

As a result, the model created on the basis of a consistent training sample will have significantly higher confidence than the previous model with conflicting and inconsistent expert assessments, and it will also have fewer assignments of objects to classes which do not match the generalized point of view of the expert group.

Step-7: Check whether the model created after the adjustment in step-6 has a higher confidence in the Van Rizbergen F-measure than the previous model, i.e. whether there are fewer false-positive solutions in this model than in the previous model.

Step-8: If there are fewer false-positive solutions in the current model than in the previous model, then go to Step-3 with the adjusted training sample, otherwise exit the algorithm.

The intelligent system "Eidos" is currently used as a software tool. This mode of the "Eidos" system (mode 3.7.9) was created simultaneously with the writing of this work. A description of this system is available on the author's (prof. E. V. Lutsenko) website at: http://lc.kubagro.ru/aidos/index.htm and therefore we consider it unnecessary to be described in this study.

4 Discussion

We will discuss a small numerical example developed specifically to test the proposed method and algorithm (methodology) in practice, as well as the suitability of the Eidos system for providing a new intellectual business service: "Generalization, weighing and coordination of group expert assessments in making management decisions". The intellectual model, on the basis of which this intellectual service will be provided, is required: under the same conditions, to issue the same forecasts and recommendations as the group of experts. For this numerical example, we will take the initial data and their expert assessment from the works of [Lutsenko, 2010–2011] as a basis. Let's consider the original model and compare it with the model created as a result of the listed above iterative algorithm for generalizing, weighing, and matching group expert estimates.

Model-1: "Initial expert assessments, generalized classes". We perform the 1st step of the algorithm proposed in this article and, following the instructions of the system, install the Eidos intelligent cloud application No. 290 in the application manager of the Eidos system (mode 1.3). Next, we will execute synthesis and verification of models in the 3.5 mode based on the forecasts of a group of experts. We can estimate the reliability of the model in the 3.4 mode. The reliability of the most reliable INF3 (chi-square) model according to the Van Rizbergen F-measure was 0.867 with a maximum of 1,000, which is a good result for medical intelligent applications. However, for patients with id: K0006 IB0001, K0026 IB0001, K0032 IB0001, K0049 IB17037, K0051 IB17105, K0021 IB7837, experts made false positive forecasts for the duration of the postoperative recovery period. That is why the erroneous point of view made by these experts should either be corrected, or should not be used at all to form a model that meets the requirement under the same conditions to make the same forecasts as the group of experts. This means that at least one more model should be created based on the adjusted point of view of experts.

Model-2: "Generalized, weighted and aligned (adjusted) group expert assessments". To correct false-positive expert forecasts of the duration of the postoperative recovery period, we will create the 2nd model in the mode 3.7.9 of the Eidos system. The screenshot of the form in this mode is shown in Fig. 2.

As a result of the operation in this mode for two iterations, we have carried out the generalization, weighting and coordination (adjustment) of group expert assessments for patients, whose IDs are shown in Table 1.

After this mode was activated in the Eidos system 3.5 mode, all 10 models were synthesized and verified. As a result, the reliability of the most reliable INF1 model (a measure of the amount of information according to A. Harkevich) by Van Rizbergen's F-measure was 0.950 with a maximum of 1,000, which is an excellent result for medical intelligent applications. Thus, as a result of the implementation of the proposed algorithm, the reliability of the model increased by 9.57%.

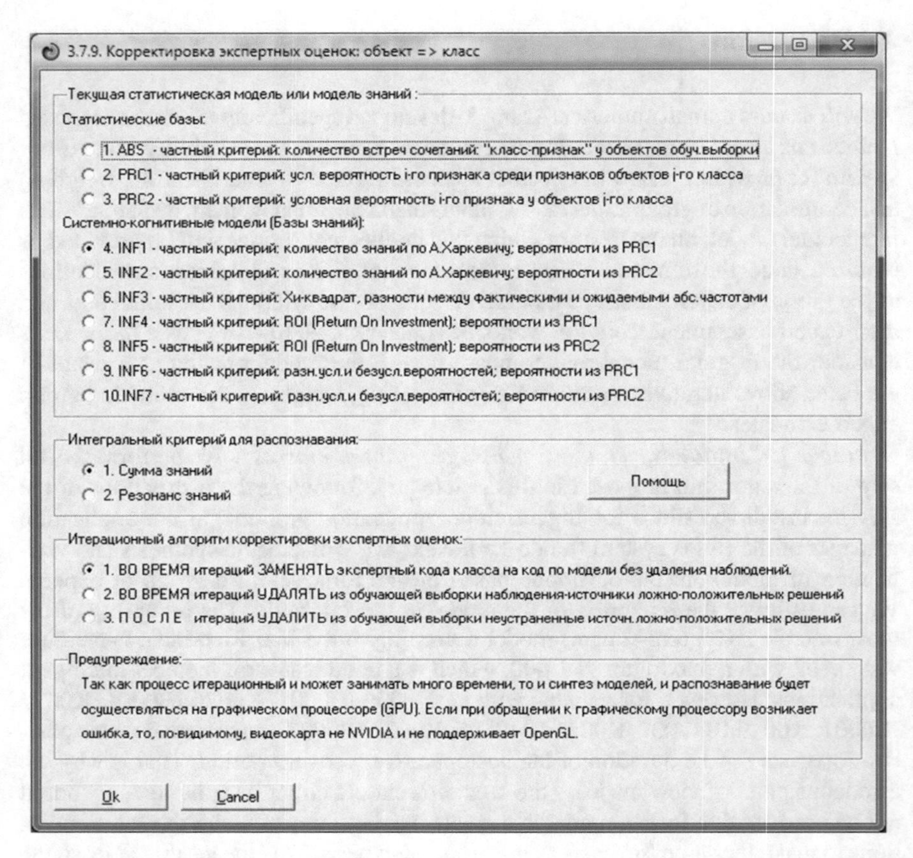

Fig. 2 On-screen form for setting the parameters in the mode: "Generalization, weighing and coordination of group expert assessments when making management decisions". *Source* ("Compiled by the authors")

5 Conclusions

Numerical experiments conducted to test the proposed method and algorithm (methodology) in practice, as well as the suitability of the Eidos system for providing a new intellectual business service, which is "Generalization, weighing and coordination of group expert assessments in making management decisions", have convincingly shown that the proposed algorithm and the tools actually provide this service to everyone who needs it, free of charge.

However, the limitation on the scope of this work does not allow us to give any detailed description of this numerical example. This is planned to be done in future publications.

Table 1 Report on the results of the mode of generalization, weighing and coordination (adjustment) of group expert assessments

Iteration number	Patient code	Patient ID	Expert solution		System solution based on the model		
			Forecast code	Forecast of the duration of recovery after surgery	System confidence in the forecast	Forecast code	Forecast of the duration of recovery after surgery
1	4	К0006 ИБ0001	1	1/3–0 days	37,452	2	2/3–3 days
1	18	К0026 ИБ0001	3	3/3–7 days	39,680	2	2/3–3 days
1	21	К0032 ИБ0001	3	3/3–7 days	16,812	2	2/3–3 days
1	29	К0049 ИБ17037	3	3/3–7 days	26,599	1	1/3–0 days
1	31	К0051 ИБ17105	3	3/3–7 days	20,979	1	1/3–0 days
2	13	К0021 ИБ7837	2	2/3–3 days	25,985	1	1/3–0 days

In the public scientific literature, there are no analogues of the algorithm and tools proposed in this work, which allows us to conclude that the solutions proposed in this study are of a high level of scientific novelty.

The main result of the article is an effective method that actually works, including an algorithm (methodology) and available software tools (open source software), which provides the provision of a new popular and relevant intellectual business service: "Generalization, weighing and coordination of group expert assessments". The proposed method is based on an automated system-cognitive analysis and its software tools, which are currently presented in the Eidos intelligent system. Using a real numerical example, the proposed method has provided an increase in the reliability of the automated forecast of the duration of the postoperative recovery period by 9.57%. Thus, the goal of the work has been achieved.

Acknowledgements The research described in this work was carried out with the financial support of the Russian Foundation for Basic Research (project No. 20-010-00076).

References

1. Mogilko, D.Y., Ilin, I.V., Iliashenko, V.M., Svetunkov, S.G.: BI Capabilities in a Digital Enterprise Business Process Management System. Lecture Notes in Networks and Systems **95**, 701–708 (2020)

2. Izotov, A., Rostova, O., Dubgorn, A. The Application of the Real Options Method for the Evaluation of High-Rise Construction Projects. E3S Web of Conferences, vol. 33, p. 03008 (2018)
3. Bril, A., Kalinina, O., Ilin, I.: Small innovative company's valuation within venture capital financing of projects in the construction industry. In: MATEC Web of Conferences, vol. 106, p. 08010 (2017)
4. Alesinskaya, T.V., Arutyunova, D.V., Orlova, V.G., Ilin, I.V., Shirokova, S.V.: Conception BSC for investment support of port and industrial complexes. Acad. Strat. Manag. J. **16**(Specialissue1), 10–20 (2017)
5. On Purposeful Systems, https://gtmarket.ru/library/basis/7083. Accessed 16 April 2021
6. The Principles of Scientific Management, https://gtmarket.ru/library/basis/3631, 16 April 2021
7. Lutsenko, E.V., Sergeeva, E.V.: Forecasting the duration of the postoperative recovery period by the method of cardio-respiratory synchronism (CRS) with the use of the ASC-analysis (part 1). Sci. J. KubGAU, 142–178 (2010)
8. Lutsenko, E.V.: Conceptual principles of the system (emergent) information theory and its application for the cognitive modeling of the active objects (entities). In: 2002 IEEE International Conference on Artificial Intelligence Systems, ICAIS 2002, pp. 268–269. (2002)

Formation of a Business Model for a Geographically Distributed Medical Organization

Oksana Iliashenko, Ekaterina Lukyanchenko, Dmitrii Karaptan, and Carlos A. Julia

Abstract The article describes the creation of a business model for a geographically distributed medical organization. Based on the reference model, the key segments of the organization's activities were analyzed and an adapted business model was drawn up. When carrying out activities in conditions of territorial distribution, a medical organization is faced with certain features described in this study.

Keywords Business model · Reference model · Geographically distributed medical organization

1 Introduction

The healthcare sector has been and remains one of the key areas in the life of society. Currently, it is actively developing due to the growing demand for medical services, digitalization of healthcare and a rapid increase in the number of elderly people. In addition, the state is facing an acute solution to the problem of lack of medical personnel in rural areas, far from large cities, as well as improving the quality and availability of medical care [1]. According to the forecasts of the World Health Organization, the number of people over 60 years old will exceed 2 billion by 2050 [2]. The acceleration of the aging of the population introduces features into the strategies for the further development of the social sphere of states. First of all, medical organizations, medical workers, insurance funds and the healthcare sector in general will face difficulties.

According to a study by KPMG [3], despite the unfavorable economic conditions on the Russian market in the healthcare segment, there is an increase in the number of private medical centers operating under compulsory health insurance (CHI) not only in Moscow, St. Petersburg and Kazan, but also in such subjects like Udmurtia,

O. Iliashenko (✉) · E. Lukyanchenko · D. Karaptan
Peter the Great St. Petersburg Polytechnic University, St. Petersburg, Russia

C. A. Julia
Colegio Mayor Nuestra Señora del Rosario, Bogotá, Colombia

C. Jahn et al. (eds.), *Algorithms and Solutions Based on Computer Technology*, Lecture Notes in Networks and Systems 387,
https://doi.org/10.1007/978-3-030-93872-7_18

Chuvashia, Penza region and others. The demand for the provision of medical services in private medical centers is observed both in large Russian cities and in the regions, which is a key factor in the territorial distribution of medical organizations on the territory of the Russian Federation and the conduction of activities according to the franchise model, or the opening of branches by large medical organizations [4].

To successfully manage a network of geographically distributed medical organizations, the management needs to understand the strategy for further development and work out the business model of the organization. In this article, the authors propose a business model for a geographically distributed medical organization based on a reference business model.

A feature of the formed business model is the territorial distribution of branches of the medical company, for which the model is being created, in various regions, cities of the Russian Federation [5]. This model is relevant for medical organizations in Russia due to the length of the territory of the state, which causes differences in the quality and quantity of medical care provided to the population.

2 Materials and Methods

The main research method used in this article is the method of forming an organization's business model based on the framework proposed by A. Osterwalder and I. Pigneur. According to [6], there are many approaches to building a business model: Henry Chesbrough and Rosenbaum's business model, Verna Ellie's value network business model, Patrick Stalker's business model in the digital economy, A. Osterwalder's and I. Pigneur's business model canvas [7], the business model of Chan Kim and R. Moborne and many others.

Most of these models consider the following 5 critical components [8]:

1. An offer that brings value to the client;
2. Interaction with clients;
3. Key competencies of the company;
4. Competitive advantages of the company;
5. Financial efficiency of the company.

In this study, the choice was made in favor of using the model of Osterwalder and Pigneur [9] due to the clear segmentation of its key parameters, instrumental support and wide practical application. A. Osterwalder segments the business model into 9 structural blocks: key partners, key activities, key resources, value propositions, channels, customer segments, revenue streams, cost structure. Osterwalder's approach to shaping the business model is used in the TOGAF standard and is instrumented in the Archi software.

3 Results

According to a study [10], an imbalance in the geographical distribution of the health workforce is observed in many countries and can be caused by a number of factors:

1. Social factors, including educational level, age, gender, marital status, plans and prospects, goals and beliefs. Thus, women are less likely than men to accept job offers in remote regions; younger healthcare workers are more likely to be willing to relocate, including to rural or remote areas. The locality in which a person grew up and was brought up also has a great influence, since cadres who grew up in remote regions are more likely to return to work there.
2. Factors of the organizational environment: career opportunities, salary scales, practices of working with personnel also play an important role in deciding on a place of work [11, 12].
3. Factors associated with the education and health care system affect both the choice of educational institution for the development of the medical profession, and the subsequent continuation of education, internship and choice of place to work. The study shows that graduates of medical schools located in small towns are more likely to agree to work in remote, rural areas, and also more often choose general medical practice as their specialization [13].
4. Institutional factors are associated with the decentralization of health care, which leads to difficulties in hiring medical personnel in rural areas, since most doctors still prefer to work in urban areas. Despite the adoption of a number of measures to improve working conditions in rural health facilities, decentralization leads to a reduction in local government budgets and the inability to offer competitive salaries to doctors practicing in small towns and villages [14].

In connection with the above factors, the territorial distribution of medical organizations is becoming an urgent solution to the problem of a shortage of medical personnel and an improvement in the quality of medical care in remote regions of the Russian Federation. Territorial distribution implies the presence of a central medical organization and its branches located in various cities and subjects of the state. The management strategy for such medical organizations is to use the best business process management practices, which allows to successfully open branches of the clinic, replicating the main business processes, measuring their effectiveness and identifying bottlenecks for their further elimination.

The developed business model takes into account the presence of a central multidisciplinary medical organization and several of its branches, in which performance indicators are monitored. The study is based on the methodology proposed by A. Osterwalder and I. Pigneur for the development of an innovative business model of an organization, as well as on a reference business model of a medical organization that implements the Smart Hospital concept proposed by the authors in [15, 16] (see Fig. 1).

The specificity of the business model of a geographically distributed medical organization is due to the following features:

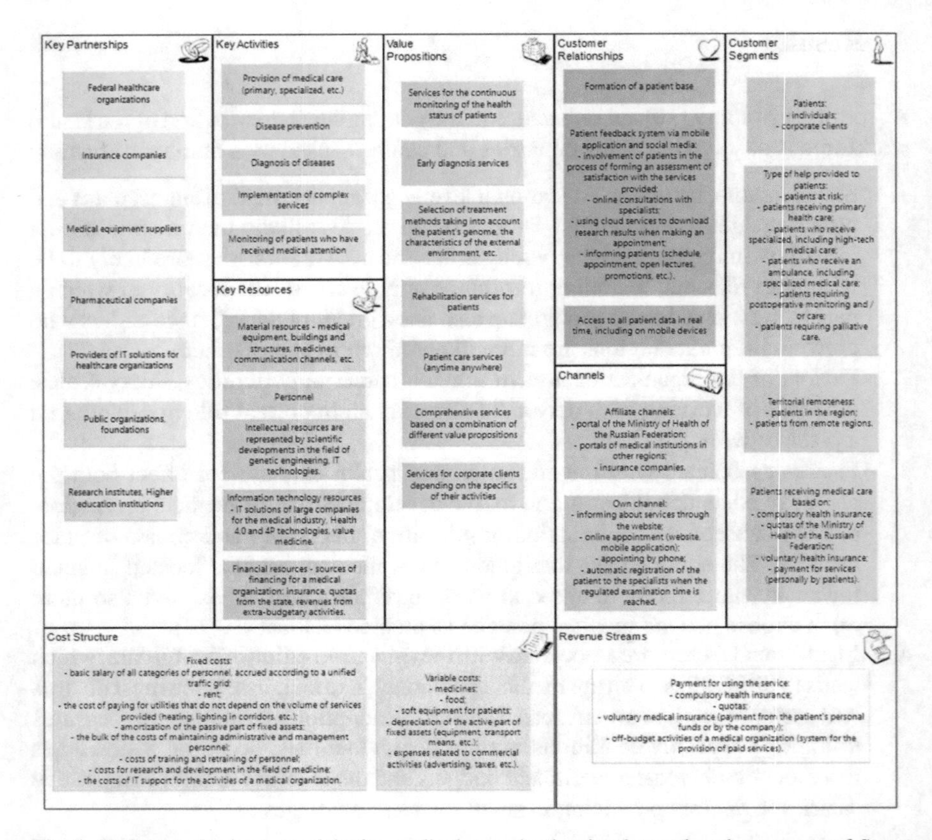

Fig. 1 Reference business model of a medical organization implementing the concept of Smart Hospital

- A geographically distributed medical organization is understood as a network of multidisciplinary medical organizations with a central, main organization that manages the strategic development of the network, financial and marketing activities, quality management, monitoring of performance indicators of a medical organization and its branches, etc.
- In the branches of a geographically distributed medical organization, the best practices for managing business processes, services, patient flows are being introduced, while the main business processes related to the provision of medical care to patients are replicated from previously opened branches and adapted taking into account territorial characteristics. This leads to a reduction in the time required to open a new branch, and also guarantees the efficiency of the branch.
- One of the key activities of a geographically distributed medical organization is the conduction of professional preventive medical examinations of employees of large industrial and oil and gas enterprises working in remote areas.

As a result of the research carried out and taking into account the specifics of a geographically distributed medical organization, the following business model was developed:

1. Consumer segments: both the central multidisciplinary medical organization and its branches work with patients who can be classified according to various criteria [17]:
- By the type of medical care provided (primary health care, specialized medical care, emergency medical care, postoperative health monitoring, etc.);
- By the source of financing for the provided medical care (compulsory health insurance, quotas of the Ministry of Health of the Russian Federation, voluntary health insurance, payment for services by the patient, etc.);
- By the patient's legal status (corporate clients and individuals).
2. Value propositions offered by a network of geographically distributed medical organizations include [18]:
- Services for continuous monitoring of patients' health status;
- Rehabilitation and patient care services;
- Services for the early diagnosis of diseases;
- Selection of treatment methods taking into account the patient's genome, characteristics of the external environment, etc.;
- Comprehensive services based on a combination of different value propositions, for example, in the provision of early diagnosis services, the patient receives services related to treatment and rehabilitation if necessary;
- Services for corporate clients, depending on the specifics of their activities (for example, conducting specialized preventive examinations for employees of oil and gas companies working on remote drilling rigs, etc.);
- Services based on telemedicine technologies.
3. Key resources of a geographically distributed medical organization are represented by the following types of resources:
- Material resources—medical equipment, buildings and structures, medicines, communication channels, transport and engineering communications.
- Personnel—human resources that support the activities of a medical organization: doctors, nurses, management personnel, employees of IT departments, HR, accounting, etc.
- Intellectual resources are represented by scientific developments in the field of genetic engineering, IT technologies.
- Information technology resources—IT solutions of large companies for the medical industry, Health 4.0 and 4P technologies, value medicine [19]. For geographically distributed medical organizations an important aspect is the possibility of a single information exchange between the main organization and branches through a single data exchange platform deployed on the basis of cloud IT services.
- Financial resources—sources of financing for a medical organization: insurance, quotas from the state, revenues from extra-budgetary activities.

4. Sales channels through which the value propositions of a medical organization reach the consumer are represented by an information network with 2 types of channels: partner and own. Partner channels include the portal of the Ministry of Health of the Russian Federation, portals of medical organizations in other regions, and insurance companies. The own channel includes the website of a medical organization, which provides information about the services provided both in the central organization and in the branches, the possibility of online appointment [20].
5. The relationship with the client is carried out through a unified information system for data exchange, deployed on the basis of cloud IT services on the one hand and smartphones, smart watches, sensors, as well as mobile applications for patients, on the other hand. A personalized approach to each patient based on digital data aggregated from various sources about him is the key to the successful implementation of value-oriented medicine [21].
6. The key activities of the considered medical organization are:

- Diagnostics of diseases;
- Provision of medical care, including high-tech and specialized;
- Disease prevention;
- Rehabilitation and recovery;
- Customer service;
- Scientific and research activities;
- Educational activities;
- Marketing activities;
- Conducting professional medical examinations for corporate clients—large industrial companies operating in remote areas.

7. Streams of Income:

- Individuals—within the framework of compulsory health insurance, voluntary health insurance, quotas or on the basis of the provision of paid services;
- Legal entities—within the framework of voluntary health insurance or on the basis of the provision of paid services.

8. Key Partners Are:

- State bodies of the health care system of the federal, regional and local levels;
- Government medical centers and commercial clinics;
- Manufacturing and transport companies;
- Insurance companies;
- Scientific and research centers;
- Educational institutions;
- Pharmaceutical companies;
- Providers of medical equipment and IT solutions.

9. The cost structure of a geographically distributed medical organization is based on two components: fixed and variable costs. Fixed costs remain unchanged

when the volume of services provided changes. These include: payroll fund, training and advanced training of employees, maintenance of real estate, utility bills, depreciation, etc. Variable costs are a dynamic indicator and vary depending on the volume of services provided. These include: consumables and the cost of food, medicines, the cost of implementing commercial projects [22].

As a result of the description of the main 9 blocks of the business model of a geographically distributed medical organization on the basis of the reference model, a business model was formed, shown in Fig. 2. The elements of the business model developed for a geographically distributed medical organization are highlighted in white.

The created business model shows the key components of the successful work of a geographically distributed medical organization. The model examines in detail the main groups of clients (patients), as well as the main types of care, main value propositions and key partners of the medical organization.

The developed model can be applied in practice for the strategic planning of the activities of a multidisciplinary high-tech medical organization that already has or plans to open branches.

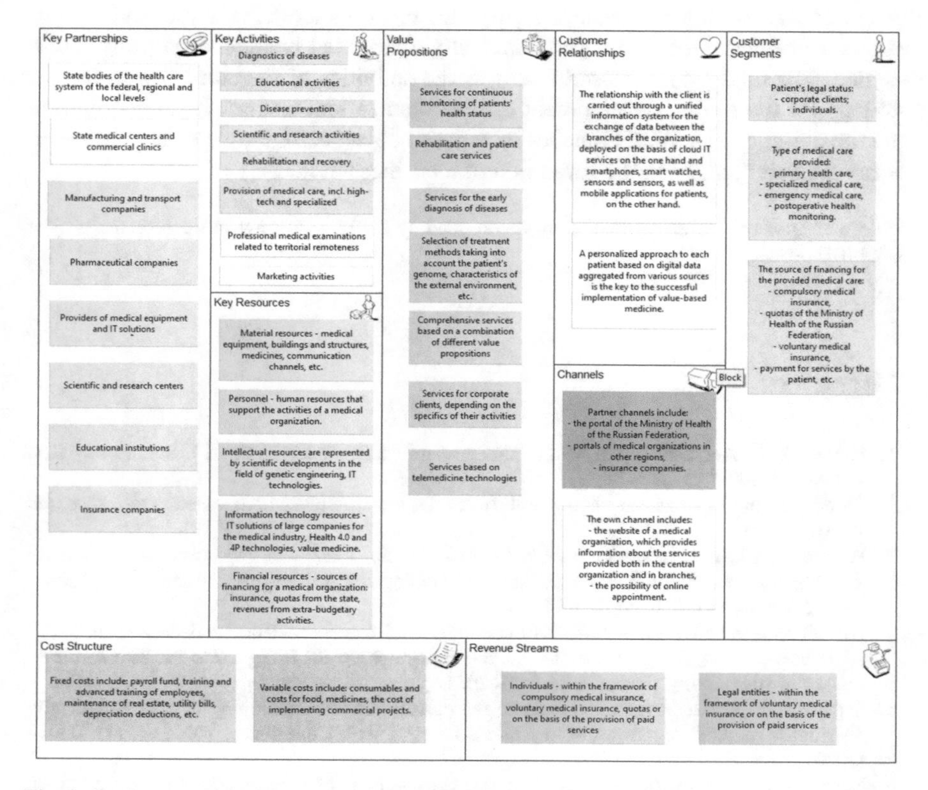

Fig. 2 Business model of a geographically distributed medical organization

4 Discussion

As a result of the study, the reference business model of a medical organization that implements the Smart Hospital concept was supplemented to reflect the specifics of the activities of a geographically distributed medical organization.

The formed business model can be adapted for a specific geographically distributed medical organization and used for the further development of the organization, the introduction of innovative technologies and modern medical trends and concepts.

It should be noted that the developed model assumes the use of new technologies by a medical organization in the process of carrying out activities, for example, the Internet of Things for remote monitoring, Big Data technologies, cloud computing, Machine Learning and Artificial Intelligence [15].

5 Conclusions

Based on the previously developed reference business model of a medical organization, a business model of a geographically distributed medical organization was created. The model can be used to assess the activities of the company, to form a strategy for the further development of the organization. Geographical distribution and the presence of several branches in various subjects of the state were considered as fundamental factors in the formation of a business model.

Acknowledgements The reported study was funded by RFBR according to the research project № 20-010-00955.

References

1. Panova, L.V.: Accessibility of health care: russia in a European context. J. Soc. Policy Res. **17**(2), 177–190 (2019)
2. World Health Organization. Aging and life cycle, https://www.who.int/ageing/ru/. Accessed 17 April 2021
3. Private Healthcare Market in Russia: Outlook for 2017–2019, https://assets.kpmg/content/dam/kpmg/ru/pdf/2017/03/en-en-research-on-development-of-the-private-medical-services-market.pdf. Accessed 26 April 2021
4. KPMG's Views on Development of Commercial Health Care in the Russian Federation in 2017–2019, https://home.kpmg/ru/en/home/media/press-releases/2017/03/private-medical-services-market-in-russia.html. Accessed 26 April 2021
5. Dubgorn, A., Svetunkov, S., Borremans, A.: Features of the functioning of a geographically distributed medical organization in Russia. In: E3S Web Conference, vol. 217, No. 06014 (2020)
6. Morris, M., Schindehutte, M., Allen, J.: The entrepreneur's business model: toward a unified perspective. J. Bus. Res. **58**(6), 726–735 (2005)

7. Osterwalder, A.: Building Business Models: Handbook of Strategist and Innovator. Alpina Publisher, Moscow (2016)
8. DaSilva, M.C., Trkman, P.: Business model: what it is and what it is not. Long Range Plan. **47**(6), 379–389 (2006)
9. Osterwalder, A.: Development of value propositions: how to create goods and services that consumers want to buy. Alpina Publisher, Moscow (2020)
10. Dussault, G., Franceschini, M.C.: Not enough there, too many here: understanding geographical imbalances in the distribution of the health workforce. Hum Resour Health **4**, 12 (2006). https://doi.org/10.1186/1478-4491-4-12
11. Iljashenko, O., Bagaeva, I., Levina, A.: Strategy for establishment of personnel KPI at health care organization digital transformation. In: IOP Conference Series: Materials Science and Engineering, vol. 497, p. 012029 (2019).https://doi.org/10.1088/1757-899X/497/1/012029
12. Kovaleva, I.P., Shuliko, E.V., Strizhak, M.S.: Russian peculiarities of formation of motivative and incentive systems of payment in non-governmental healthcare institutions. Probl. Soc. Hygiene, Public Health Hist. Med. **28**(1), 106–113 (2020)
13. Liu, S., Qin, Y., Xu, Y.: Inequality and influencing factors of spatial accessibility of medical facilities in rural areas of China: a case study of Henan Province. Int. J. Environ. Res. Public Health 2019 **16**, 1833 (2019). https://doi.org/10.3390/ijerph16101833
14. Belinskaya, I.B., Loskutova, M.V.: The specifics of financing the healthcare sector at the regional level. Socio-Econ. Phenom. Process. **14** (2(106)), 73–80 (2019)
15. Ilin, I., Levina, A., Lepekhin, A., Kalyazina, S.: Business requirements to the IT architecture: a case of a healthcare organization. In: Murgul, V., Pasetti, M. (eds.) International Scientific Conference Energy Management of Municipal Facilities and Sustainable Energy Technologies EMMFT 2018. pp. 287–294. Springer International Publishing, Cham (2019). https://doi.org/10.1007/978-3-030-19868-8_29
16. Iliashenko, O.Y., Iliashenko, V.M., Dubgorn, A.: IT-architecture development approach in implementing BI-systems in medicine. In: Arseniev, D.G., Overmeyer, L., Kälviäinen, H., and Katalinić, B. (eds.) Cyber-Physical Systems and Control. pp. 692–700. Springer International Publishing, Cham (2020). https://doi.org/10.1007/978-3-030-34983-7_68
17. Ilin, I.V., Levina, A.I., Lepekhin, A.A.: Reference model of service-oriented IT architecture of a healthcare organization. In: Arseniev, D.G., Overmeyer, L., Kälviäinen, H., and Katalinić, B. (eds.) Cyber-Physical Systems and Control. pp. 681–691. Springer International Publishing, Cham (2020). https://doi.org/10.1007/978-3-030-34983-7_67
18. Ilin, I.V., Lepekhin, A.A., Ershova, A.S., Borremans, A.D.: IT and technological architecture of healthcare organization. In: IOP Conference Series: Materials Science and Engineering, vol. 1001, p. 012141 (2020). https://doi.org/10.1088/1757-899X/1001/1/012141
19. Ilin, I.V., Iliashenko, O.Y., Iliashenko, V.M.: Architectural approach to the digital transformation of the modern medical organization. In: Proceedings of the 33rd International Business Information Management Association Conference, IBIMA 2019: Education Excellence and Innovation Management Through Vision 2020, pp. 5058–5067 (2020)
20. Shumakova, O.V., Kryukova, O.N.: New transaction costs and global digitalization. Human Science: Humanitarian Research. T. **14**(3), 189–197 (2020). https://doi.org/10.17238/issn1998-5320.2020.14.3.23
21. Chesbrough, H.: Business model innovation: it's not just about technology anymore. Strat. Leader. **35**(6), 12–17 (2007). https://doi.org/10.1108/10878570710833714
22. Dubgorn, A., Ilin, I., Levina, A., Borremans, A.: Reference model of healthcare company functional structure. Presented at the (2019)

Self-Organization of a Group of Drones in the Conditions of Absence of Data on the Exact Location

Ruben Girgidov and Timur Khakimov

Abstract Self-organization of a group of drones in the conditions of absence of data on the exact location and absolute velocity is an infrequent topic for scrutiny as a basic problem, though being of great relevance in the absence of navigation (for instance, in extraterrestrial or undersea explorations). Utilization of control rules based on the analysis of schooling behavior of fish, birds and other biological objects [1] enables implementation of self-organized groups but is not limited to it. The crystal-based analogy, that we adopted, proved the ability of drones to self-organize as a swarm even with low requirements to systems of intercommunication and navigation.

Keywords Swarm technologies · Drone · Control system · Self-organization · Optimization

1 Introduction

Scientific literature covers two main control principles for groups of drones: centralized and decentralized [2].

Centralized approach implies the processing centre, generally beyond the group's operating zone. Centre allows the calculation of every single drone's trajectory and transmits control commands, or interacts with one specific drone (a so-called leader) which in turn coordinates his followers within the group (hierarchical structure). Such a strategy gives certain advantages: constant control over the situation, real-time engagement coordination, rapid intervention capability in case of emergency. However, the disadvantages arise self-evidently from advantages:

- dependency on the centre makes the group vulnerable to disruption of channels of communication. Malfunction of the centre itself (or of the leading drone in case of hierarchical structure) can totally disable it

R. Girgidov (✉) · T. Khakimov
Peter the Great St. Petersburg Polytechnic University, Polytechnicheskaya str. 29, 194064 St. Petersburg, Russian Federation
e-mail: ruben@betria.com

C. Jahn et al. (eds.), *Algorithms and Solutions Based on Computer Technology*, Lecture Notes in Networks and Systems 387,
https://doi.org/10.1007/978-3-030-93872-7_19

- restriction of the flight range due to the limited range of communication or prolonged time of signal transmission. For example, for modern space vehicles exploring the surface of Mars, signal delay is approximately 13 min and the channel itself is available for only 8 min a day.
- heavy computational burden on the centre reduces flexibility, when it comes to resizing the group.
- higher requirements to the uplink. This problem is negligible for groups of under 100 drones, but for those containing more, becomes especially acute. In our research we considered groups of up to 1000 units.

As for the decentralized approach, every single drone is an independent entity (agent) operating on the basis of a built-in program and information from the immediate surroundings (from its own sensors or nearest neighbors) [3]. Apart from levelling weak points of the centralized approach such strategy enables even a straying drone to continue his mission while trying to establish contact with the rest of his kind. Besides that, distribution of calculations lowers the cost of the equipment used and generally simplifies the structural design. Thus, decentralized control strategy is more reliable and scalable, but more complex in terms of automation.

Within the decentralized approach, three modes of movement are presented:

1. In formation—which means that the group moves keeping the necessary structure—wedge, row, grid, convoy, etc. [4] Such mode implies two-tier hierarchy: leader (or leaders) and followers. Main disadvantage is similar to that of centralized control—vulnerability due to dependency on leading drones.
2. In swarm—self-organized movement without the need to maintain solid structure. Data exchange between units is optional. Such mode is more flexible and resistant to abnormal situations.
3. Combined (mixed) mode brings together concepts of both of the above [5]. In the basic mode, the group operates autonomously, but operator can, if necessary, intervene and adjust mission objectives or abort it in case of emergency. However, he does not program behavior of every single drone separately, but gives general commands while leaving it up to a swarm to handle it.

Our goal was to develop a program that controls drones (as independent agents) and meets certain requirements:

1. We should be able to stabilize the group of drones both in the presence of a rally point and its absence (when the group meets by itself);
2. Minimized complexity and accuracy standards of sensors;
3. Research self-organized group depending on the presence or absence of the central rally point;
4. Zero dependency from leaders.

There are also some additional features, e.g.:

- ability to operate without data exchange within the group.

We define the swarm as a group of drones (consisting of more than 6 members), interacting with each other and having the same control program as well as a common goal. It must be noted that the swarm can consist of different types of drones, but it is out of our scope.

State of the swarm will be deemed stable if there is no relative motion between agents and as a result the swarm structure is preserved. For instance, in a situation when swarm is similar to a crystal lattice.

Finally, the state will be referred to as quasi stable, if there is a possibility of relative motion, though the structure of the swarm (constant boundary) is preserved.

2 Materials and Methods

Traditionally, when considering control problems, object of control is separated from control system. We shall also make use of this approach (Fig. 1).

Drone in the model is a material point of mass m moving in two-dimensional space defined by x,y coordinates. The point is affected by viscous damping force:

$$F_{tr} = -\mu V, \tag{1}$$

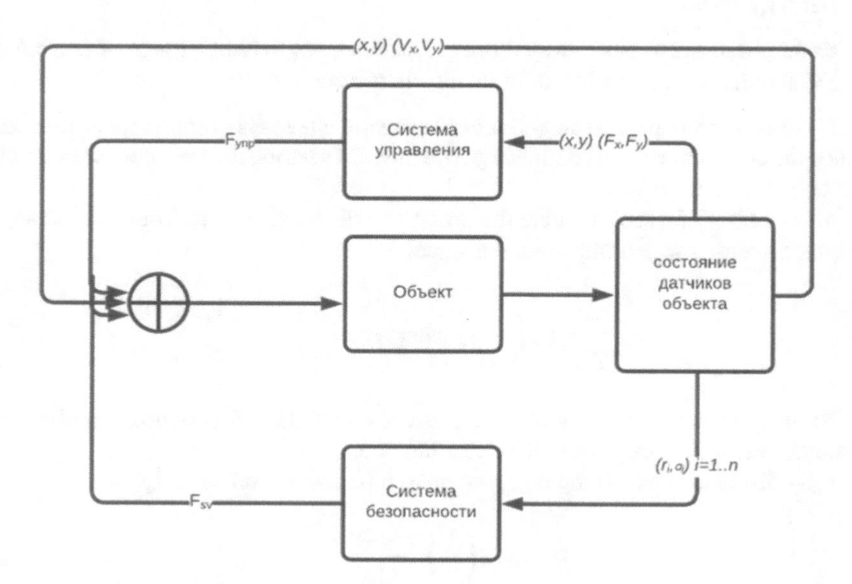

Fig. 1 Scheme of the model

where **V** is a velocity vector, and μ—is damping coefficient. It is assumed that at any moment a drone as a material point can be affected by a control force acting in an arbitrary direction.

The drone is affected by limitations:

$|V| < V_{\max}$—velocity can not be above some maximum value.

$|a| < a_{\max}$—acceleration can not be above maximum, which means that the force acting on the drone is also limited.

In situations when velocity and acceleration exceed maximum values, they remain constant, but this does not affect movement direction.

These empirical limitations are derived from the physical ones. For example, velocity and acceleration of standard civilian-pattern quadcopters does not exceed $20\mathrm{M}/s$, and $14\mathrm{M}/s^2$ respectively. They can also be derived from the model where real viscous damping, engine power, etc., are satisfied. However, we are interested not in physically accurate modelling, but in that sufficient for the implementation and research of control programs and object itself.

Sensors that we are supposed to make use of are range finders (that determine distance to the nearest aim) and optoelectronic sensors determining angle within a sector. Thus, every drone receives coordinates of its neighbors in polar coordinates where it acts as a starting point.

Let's consider individual parts of our model:

Security system:

- enables a drone to push away those neighbors being within a range of detection. Detection accuracy includes following characteristics:

r_i—distance to a nearest neighbor inside a sector where *i* is a drone's index number. Since the cost of a sensor is defined by a quantity of sectors, the less there are sectors, the cheaper a device is.

a_i—number of a sector where the drone was detected or in fact number of range detector's sensor which transmitted a signal.

$$F_{sv} = \sum_i v_i \mathrm{f}_{sv}\left(\frac{1}{r_i}\right), \text{ where } v_i = \begin{cases} v, \text{ if } r_i \le r_{sv} \\ 0, \text{ if } r_i \le r_{sv} \end{cases} \tag{2}$$

Security system's main goal is to prevent collisions. Direction of repulsion is defined by a sector reversed to the detecting one.

r_{sv}—distance at which the drone starts to push away from its neighbor.

$$\mathrm{f}_{sv}\left(\frac{1}{r}\right) = \frac{-\mathrm{r}}{|\mathrm{r}|^2} \tag{3}$$

This force of repulsion is proportional to the distance to the neighbor.

Thus, it is a superposition of repulsion forces that defines a security system's impact.

Control system:

– ensures the fulfillment of the task assigned to the drone.

In the first case the task was to make drones find themselves at the minimal distance to the rally point.

To achieve this, we have chosen a feedback function working in accordance with the law (4) and depending on the proximity to the aim.

$$\mathrm{F}_{upr} = k\mathrm{R} \tag{4}$$

After analyzing the structure of the group, similarities to crystal lattice were found, thereby suggesting that random "thermal" vibrations can speed assemblage of the swarm. Random vibrations will be defined as the impact of a randomly directed force in the form of a discrete-time random process (white noise) $\xi(t)$ in a two-dimensional space $(\mathbb{R}, \mathbb{R})$.

$$F_{rnd} = x \cdot \Xi(\mathrm{t}), \text{ where } \Xi(\mathrm{t}) = \frac{\xi(\mathrm{t})}{|\xi(\mathrm{t})|} \tag{5}$$

where x is a weighting coefficient of a random process.

Analogy with a crystallization process was confirmed by the reduction of assembly time nearly twice (Figs. 2 and 3). Thus all follow-up experiments were conducted using random vibrations.

As for the second case (when the rally point is not determined), the whole swarm can be stabilized in an arbitrary point.

In the absence of a central rally point, as a control force we used an equivalent of potential force acting between atoms. In general it is:

$$\mathrm{F}_{upr} = \sum_i \lambda_i \mathrm{f}_{ac}\left(\frac{1}{R_i}\right), \text{ where } \lambda_i = \begin{cases} \lambda, \text{ if } \mathrm{r}_{sv} \le R_i \le R_{as} \\ 0, \text{ if } R > R_{as} \text{ or } r_{sv} > R_i \end{cases} \tag{5}$$

f_{ac}—attractive force between the drone with number i and the current one,

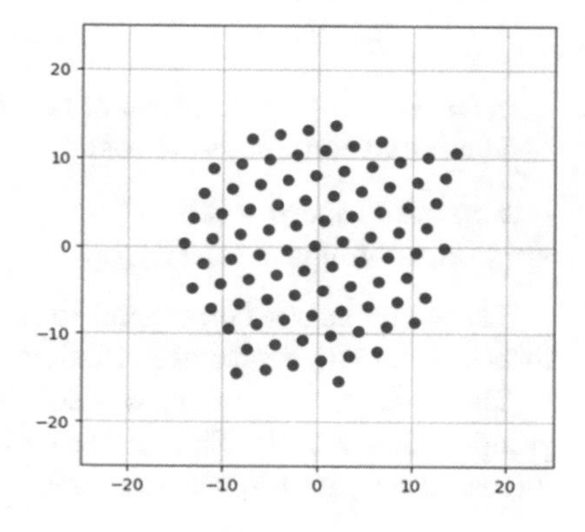

Fig. 2 Swarm structure when the detection field is divided by eight sectors—with vibrations (assembly time—95 s)

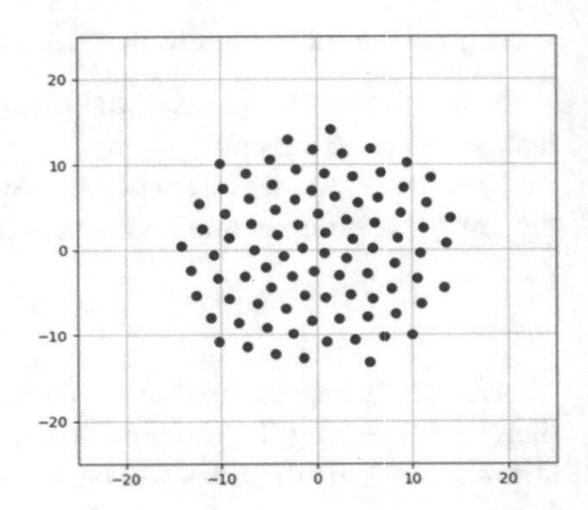

Fig. 3 Swarm structure when the detection field is divided by eight sectors—without vibrations (assembly time—180 s)

$$f_{ac}\left(\frac{1}{R}\right) = \frac{R}{(|R| - r_{sv})^2} \tag{6}$$

Thus an analogue of the Lennard–Jones potential was used [6], that performed steadily and enabled to combine security and control modules.

During research the following drone model with various types of interaction and composition of sensors was tested:

- coordinate setting by sector which means that the drone receives relative coordinates in the polar system with the following limitation: within a sector drone can detect only a single neighbor. Angle determination within a sector is precise. Area around the drone can be divided by several sectors (from 3 to 8).

Thus, every drone in the swarm moves according to the equation:

$$\begin{gathered} m\mathrm{a}_i = \mathrm{F}^i_{upr} + \mathrm{F}^i_{sv} + \mathrm{F}^i_{tr} + \mathrm{F}^i_{rnd},\ \text{where}\ i = \overline{1 \ldots N} \\ \mathrm{V}^i(0) = \left. \begin{matrix} v^i_x \\ v^i_y \end{matrix} \right|_{t-0} = 0 \\ \left. \begin{matrix} x_i \\ y_i \end{matrix} \right|_{t=0} = C_i \end{gathered} \tag{7}$$

$\mathrm{a}_i = (\ddot{x}_i,\ \ddot{y}_i)$—acceleration of the respective drone (i).

The initial conditions are set as:

- zero initial speed $\mathrm{V}^i(0)$.
- initial coordinates $(x_i,\ y_i)_{t=0}$.

These conditions correspond to the case when a group of drones takes off the surface where they have been randomly positioned.

This system of equations was solved using the Euler method for all drones in the group simultaneously. This choice was guided by the necessity to adjust both control force and security force at any time.

3 Results

In the course of modelling the number of sectors was changed from from the lowest to the highest value. The quantity was defined from 3, when the drone sees no more than three neighbors (without necessity to be the closest ones, but being in different sectors) to 8 sectors respectively when it can interact with (or in other words detect) eight neighbors. Different types of configuration depending on the number of sectors are shown below.

From Figs. 4, 5, 6, 7 it is obvious that the crystalline structure is preserved regardless of the number of sectors used. However when their number is small (4–5) the "crystal" itself contains either flaws (Fig. 4) or anomalies on the perimeter (Fig. 3). It becomes a problem if the task is to uniformly cover some territory with the swarm, but does not deny the swarm's structural stability. At the same time assembly speed depends on the precision and completeness of measurements and in our experiments it varies from 350 s for 3 sectors to 95 s for 8 sectors.

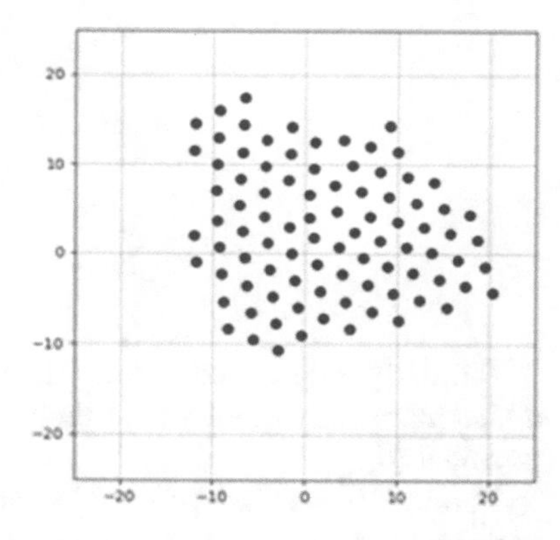

Fig. 4 Swarm structure when the detection field is divided by three sectors (assembly time—350 s)

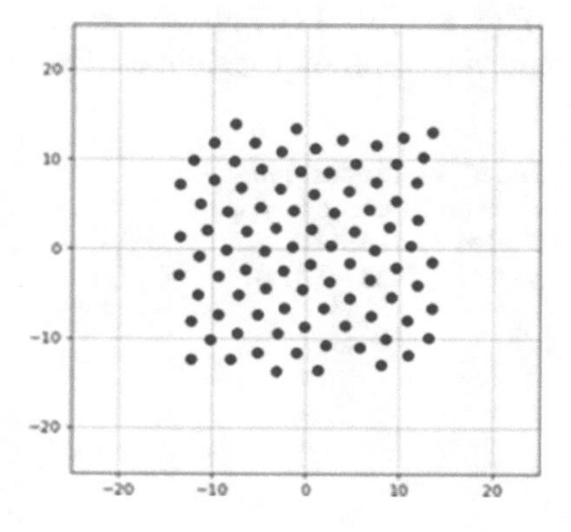

Fig. 5 Swarm structure when the detection field is divided by four sectors (assembly time—300 s)

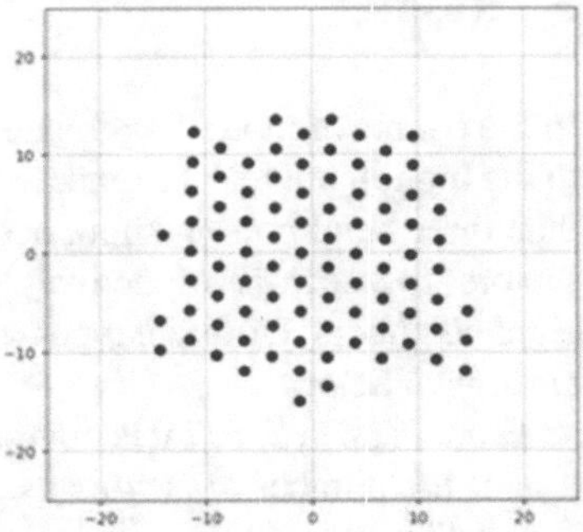

Fig. 6 Swarm structure when the detection field is divided by six sectors (assembly time—120 s)

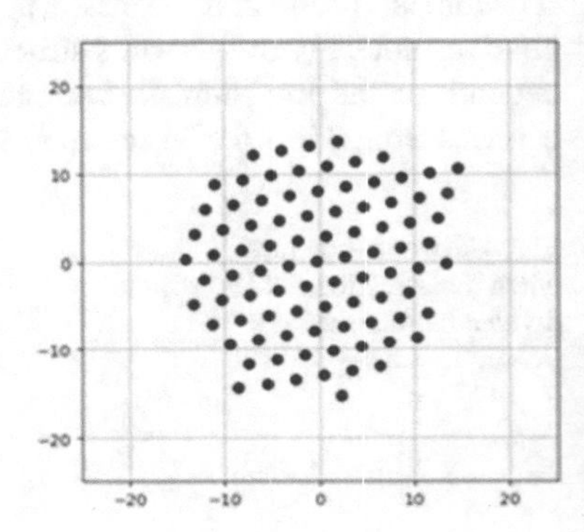

Fig. 7 Swarm structure when the detection field is divided by eight sectors (assembly time—95 s)

Similarly, we have considered the model for constructing a stable swarm without a stationary rally point where the control force was based on the Lennard–Jones potential (6).

Perhaps, those "independent" swarms alienated from centralized control can go with low-tier sensors as it can be seen from the data received during experiments (Figs. 8, 9, 10, 11). Enhancement of sensors' quality does not cause any significant growth of swarm capabilities. For instance, for a swarm with 3 sectors, a stable

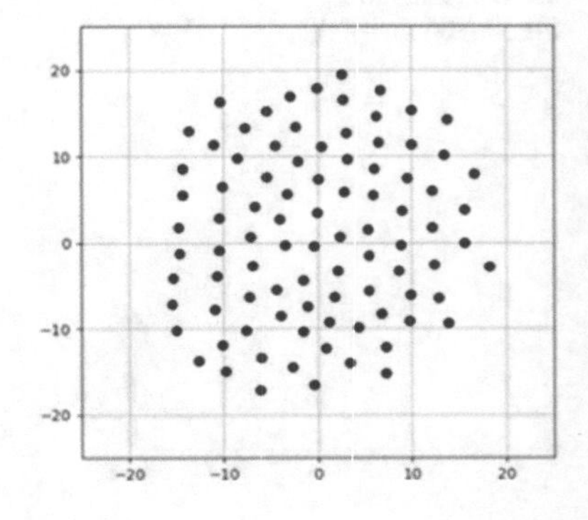

Fig. 8 Swarm structure when the detection field is divided by 3 sectors (assembly time—170 s)

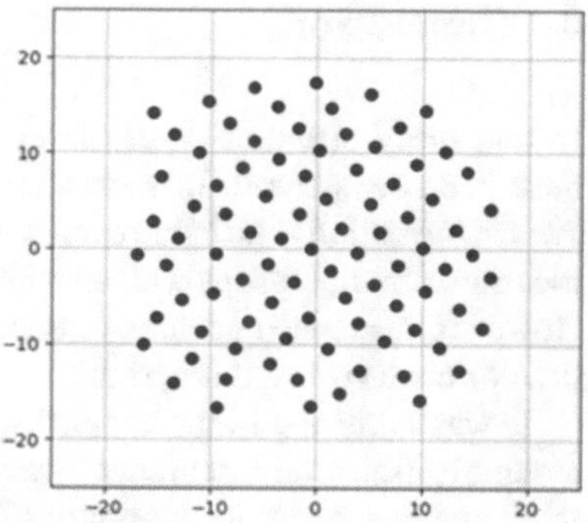

Fig. 9 Swarm structure when the detection field is divided by 4 sectors (assembly time—110 s)

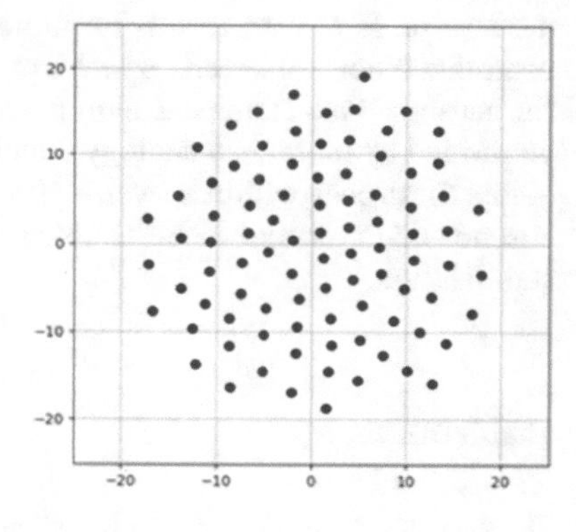

Fig. 10 Swarm structure when the detection field is divided by 6 sectors (assembly time—100 s)

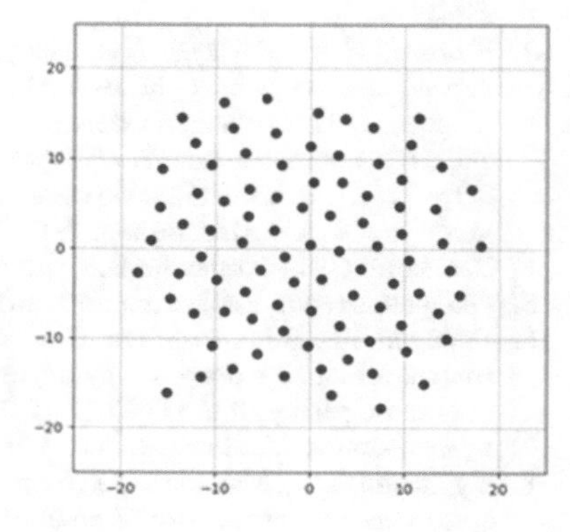

Fig. 11 Swarm structure when the detection field is divided by 8 sectors (assembly time—92 s)

assembly process takes 350 s in the presence of a central rally point and only 170 s without it (that is to say twice as fast), but with division by 8 sectors the difference is only 3%, which means that assembly time shortened from 95 to 92 s.

4 Conclusions

In this study different types of sensor array's composition and control algorithms have been considered. It was found that in the presence of a viscous friction force, swarm formation is possible even without data exchange and information about exact location of neighbors as well as with sufficiently large response time (of up to 0.1 s). However, decreased accuracy consequently resulted in an increase of stabilization time or even swarm disruption.

It was confirmed that simplification of the drone's composition entails slower assembly times and degraded area coverage. Thus this problem moves from the mathematical realm to the realm of financial reasonableness of implementation of different drone compositions with an attachment to the specific task assigned to the swarm. However, results of numerical experiments show that minimization of centralized control speeds up swarm assembly time and reduces the requirements for sensors. This statement requires additional research, although being indirectly confirmed by biological analogy which is very important for understanding requirements for drones within a swarm. The experiments also confirmed that existence or absence of the leader or hierarchical structure is not really important for swarm's stabilization.

References

1. Reynolds, C., Flocks, W.: Herds, and schools: a distributed behavioral model. Comput. Graph. **21**(4) (1987)
2. Каляев, И.А., Гайдук, А.Р., Капустян, С.Г.: Методы и модели коллективного управления в группах роботов. М.: Физматлит, 2009. 280 с
3. Морозова, Н.С.: Формирование строя и движение строем для мультиагентной системы с динамическим выбором структуры строя и положения агента в строю // Труды XII Всероссийского совещания по проблемам управления ВСПУ-2014. М.: ИПУ им. В. А. Трапезникова РАН, 2014. С. 3822–3833
4. Зенкевич С.Л., Галустян Н.К.: Децентрализованное управление группой
5. квадрокоптеров // Мехатроника, автоматизация, управление. 2016. №11
6. Городецкий, В.И., Карсаев, О.В., Самойлов, В.В., Серебряков, С.В.: Прикладные многоагентные системы группового управления // Искусственный интеллект и принятие решений. 2009. № 2. с.3–24
7. Lennard-Jones, J.E.: Proc. R. Soc. A **106**, 463 (1924)
8. Бурдаков, С.Ф., Марков, А.О.: Управление квадрокоптером при полетах с малыми и средними перегрузками; М-во образования и науки Российской Федерации, Санкт-Петербургский политехнический ун-т Петра Великого. - Санкт-Петербург : Изд-во Политехнического ун-та, 2016. - 250 с. : ил.; 21 см.; ISBN 978–5–7422–5059–3
9. Бурдаков, С.Ф., Сизов, П.А.: "Алгоритмы управления движением мобильного робота в задаче преследования", Научно-технические ведомости СПбГПУ. Информатика. Телекоммуникации. Управление, 2014, № 6(210), 49–58
10. Awasthi, S., Balusamy, B., Porkodi, V.: Artificial intelligence supervised swarm UAVs for reconnaissance. Data Sci. Anal. **1229** (2020) ISBN : 978-981-15-5826-9

11. Stolfi, D.H ; Brust, M.R., Danoy, G., Bouvry, P.: Optimizing the performance of an unpredictable UAV swarm for intruder detection. Optim. Learn. **1173,** 37–48 (2020). ISBN : 978-3-030-41912-7
12. Первозванский, А.А.: Курс теории автоматического управления: Учеб. пособ - М.: Наука. Гл. ред. физ.-мат. лит., 1986. - 616 с.
13. Svetunkov, S.G., Chanysheva, A.F.: Properties of complex dispersion and standard deviation. Espacios **38**(33), 25 (2017)
14. Svetunkov, S.G., Chanysheva, A.F.: Analyses of erdenet corporation performance using the exponential production function of complex variables. Gornyi Zhurnal, (1), 26–31 (2017)
15. Svetunkov, S.G.: The possibility using the power production function of complex variable for economic forecasting. Econ. Region **12**(3), 966–976 (2016)
16. Cooley, R., Wolf, S., Borowczak, M.: Secure and decentralized swarm behavior with autonomous agents for smart cities. In: 2018 IEEE International Smart Cities Conference (ISC2), pp. 1–8 (2018). https://doi.org/10.1109/ISC2.2018.8656939
17. Dai, F., Chen, M., Wei, X., Wang, H.: Swarm intelligence-inspired autonomous flocking control in UAV networks. IEEE Access **7**, 61786–61796 (2019). https://doi.org/10.1109/ACCESS.2019.2916004
18. Zhu, X., Liu, Z., Yang, J.: Model of collaborative UAV swarm toward coordination and control mechanisms study. Proc. Comput. Sci. **51**, 493–502 (2015). ISSN 1877-0509
19. Hereford, J.M.: A distributed particle swarm optimization algorithm for swarm robotic applications. IEEE International Conference on Evolutionary Computation **2006**, 1678–1685 (2006). https://doi.org/10.1109/CEC.2006.1688510
20. Rosalie, M., Danoy, G., Chaumette, S., Bouvry, P.: Chaos-enhanced mobility models for multilevel swarms of UAVs. Swarm Evolut. Comput. **41**, 36–48 (2018). ISSN 2210-6502
21. Brust, M.R., Zurad, M., Hentges, L., Gomes, L., Danoy, G., Bouvry, P.: Target tracking optimization of UAV swarms based on dual-pheromone clustering. In: 2017 3rd IEEE International Conference on Cybernetics (CYBCONF), pp. 1–8 (2017). https://doi.org/10.1109/CYBConf.2017.7985815

Smart Hospital Architecture: IT and Digital Aspects

Anastasia Levina, Victoria M. Iliashenko, Sofia Kalyazina, and Ed Overes

Abstract The paper proposes to consider the IT and Digital concepts of Smart Hospital. Today, smart clinics are a healthcare facility that uses the latest digital technologies to improve the quality of healthcare services, simplify healthcare workflows, reduce the burden on clinic staff, and ultimately increase patient satisfaction. Therefore, the relevance of this article is clear. The analysis of world experience in terms of digital technologies and making smart hospitals is presented. A detailed analysis of modern digital technologies used in medical institutions has been carried out. The architecture of Smart Hospital was proposed. The purpose of research is formation of the top level of IT architecture of a smart hospital and presentation of requirements for complex architecture of smart hospital. The methodological basis of the paper is the analysis of open sources.

Keywords Smart hospital · Digital technologies · Reference model · IT-architecture

1 Introduction

In medicine, as in other industries, digitalization has penetrated, and it is here that it works most noticeably: the effectiveness of treatment is growing even for the most severe diseases, diagnostics predicts problems at the very beginning or even before they arise. In addition, service has received a special role in healthcare today. Smart hospitals are being created all over the world, or old hospitals are being transformed into them. In such a "smart" hospital, the patient is comfortable, calm [1].

Smart hospitals, as they are commonly called, are built on technologies that automate not only many "manual" tasks, but also communication and interaction with

A. Levina · V. M. Iliashenko (✉) · S. Kalyazina
Peter the Great St.Petersburg Polytechnic University, Polytechnicheskaya, 29, St.Petersburg 195251, Russia
e-mail: vmi1206@yandex.ru

E. Overes
Hogeschool Zuyd, Heerlen, Netherlands

C. Jahn et al. (eds.), *Algorithms and Solutions Based on Computer Technology*, Lecture Notes in Networks and Systems 387,
https://doi.org/10.1007/978-3-030-93872-7_20

patients, that is, they digitize most of the processes. These hospitals are generally paperless. They usually purchase and/or develop modern equipment, work professional doctors and scientists, conduct clinical research [2].

"IoT" (Internet of Things)—technology, permissive or implicit possibility connecting to the global data network of any devices that are not originally IT equipment (servers, personal computers, smartphones). "Big Data"—the term adopted for descriptions of modern volumes of information related with a digital society, digital economy. Term "Big data" characterizes data sets with possible exponential growth that is too large, too unformatted, or too unstructured for analysis by traditional methods [3, 4].

The main goal of this article is making an overview of IT and Digital concepts of Smart Hospital; to make an overview of the world's leading clinics using advanced digital technologies, consider their structure and the interaction of the main components of the hospital. Moreover, the article suggests the architecture of the upper level of interaction of the components of a smart clinic.

As a result of the research, the authors of the article want to get a set of requirements to smart hospital. Moreover, the result of research is formation of the top level of IT architecture of a smart hospital and presentation of requirements for complex architecture of Smart hospital, which takes all blocks as building, digital technologies, smart wards into account.

2 Materials and Methods

2.1 Johns Hopkins Hospital

Johns Hopkins Hospital is the most famous medical institution in the United States, which is at the same time a training base for the Johns Hopkins Medicine educational complex, Baltimore, Maryland, USA [5] (Figs. 1 and 2).

In terms of innovation, the clinic can be distinguished:

- Enhanced filtered air circulation system to prevent the spread of germs and reduce respiratory complications.
- The latest surgical technologies in 33 operating rooms, including intraoperative MRI. Innovative patient ward layout to ensure the best possible care and collaboration between your medical team.
- A real-time positioning system that instantly tracks equipment anywhere in the hospital and can locate staff in inpatient wards.
- Radiological imaging packages with the most advanced diagnostic and radiological services in the country.
- An automated underground system that delivers materials and waste through a tunnel and to a hospital loading bay, a quarter mile from patient service areas.
- High-tech computer systems to help your healthcare team respond to problems.

The Hopkins Hospital Health System includes the following units (Fig. 2):

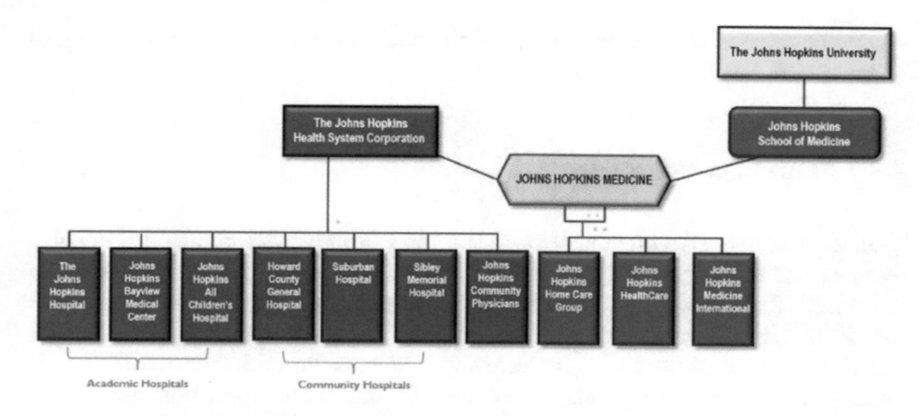

Fig. 1 Johns Hopkins medicine organizational chart

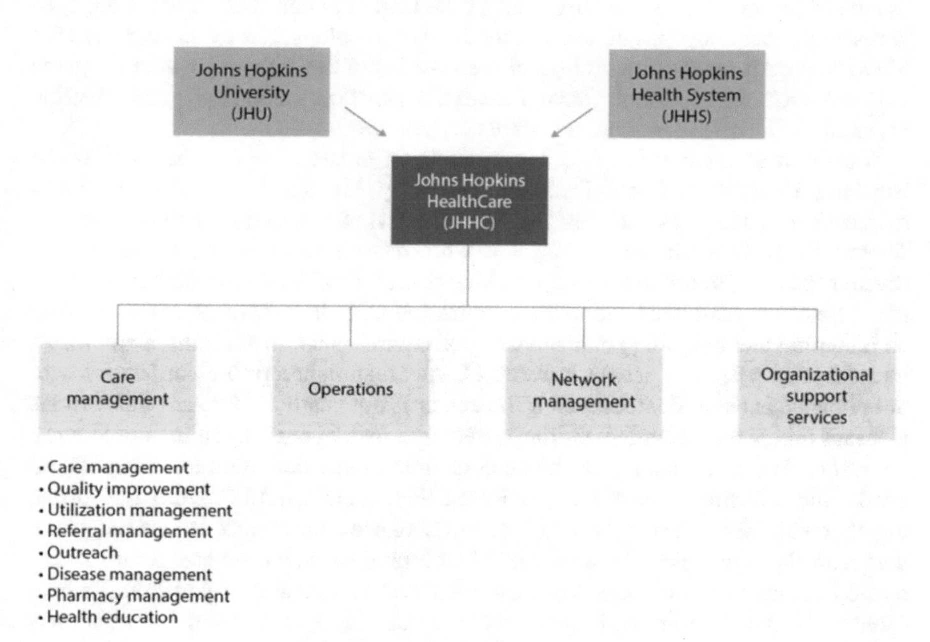

Fig. 2 Johns Hopkins healthcare: interaction with other blocks

- Johns Hopkins University is necessary for laboratory and fundamental research, staff formation.
- Johns Hopkins Health system which is necessary for which is necessary for carrying out the key functions of the clinic—medical activity.
- Care management includes supervision and educational activities conducted by healthcare professionals to help patients with chronic conditions.

– The operations block includes one of the key processes of the clinic—surgical treatment.

2.2 Breakthrough R&D

"Smart" hospitals also often appear on the basis of university clinics that conduct research and train doctors. For example, Stanford Medical Complex is famous for the world's first heart, lung transplants and other surgeries. The hospital also has the world's first hybrid ward that diagnoses and treats vascular disease using stereoscopic digital subtraction angiography. In Germany, Asklepios Barmbek has such a scientific fame, a clinic where doctors take on the most difficult cases in gynecology, urology, and neurology. Sometimes smart clinics have narrow specializations. For example, the world-famous Mayo Cancer Center in the United States. This clinic is engaged in research, including unique ones, in the field of oncology. For example, it was Dr. Mayo who performed the neurological examination of the rugby team after the game with the V-Go remote robot. Indian research center Fortis also pays great attention to research, in particular, there is a stem cell laboratory [6, 7].

Ideally, a smart hospital is a combination of smart components, such as the Bundang Hospital of Seoul National University. Bundan is a multidisciplinary medical center that uses the latest technology and methods to treat serious illnesses. There is the BestCare information system with an electronic dossier, a biometric data transfer system, "smart" systems for making clinical decisions and resource management. BestCare maintains digital communication with the patient, stores all the data on incoming and outgoing people in a digital environment, in addition, it automates internal processes. The patient is aware of what manipulations are carried out with him, how he is being treated, even if the doctor is not nearby. There are also various pleasant things that complement the system and make your stay in the clinic more enjoyable. For example, a smart bed can maintain individual light and temperature levels. Such hospitals exist not only in Korea. For example, in El Camino, California, together with Silicon Valley-based IT startup Lockheed, they created a similar system that controls all processes in the clinic. The hospital has achieved particular success in the treatment of oncology. Also, a similar system operates at Anadolu Medical Center in Turkey. The institution does not have a single paper card—all information is stored in computers and smartphones, and telemedicine is the main way of communicating with patients. In the Palomar medical center, together with the "smart" system of organization of the clinic, biometric identification works, which further increases the convenience of patients and doctors [8].

Digitalization has become the basis of many modern clinics, including the oldest ones, which simply invested in the modernization of medical facilities. For example, the Thai hospital Bumrungrad, established in 1980 in Bangkok. Today it is the largest clinic in Asia, which every year purchases the latest equipment—digital PET-CT scanners, drug dosing and distribution systems, mammography. Ramkamhang

Hospital, also a Thai medical institution, has digitized the entire process of interaction with patients: at the reception, everyone is given a digital hospital card—a guide card, which stores all information about him. Doctors only need to scan the card code, and they see the medical history on a computer or tablet. All diagnoses are stored electronically, and prescriptions are sent to a digital pharmaceutical system that dispenses the required amount of drugs by itself [9].

There are also relatively new medical organizations that have begun to emerge on the basis of modern IT solutions, which is called "from scratch." In 2010, Vail Hospital was opened in the small town of Hensol in the UK. The hospital has digital operating theaters, the design and layout of which changes depending on the operation. In 2013, the Nuffield Health Clinic was launched in Bristol. It is believed to be the first fully digital hospital in England. Such technologies include a digital communication system and the latest technology (MR, CT scanners, PET, 3D mammography, equipment for endoscopy). Of course, such high-tech clinics require large investments—tens of millions of dollars.

2.3 *SunCom Smart Clinic*

The Digital Clinic provides real-time communication between doctors. Communication can take place through a secure messenger, as well as using audio and video communication channels using telemedicine complexes installed in operating units. The combination of Avaya technologies and Sun Com developments allow for emergency consultations from the doctor's office and from the operating room. In addition, the Digital Clinic increases the availability of medical personnel and the efficiency of decision-making by displaying the employment and location of employees in a medical institution. A complete record of what is happening in the operating room and the indications of medical devices is also available. This provides objectification of the course of operations, including the possibility of using the accumulated content for the analysis and training of the clinic staff [10].

"Digital Clinic" solves not only the internal tasks of the clinic, but is also a convenient tool for patient communication. The solution improves the quality of service and provides additional opportunities to contact the clinic remotely.

In particular, it is possible to receive research results and provide remote rehabilitation and monitoring of the patient's health status. Also, this solution provides integration with e-mail and the patient's busy calendar.

Grand Medica is considered the largest smart hospital in Siberia and the Far East region. On 19,000 m^2 contains special equipment and a highly efficient EcoStruxure platform from Schneider Electric, which allows you to manage the devices connected to it, carrying out monitoring and local control. The technology is flexible: as the facility grows and develops, it can be easily scaled, modernizing the clinic to further improve the productivity and quality of patient care.

Remote management of power grids and medical equipment has reduced operating costs by 20%, according to Grand Medica. Heat consumption has dropped by almost a third. Thanks to Schneider Electric product compatibility, the medical center saved approximately $ 176,000 in capital expenditure [11].

At the heart of the Grand Medica Smart Hospital is Schneider Electric's EcoStruxure healthcare platform. It combines three layers of unique IoT solutions: connected devices, local management and data collection, analytics and services. These solutions support hospitals and clinics at all levels—from emergency departments to executive offices.

3 Results

3.1 Integration of Digital Platforms in Medical Centers

There is a growing awareness in the healthcare system of the need to involve patients in the treatment process. An important aspect for restructuring the industry is the ability to leverage the continuous flow of health data from wearable devices, sensors and other similar systems and devices. Such technology, in particular remote monitoring systems and mobile applications, will be able to provide a closer contact between patients and their doctor and open up new opportunities for patients to exchange data with doctors. This makes it possible for doctors to gain a more complete picture of patients, especially how patients behave in between appointments at the clinic, which can significantly improve the results of treatment of chronic diseases [12]. Note that the patient's personal data is not only clinical data, but also data that determine the social lifestyle, which allows the doctor to get a holistic view of the patient. But today, doctors tend to resist adopting such technologies and using them in their diagnostic and treatment decisions. There are many reasons: information and emotional overload, incomplete or unreliable data, privacy and security issues, and so on. Nevertheless, this technology must be present in the clinical workflow as it is critical to the success of treatment, given the workload that most physicians face today. But it should not increase the burden on doctors, but, on the contrary, provide them with information in a revised form, ready for clinical use. To do this, these data must be collected from different sources, accumulated and analyzed.

Today there is no easy way to move data between systems, such technologies are just being created. Data interoperability remains an issue as much of EHR data is tied to manufacturers' own platforms. But the implementation of information exchange standards such as Fast Healthcare Interoperability Resources (FHIR) is expected to improve data interoperability in the near term. Collecting patient data will help develop predictive analytics models based on advanced algorithms, artificial intelligence and historical data. Moving from inpatient care to remote patient care is opening up new opportunities that can help identify, treat, and prevent disease. But for this, it is necessary to expand the capabilities of collecting and analyzing data using complex algorithms. To do this, wearable and home sensors and systems must be combined into a simplified and easily integrated platform so that developers can focus on the application. Today, such platforms are already being developed by several developers who create systems that can take information from many different sensors with the ability to transmit data for almost any type of application. These solutions separate wearable sensors, network, and data infrastructure from end-use applications, allowing developers to focus on their specific area, be it detecting congestive heart failure, monitoring infections during chemotherapy, or monitoring bedside deterioration in bedridden patients. Sensors are abstracted from the application using a common protocol, making it easy to integrate multiple data sources at once. In addition, the provision of reliable data transmission, for example, in the event of a network failure, between the sensor and the application is carried out within the framework of the protocol, instead of requiring the developer to install the application itself.

Validic

This integration system, which was launched back in 2013, allows you to access and improve the quality of data generated by patients and medical organizations on a daily basis. The platform and associated mobile solutions provide continuous access to personal health data from more than 350 medical devices, mobile applications and wearable sensors in the home. The platform can transmit all these data in the form of a single stream of information to any medical information system. In this case, the data is checked for correctness before being transferred to a medical or other system [13].

Validic has a rich set of tools for administering and maintaining services, improving the quality of critical data, registering and maintaining users, and managing device connections. Patients receive periodic reports in the form of so-called a dashboard containing their health data. Physicians, in turn, receive more complete information on their dashboard with expanded interpretations of the same data, which allows for a richer discussion with patients about their treatment. Providing information to the patient in such a visual form was noted by experts as the most important incentive for patients to take care of their own health. Connecting to the platform is very easy and takes only a few hours as data from all mHealth applications is standardized within Validic. At the same time, all security requirements for the storage and transfer of protected health information are taken into account, which comply with the "Safe Harbor" de-identification standard in accordance with the US Health Insurance Portability and Accountability Act (HIPAA). The system is used today in more than 50 countries.

Capsule Tech

This platform is a secure, comprehensive, technology-independent cloud service that connects medical devices (more than 875 different types of sensors and devices) for the free exchange of data between users of such devices and medical institutions. This service has received approval for use from the FDA and, accordingly, can be used by medical institutions. The Qualcomm 2net network is based on open standards, and the platform itself is integrated with electronic health record systems from 50 different manufacturers, as well as a number of other technology companies. 2net consists of two main modules.

The first is 2net Core, which interacts with sensors and devices for collecting health data and collects information from them, performing the function of a "library" of data, transferring them to the second module—2net Application, which contains the user interface. The modules are self-contained, which allows customers to create their own applications based on 2net Core, with their own set of capabilities and functions. An add-on to the Clinical Observation platform can provide clinicians with contextual information about a patient's condition in real time, which can facilitate early intervention, improve patient safety and improve clinical outcomes. The system can analyze data and provide actionable guidance to help clinicians make informed decisions quickly. The 2net SDK is available to developers of mobile applications on the Android platform. Developers will be able to offer automated connectivity to the open ecosystem of biometric sensors supported by 2net through their applications [14].

TactioRPM

TactioRPM remote patient monitoring platform, developed in Canada, includes a mobile application, browser tools, a secure cloud service, and a vendor-agnostic medical device integration system. In total, the system currently supports more than 150 medical devices for various purposes (thermometers, pressure meters, glucometers, scales, oximeters, heart rate meters and activity trackers). The TactioRPM platform brings together mobile patient apps, clinical portals, integrated healthcare systems (including Garmin, A&D Medical, BÜHLMANN Laboratories, Fitbit, Roche, Nonin, Omron, MIR & Welch Allyn), patient profiles, digital training programs. The system allows doctors to remotely view the data generated by the patient's devices and devices, provide the latter with educational materials and involve patients in new mobile-based relationships with medical professionals. In addition, TactioRPM provides a rich set of APIs for medical organizations, which allows the latter to organize additional data integration, automate their processes and connect special applications [15].

Having analyzed the current world experience in creating smart solutions for clinics, the authors of the article propose the following scheme for the interaction of the key blocks of a smart hospital (Fig. 3).

We can also highlight the key points that a smart clinic should include:

1. Intelligent building

Through the introduction of an automation system, it is possible to maintain a favorable climate in all rooms of the building, conveniently control lighting, and ensure visitor awareness. Thanks to the introduction of fire and security systems, video surveillance and access control systems, the level of security is increased. The output of information on the operation of all engineering systems to a single dispatch center allows to reduce the number of maintenance personnel. The automated system is created in order to provide centralized monitoring and control of critical systems, increase the comfort and safety of operation, and reduce the cost of maintaining

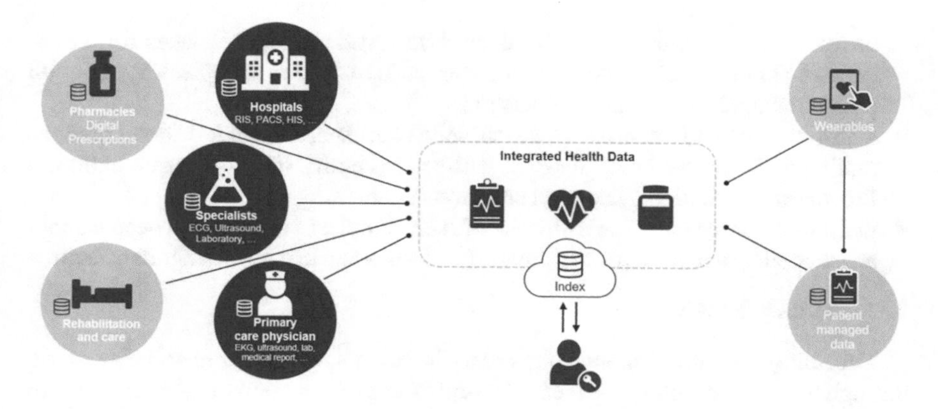

Fig. 3 Smart hospital data integration architecture

engineering equipment and consumed energy resources. The purpose of the system is a high level of awareness of the operation of systems, critical parameters of the facility, continuous quality control with a high level of reliability of data on the state of the facility and the ability to respond quickly in emergency situations.

2. Integrated automation and dispatching system for building engineering systems (BMS)

It is envisaged to install a decentralized control system, which provides a high level of reliability, fault tolerance and autonomy. The main information buses of the control system are KNX and Ethernet TCP/IP buses. Automation cabinets are distributed throughout the building, which ensures high safety, reliability and fault tolerance of the system. To ensure the uninterrupted operation of terminal and panel equipment, uninterruptible power supply units are used. Provides software and hardware for self-diagnostics and equipment condition monitoring. Provides the ability to send SMS and e-mail messages in emergency situations, with the ability to set mailing groups depending on the type of accident.

A BMS system is envisaged with a remote operator's workplace and access via a web browser. The BMS system allows monitoring and control of all engineering systems from a single dispatching point or remotely from mobile devices. Information on the BMS screen is displayed in the form of tables and mnemonic diagrams, including floor plans, individual rooms, general screens with the most important information.

3. Integrated building security system

To ensure sufficient safety, the following items must be installed:

- access control system (ACS) to restrict access to an object or to individual rooms. Fencing means (electromechanical and electromagnetic locks on doors, barriers,

turnstiles, etc.) are used, as well as an access system, input devices for identification (fingerprint scanner, iris scanner, RFID tag) and a control device with protection against unauthorized access;
- alarm system and security video surveillance. These include video cameras, motion and window break sensors, perimeter security sensors, "panic buttons", fire alarms, aspiration systems, a situation center;
- fire protection system. The main task of such complexes is to inform people about an emergency and assist in coordinating actions when leaving a dangerous facility.

4. Multimedia systems

The ability to connect mobile devices to the room's audio system and play music through the built-in ceiling speakers. Possibility of playing Internet radio in the room is supposed.

5. Smart-ward

- Lighting control. Each room provides for light control via relays/dimmers using the KNX protocol. The light is controlled using an LCD wall panel, keypads and mobile devices. The ability to create lighting scenarios is provided. Information about the operation of the lighting system is transmitted to a single control center. It is possible to control the light from the control room.
- Climate control. Provide for the installation of a climate control system to maintain the specified air temperature in the ward. The climate control system regulates the operation of local air conditioners, underfloor heating and heating convectors.
- Safety. Integration of comfort systems with security systems is envisaged. On a signal from the ACS system, a signal is received about the presence of people. Depending on who is in the ward, the patient or the service staff, they will have different levels of access to the systems. If full access is provided for patients, then access to certain subsystems can be restricted for service personnel.

3.2 Methodology for the Formation of an Integrated Architecture of a Medical Organization

Building an architectural model—an architectural process—follows certain principles. The most important of them are:

- The principle of gradual detailing. One should start by looking at the enterprise from a great height (the "owner" perspective in the Zachman model or the business architecture in the TOGAF model, etc.), rather than with a detailed description of one or more underlying elements. This is due to the fact that, describing any of the underlying elements of the general enterprise model, we will not get a single systemic view of the enterprise, which its architectural model is designed to give. The detailing can be deepened as needed, but this should be determined by how the architectural model is used.

- The principle of consistency of layers. If the basis is the most popular view of enterprise architecture as a "layer cake" (four levels in the TOGAF model or "perspective" in the Zachman model) [16, 17], it is necessary to achieve consistency of the layers. After all, the goal of building an architectural model is to get a unified and interconnected picture of the enterprise. Therefore, you should not describe the layers separately, entrusting this work to various departments. It is better to single out a process, or, at first—a project—describing the enterprise architecture, and create a single team.
- The principle of the independence of layers. At the same time, the layers (levels, "perspectives") must be independent. It is possible to select any necessary layers or levels (for example, an integration architecture that defines the principles of interaction and integration of applications, data and business processes in a distributed company environment in the Frameworx model (formerly NGOSS), or a security architecture). Therefore, when isolating them, the following conditions are applied: • if the lower layer is out of order, the upper one cannot work; • inoperability of the upper layer does not affect the performance of the lower one; • the performance of elements within one layer may or may not affect the performance of other elements of the same layer [18].
- The principle of completeness. The enterprise architecture model should describe the enterprise with the required completeness. At the same time, one should not get carried away with a large number of architectures and levels, as this complicates the model. The number of architectures and levels should be determined by specific tasks for the solution of which the architectural model is built.
- The principle of consistency. Elements of the architectural model should not contradict each other.
- The principle of no duplication. Elements of the architectural model should not duplicate each other [19, 20].
- The principle of continuous transformation of the current enterprise architecture. It should not be forgotten that any enterprise is in constant development. This means that its architectural model is useful only when it is relevant and constantly brought in line with the real state of the enterprise.

Based on the global experience of introducing digital technologies in medical institutions and describing the key steps in building an integrated enterprise architecture, the authors propose the following development roadmap for creating a smart clinic [21]. It includes 9 stages that cover both technical and financial aspects of the project. Particular attention should be paid to the formation of the IT landscape, service architecture and information systems integration (Fig. 4).

4 Discussion

The results of such studies will serve as a basis for creation complex IT-architecture of Smart Hospital considering all the technological features of modern trends. Based on

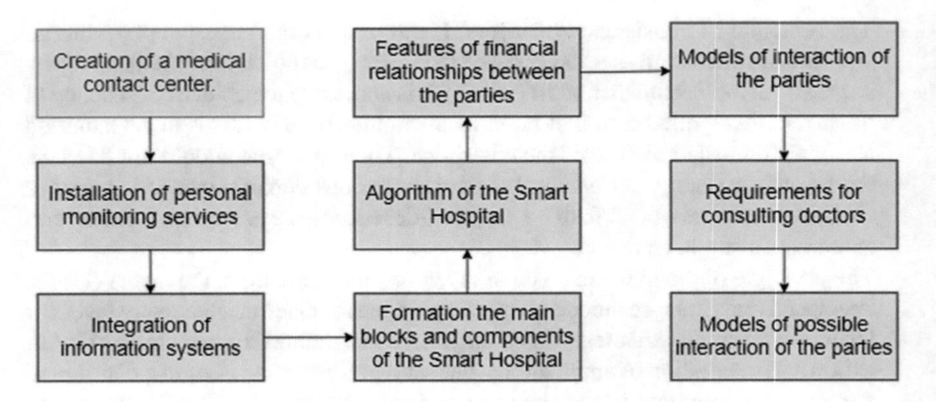

Fig. 4 Stages of the smart hospital concept creation

the described requirements for the integrated architecture of a medical organization that implements the principles of value-based and personalized medicine, the authors plan to form a comprehensive architecture of a digital clinic within the framework of scientific research based on the federal medical center.

5 Conclusions

In this paper, the IT and Digital concepts of Smart Hospital were considered. The analysis of world experience in terms of digital technologies and making smart hospitals was presented.

An in-depth analysis of the world's experience in creating smart clinics and digital platforms that support the development of the organization's digital resources was carried out.

Based on it, the authors suggested top level of IT architecture of a smart hospital and presentation of requirements for complex architecture of smart hospital. As a result, the paper suggests steps to form Smart Hospital considering different aspects as technical and financial.

Acknowledgements The reported study was funded by RFBR according to the research project № 19-010-00579.

References

1. Ilin, I., Levina, A., Borremans, A., Kalyazina, S.: Enterprise architecture modeling in digital transformation era. In: Advances in Intelligent Systems and Computing, 1259 AISC, pp. 124–142 (2021)

2. Digital business: 3 requirements to make it real. https://inform.tmforum.org/features-and-analysis/2017/01/digital-business-3-requirements-make-real/. Last accessed 02 February 2021
3. Ilin, I.V., Iliashenko, O. Yu., Iliashenko, V.M.: Architectural approach to the digital transformation of the modern medical organization. In: Proceedings of the 33rd International Business Information Management Association Conference, IBIMA 2019: Education Excellence and Innovation Management through Vision 2020. IEEE, pp. 5058–5067 (2019)
4. Dubgorn A., Ilin I., Levina A., Borremans A.: Reference model of healthcare company functional structure. In: Proceedings of the 33rd International Business Information Management Association Conference, IBIMA 2019: Education Excellence and Innovation Management through Vision 2020, vol. 33. IEEE, pp. 5129–5137 (2019)
5. Johns Hopkins University Organizational Chart—Organizational Structure Finance And Administration. https://mumbai.org/johns-hopkins-university. https://sharda.organizational-chart/. Last accessed 03 May 2021
6. Smart medicine: what makes a hospital "smart". https://rb.ru/opinion/smart-medicine/. Last accessed 03 May 2021.
7. Digital medicine in Russia: how new technologies are applied in practice. https://supermed.pro/digital-med.html. Last accessed 03 May 2021
8. Smart hospitals and digital transformation. https://www.unity.de/en/industry-solutions/healthcare/smart-hospitals-and-digitalization/. Last accessed 03 May 2021
9. What is a smart hospital? https://healthcareglobal.com/hospitals/what-smart-hospital. Last accessed 03 May 2021
10. Abdumanonov A.A., Karabaev M.K.: Computerization Medical Institutions for the Organization and Optimization of Clinical Processes (2016)
11. GrandMedica. https://gm.clinic. Last accessed 03 May 2021
12. Mirsaidova M.H.: The use of modern information and communication technologies into diagnostic practice. Med. Sci. **3** (2020)
13. Digital platforms for collecting medical data. https://evercare.ru/tsifrovye-platformy-dlya-sbora-meditsinskikh-danny. Last accessed 03 May 2021
14. Medical device integration. https://capsuletech.com/. Last accessed 06 May 2021
15. Digital platforms for collecting medical data. https://evercare.ru/tsifrovye-platformy-dlya-sbora-meditsinskikh-danny. Last accessed 06 May 2021
16. Bui, Q.N.: Evaluate enterprise architecture frameworks using essential elements. Commun. Assoc. Inf. Syst. **41**(1):121–149 (2018). https://doi.org/10.17705/1CAIS.04106
17. Mogilko, D.Y., Ilin, I.V., Iliashenko, V.M., Svetunkov, S.G.: BI capabilities in a digital enterprise business process management system. Lect. Notes Netw. Syst. **95**, 701–708 (2020)
18. Moreira, M.W.L., Rodrigues, J.J.P.C., Kumar, N., Saleem, K., Illin, I.V.: Postpartum depression prediction through pregnancy data analysis for emotion-aware smart systems. Inf. Fusion **47**, 23–31 (2019)
19. Anisiforov, A., Dubgorn, A., Lepekhin, A.: Organizational and economic changes in the development of enterprise architecture. In: E3S Web of Conferences, vol. 110. IEEE, p. 02051 (2019)
20. TOGAF, sewiki.ru/TOGAF. Last accessed 03 May 2021
21. IT-director book. https://book4cio.ru. Last accessed 06 May 2021

Models for Optimizing the Supply Volume in Conditions of Uncertainty in Demand, Taking into Account the Risks of Accrual of Penalties, Loss of Customers and Additional Costs

Oleg Kosorukov, Sergey Maslov, Olga Sviridova, and Karmak Bagisbayevc

Abstract The article describes stochastic optimization models of enterprise inventory management, taking into account the uncertainty of demand, which, according to most authors, is the main carrier of uncertainty in inventory management systems. As distribution laws describing this uncertainty, a triangular distribution is considered, which can be obtained in practice by expert assessment methods, for example, in the absence of a sufficient volume of statistical data. Formalization of the problems of finding the optimal delivery volume and their solution by analytical methods for goods of two types, namely, with a long shelf life and with a limited shelf life, are given. Four types of costs are taken into account, namely, storage costs (arising from the formation of unrealized surplus goods), costs of losing customers and calculating penalties (arising when it is impossible to fully meet the demand of a given period), as well as costs associated with the transportation, disposal or destruction of expired goods. which arise in the event of the formation of an unrealizable balance for goods with a limited shelf life.

Keywords Stocks · Demand · Volume of supply · Penalties · Utilization · Storage · Loss of customers

O. Kosorukov (✉) · S. Maslov
Lomonosov Moscow State University, National Research University Higher School of Economics, Moscow, Russia
e-mail: kosorukovoa@my.msu.ru

O. Sviridova
Financial University Under the Government of Russian Federation, Moscow, Russia

K. Bagisbayevc
Suleyman Demirel University, Almaty, Kazakhstan

C. Jahn et al. (eds.), *Algorithms and Solutions Based on Computer Technology*, Lecture Notes in Networks and Systems 387,
https://doi.org/10.1007/978-3-030-93872-7_21

1 Introduction

Stochastic models most accurately reproduce physical processes in a real system, in which uncertainty often arises due to inaccuracy or incompleteness of data regarding the conditions for the realization of phenomena or processes. Due to the influence of uncertain factors, a risk arises, which mainly has a large impact on the result. As a consequence, when modeling a stochastic mathematical model, it is very important to take into account these circumstances, in contrast to deterministic models, in which these conditions can be ignored.

The influence of inventory management strategies on the economic performance of an enterprise is widely presented in the scientific literature [1–8]. Processes containing random or undefined parameters are the most difficult for analysis [3, 9–13]. A description of some information systems for decision-making support for inventory management in conditions of uncertainty can be found in works [14–17].

In this article, we assume that the delivery of goods occurs periodically at certain points in time exactly at the appointed time, i.e. there is no uncertainty in delivery time. The volume of goods supplied at the beginning of each period is considered as a variable, which is an optimizing variable in this model. The demand for a product in each period is a random variable with a known distribution law, with a probability distribution density $f(x)$ and a known set of values $x \in [0, XM]$, where XM is the maximum possible demand in the period under consideration. The problem is solved autonomously for each delivery period, and therefore the parameter XM and the distribution law $f(x)$ may depend on the period under consideration.

We assume that the profit from the sale of a unit of goods is known—v. We assume that $v \geq 0$. The total profit is proportional to the volume of goods sold. In the event that demand turns out to be higher than the volume of goods available in a given period, various financial risks may arise, expressed in contractual penalties, fines, other penalties, reputational risks or risks of losing a client, and in the case of a product with a limited shelf life costs of export, destruction or disposal of expired goods. The risks of losing a customer can also be quantified in financial terms. Logistic regression models of this kind that estimate the likelihood of a client leaving are presented, for example, in [18]. Using the regression coefficient of similar models with the variable "volume of unmet demand" and comparing its effect on the probability of a client leaving with data on customer trade histories, we can derive a certain specific coefficient characterizing the average financial losses of a company in case of dissatisfaction of demand in a unit volume. Let's designate this coefficient as w. We assume that $w \leq 0$.

If the goods available in a given period exceed the demand, an unused surplus of goods arises, which entails additional storage costs and financial losses from the funds "frozen" in the goods. In this case, it is advisable to estimate the costs of storing surplus goods during the period under consideration in proportion to the amount of unclaimed goods with a certain coefficient h. We assume that $h \leq 0$. Note that we only calculate the additional costs of storing goods that were stored for a period, but were not sold. We assume that the costs of storing the goods that were sold during the delivery period are taken into account in the profit of the goods v. The issues of optimizing the volume of supply under conditions of uncertainty were considered by the authors in a number of articles [19–21] with certain types of costs and without taking into account the costs of disposal and the accrual of penalties.

Next, we will consider two types of goods, namely, goods with a long shelf life and goods with a limited shelf life, which include, for example, food, perfumery and cosmetic products, medicines, household chemicals, etc. We assume that the first type of goods unused in this period can be used in the following. Regarding the second type of goods, we assume that its unused volumes in this period are subject to destruction or utilization.

2 Materials and Methods

The optimized criterion in this model is the mathematical expectation of profit from the sale of goods in the period under consideration, minus the four types of costs described above, namely, storage costs (arising from the formation of unrealized surplus of goods), costs of losing customers (occurring when it is impossible to fully satisfy demand of the given period) and the costs of calculating penalties for the dissatisfaction of demand in the period under review. We assume that penalties are calculated in proportion to the volume of unsatisfied demand in a given period and amount to m per unit of goods. We assume that $m \leq 0$. The objective function of the problem in this case is further represented by the expression (1)

$$F(Q) = \int_{0}^{Q} (vx + h(Q - x)) f(x)dx + \int_{Q}^{XM} (vQ + w(x - Q) + m(x - Q)) f(x)dx \quad (1)$$

The problem is to maximize function (1) with respect to the variable Q on an admissible set of values, namely, on the segment [0, *XM*]. Let us carry out the following transformations of function (1):

$$\begin{aligned} F(Q) =& v\int_{0}^{Q} xf(x)dx + hQ\int_{0}^{Q} f(x)dx + h\int_{0}^{Q} xf(x)dx \ + vQ\int_{Q}^{XM} f(x)dx \\ &+ w\int_{Q}^{XM} xf(x)dx - wQ\int_{Q}^{XM} f(x)dx + m\int_{Q}^{XM} xf(x)dx - mQ\int_{Q}^{XM} f(x)dx. \end{aligned}$$

To minimize the function *F (x),* we find its derivative using the well-known Leibniz formula for differentiating integral functions with variable upper limit [22].

$$\frac{dF(Q)}{dQ} = h\int_{0}^{Q} f(x)dx + v\int_{Q}^{XM} f(x)dx - w\int_{Q}^{XM} f(x)dx - m\int_{Q}^{XM} f(x)dx. \quad (2)$$

Considering that

$$\int_{Q}^{XM} f(x)dx = 1 - \int_{0}^{Q} f(x)dx, \quad (3)$$

and equating the derivative 0, we obtain the following equation with respect to the variable Q, considered for $x \in [0, XM]$:

$$\int_{0}^{Q} f(x)dx = \frac{m + w - v}{h + m + w - v} = \frac{1}{\frac{h}{m+w-v} + 1} = k_1. \quad (4)$$

From the second expansion in relation (4) and the signs of the parameters *h, w, m, v* indicated above, it follows that $0 \leq k_1 \leq 1$. Due to the fact that the function of the variable Q, which is on the left-hand side of Eq. (4), is a monotonically non-decreasing continuous function varying on the considered interval from 0 to 1, Eq. (4) has a solution by virtue of the Weierstrass theorem [22]. In the general case, finding the root of Eq. (4) is carried out by known numerical methods. Let Q^* be a root of Eq. (4). Further, we note that the derivative (2) is a non-increasing function on the segment under consideration, while

$$\frac{dF(0)}{dQ} = v - w - m \geq 0, \quad \frac{dF(XM)}{dQ} = h \leq 0.$$

Due to this, at the point Q^*, the derivative changes sign from " + " to "-", and therefore, the point Q^* is the point of the desired maximum and there is no need to calculate the value of the function F (x) at the ends of the segment.

Next, we will solve this problem under the assumption that the random quantity of demand is distributed according to the triangular distribution law on the interval [*a, b*]. Parameters *a, b, c*—are determined from statistical data, or using expert estimates, subject to the following condition: $a \leq c \leq b$, $a < b$, where *a* is the lower limit of the values of the random variable, *b* is the upper limit of the values of the random variable, *c*—mode (the value that occurs most often in the distribution). In this case, the probability density function $f(x)$ is expressed according to (5), and the probability distribution function according to (6):

$$f(x)=\begin{cases}0, & at\ x<a;\\ \frac{2(x-a)}{(b-a)(c-a)}, & at\ a\le x<c;\\ \frac{2}{(b-a)}, & at\ x=c;\\ \frac{2(b-x)}{(b-a)(b-c)}, & at\ c<x\le b;\\ 0, & at\ b<x.\end{cases} \quad (5)$$

$$\Phi(x)=\begin{cases}0, & at\ x<a;\\ \frac{(x-a)^2}{(b-a)(c-a)}, & at\ a\le x\le c;\\ 1-\frac{(b-x)^2}{(b-a)(b-c)}, & at\ c\le x\le b;\\ 0, & at\ b<x.\end{cases} \quad (6)$$

Whence the analytical solution of the problem for the case when a random quantity of demand is described by a triangular distribution is expressed by relations (7).

$$Q^*=\begin{cases}a+\sqrt{k_1(b-a)(c-a)}, & 0\le k_1\le\frac{c-a}{b-a},\\ b+\sqrt{(1-k_1)(b-a)(c-a)}, & \frac{c-a}{b-a}\le k_1\le 1,\end{cases} \quad (7)$$

In conclusion, we note that this model allows a narrowing of the problem for the case of the absence of any type of costs from those considered in the model.

If there is no storage overhead, i.e. $h = 0$, then $k_1 = 1$ and $Q^* = XM$. The economic meaning of this result is clear, since in this case the risk of additional storage costs is initially absent, the risk of losing customers and the risk of paying a penalty disappears when the maximum volume of possible demand is delivered, and unsold goods are transferred to the next period without any additional costs.

If there are no risks of losing customers, i.e. $w = 0$, then

$$k_1=\frac{v-m}{v-m-h}.$$

As you can see, $0 \le k_1 \le 1$ and Eq. (4) remains solvable. In this case, the decision is non-trivial since there remains a balance between the profit from the sale, the costs of storing the surplus and the risks of charging a penalty.

If there are no risks of accrual of interest, i.e. $m = 0$, then

$$k_1=\frac{v-w}{v-w-h}.$$

As in the previous case, $0 \le k_1 \le 1$ and Eq. (4) remains solvable. In this case, the decision is non-trivial since there is a balance between the profit from the sale, the costs of storing surplus and the risk of losing customers.

If at the same time there are no risks of calculating penalties and risks of losing customers, i.e. $w = 0$, $m = 0$ then.

As in the previous case, $0 \leq k_I \leq 1$ and Eq. (4) remains solvable. In this case, the decision is non-trivial since there remains a balance between the profit from the sale, the costs of storing surplus and the risk of losing customers.

If at the same time there are no risks of calculating penalties and risks of losing customers, i.e. $w = 0$, $m = 0$ then

$$k_1 = \frac{v}{v - h}.$$

As you can see, $0 \leq k_I \leq 1$ and Eq. (4) remains solvable. In this case, the decision is non-trivial since there remains a balance between the profit from the sale and the cost of storing the surplus.

3 Results

The optimized criterion in this model is the mathematical expectation of profit from the sale of goods in the period under consideration, minus 4 types of costs, 3 of which are described above, namely, storage costs (arising from the formation of unrealized surplus of goods), costs of losing customers and calculating penalties (arise when it is impossible to fully satisfy the demand of a given period), as well as the fourth type of costs associated with the transportation, disposal or destruction of expired goods, which arise in the event of the formation of an unrealized balance for goods with a limited shelf life. Unit costs for this type of costs will be denoted as u. We assume that $u \leq 0$.

Unlike the model discussed above, the costs of purchasing a consignment imported in the period under review are also deducted. This is explained by the fact that if in the previous model the unclaimed surplus of goods was transferred as an incoming balance to the next period and was taken into account when determining the volume of delivery of goods for the next period, then in this model it is assumed that the unrealized balance of the period under consideration is considered overdue and is not considered further, i.e. e. goes out of circulation. The objective function of the problem in this case is further represented by the expression (8)

$$F(Q) = \int_0^Q (c_2 x + h(Q - x) + u(Q - x)) f(x) dx + \int_Q^{XM} (c_2 Q + w(x - Q) + m(x - Q)) f(x) dx - c_1 Q, \quad (8)$$

where c_1—purchase price of product, c_2—the selling price of the product. We assume that

$$0 \le c_1 \le c_2.$$

The problem is to maximize function (8) with respect to the variable Q on an admissible set of values, namely, on the segment [0, XM]. Let us carry out the following transformations of function (8):

$$F(Q) = c_2 \int_0^Q x f(x)dx + hQ \int_0^Q f(x)dx + h \int_0^Q x f(x)dx$$
$$+ uQ \int_0^Q f(x)dx + u \int_0^Q x f(x)dx + c_2 Q \int_Q^{XM} f(x)dx + w \int_Q^{XM} x f(x)dx$$
$$- wQ \int_Q^{XM} f(x)dx + m \int_Q^{XM} x f(x)dx - mQ \int_Q^{XM} f(x)dx - c_1 Q. - c_1 Q.$$

As in the study of a model with long shelf life goods, to minimize the function $F(x)$, we find its derivative using the well-known Leibniz formula for differentiating integral functions with a variable upper limit [22].

$$\frac{dF(Q)}{dQ} = h \int_0^Q f(x)dx + c_2 \int_Q^{XM} f(x)dx - w \int_Q^{XM} f(x)dx + u \int_0^Q f(x)dx$$
$$- m \int_Q^{XM} f(x)dx - c_1. \quad (9)$$

Taking into account relation (3) and equating the derivative 0, we obtain the following equation with respect to the variable Q, considered for × 0 [0, XM]:

$$\int_0^Q f(x)dx = \frac{w + m + c_1 - c_2}{h + u + w + m - c_2} = \frac{1 + \frac{c_1}{w+m-c_2}}{1 + \frac{h+u}{w+m-c_2}} = k_2. \quad (10)$$

Taking into account the signs of the parameters w, h, u, m and the relation $0 \le c_1 \le c_2$, the following chain of inequalities is valid:

$$c_1 \le c_2 - w - m, w + m - c_2 < 0,$$

$$-1 \le \frac{c_1}{w + m - c_2} \le 0, \quad 0 \le 1 + \frac{c_1}{w + m - c_2} \le 1,$$

$$0 \leq \frac{h+u}{w+m-c_2}, \quad 1 \leq 1 + \frac{h+u}{w+m-c_2},$$

and, therefore, $0 \leq k_2 \leq 1$.

Due to the fact that the function of the variable Q, which is on the left-hand side of Eq. (10), is a monotonically non-decreasing continuous function varying on the considered interval from 0 to 1, Eq. (10) has a solution by virtue of the Weierstrass theorem [22]. In the general case, finding the root of Eq. (10) is carried out by known numerical methods. Let Q^* be the root of Eq. (10). Further, we note that derivative (9) is a non-increasing function on the segment under consideration, while

$$\frac{dF(0)}{dQ} = c_2 - w + m - c_1 \geq 0, \quad \frac{dF(XM)}{dQ} = h + u - c_1 \leq 0.$$

Due to this, at the point Q^*, the derivative changes sign from " + " to "-", and therefore, the point Q^* is the point of the desired maximum and there is no need to calculate the value of the function F (x) at the ends of the segment.

Next, we will solve this problem under the assumption that the random quantity of demand is distributed according to the triangular distribution law on the interval $[a, b]$. Parameters a, b, c—satisfy the following conditions: $a \leq c \leq b$, $a < b$, where a is the lower limit of the values of the random variable, b is the upper limit of the values of the random variable, c is the mode (the value that occurs most often in the distribution). In this case, the probability distribution function $f(x)$ is expressed according to relations (5), and the probability distribution function according to relations (6). Whence the analytical solution of the problem for the case when a random quantity of demand is described by a triangular distribution is expressed by relations (11).

$$Q^* = \begin{cases} a + \sqrt{k_2(b-a)(c-a)}, & 0 \leq k_2 \leq \frac{c-a}{b-a}, \\ b + \sqrt{(1-k_2)(b-a)(c-a)}, & \frac{c-a}{b-a} \leq k_2 \leq 1, \end{cases} \tag{11}$$

In conclusion, we note that this model allows for a narrowing of the problem for the case of the absence of any of the costs considered in the model. Below we will consider the cases of the absence of only one of the types of costs.

If there is no storage overhead, i.e. $h = 0$, then

$$k_2 = \frac{1 + \frac{c_1}{w+m-c_2}}{1 + \frac{u}{w+m-c_2}}.$$

As you can see, $0 \leq k_2 \leq 1$ and Eq. (10) remains solvable. In this case, the decision is non-trivial since there remains a balance between the profit from the sale of the goods, the costs of losing customers, the losses from the payment of penalties, the costs of disposal of the expired goods and the loss of the value of unrealized balances.

If there is no cost of losing customers, i.e. $w = 0$, then

$$k_2 = \frac{1 - \frac{c_1}{m - c_2}}{1 - \frac{h+u}{m - c_2}}.$$

As you can see, $0 \le k_2 \le 1$ and Eq. (10) remains solvable. In this case, the decision is non-trivial since there remains a balance between the profit from the sale of the goods, additional costs from storing unsold goods, the costs of payment of penalties, the costs of disposing of the unsold goods and the loss of the value of the unsold inventory.

If there are no costs of payment of interest, i.e. $m = 0$, then

$$k_2 = \frac{1 - \frac{c_1}{w - c_2}}{1 - \frac{h+u}{w - c_2}}.$$

As you can see, $0 \le k_2 \le 1$ and Eq. (10) remains solvable. In this case, the decision is non-trivial since there remains a balance between the profit from the sale of the goods, additional costs from storing unsold goods, the costs of losing customers, the costs of disposing of unsold goods and the loss of the value of unsold inventory.

If there are no costs for disposal of expired goods, i.e. $u = 0$, then

$$k_2 = \frac{1 - \frac{c_1}{w + m - c_2}}{1 - \frac{h}{w + m - c_2}}.$$

As you can see, $0 \le k_2 \le 1$ and Eq. (10) remains solvable. In this case, the decision is non-trivial since there remains a balance between the profit from the sale of the goods, additional costs from storing unsold goods, the costs of payment of penalties, the costs of losing customers and the loss of the value of unsold inventory balances.

Note also that even in the case when all the considered costs are absent, i.e. $h = 0$, w $= 0$, $m = 0$, $u = 0$ then

$$k_2 = 1 - \frac{c_1}{c_2}.$$

As you can see, $0 \le k_2 \le 1$ and Eq. (10) remains solvable. In this case, the decision is non-trivial since there remains a balance between the profit from the sale of the goods and the loss of the value of unrealized inventory balances.

4 Conclusions

So, the described models allow, with a known delivery time, to determine the volume of delivery of a new batch of goods, provided that various kinds of risks are minimized. In the case of a triangular distribution, these optimization problems have analytical solutions. I would like to note that in the overwhelming majority of stochastic models of inventory management it is not possible to obtain an analytical solution. In this case, the methods of simulation modeling are used, presented, for example, in the work [12]. The use of simulation models makes it difficult to find the optimal parameters of the inventory control algorithms due to the computational complexity of the problem and the need to use special software. imitation of demand, presented, for example, in the work [13].

The result, namely the type of analytical expression and its content, depend on the input parameters of the model, as well as the parameters of the triangular distribution a, b, c of the random quantity of demand.

In works [9–11], the problem was considered in a similar setting, but for the case when a random variable describing the deviation of the real time of delivery of goods to the warehouse from the expected one can be considered normally distributed. Firstly, this is not always true, and secondly, the company often does not have enough statistical data to test a sample of realizations of a random variable for its compliance with the normal distribution law. in the absence of a sufficient volume of statistical data, it can be produced expertly.

References

1. Anikin, B.A., Tyapukhin, A.P.: Commercial logistics. Prospect, Moscow (2012)
2. Anikin, B.A.: Logistics: A Tutorial. INFRA-M, Moscow (1999)
3. Brodetsky, G.L.: Methods of Stochastic Optimization Mathematical Models of Inventory Control: Textbook. REA, Moscow (2004)
4. Brodetsky, G.L.: System Analysis in Logistics. Choice in Conditions of Uncertainty. Academy, Moscow (2010)
5. Brodetsky, G.L., Gusev, D.A.: Economic and Mathematical Methods and Models in Logistics. Optimization procedures. Academy, Moscow (2012)
6. Prosvetov, G.I.: Mathematical Methods in Logistics. Problems and Solutions. Alfa-Press, Moscow (2008)
7. Prosvetov, G.I.: Inventory Management. Problems and Solutions. Alfa-Press, Moscow (2009)
8. Tsvirinko, I.A.: Methodology, Methods and Models of Management of Logistic Business Processes. SPbGIEU, SPb (2003)
9. Kosorukov, O.A., Sviridova, O.A.: Cost minimization model in inventory management systems. Bulletin of the Russian Economic Academy named after G.V. Plekhanov. **6** (30), 94–102, (2009)
10. Kosorukov, O.A., Sviridova, O.A.: Cost minimization model in inventory management systems. Bulletin of the Russian Economic Academy named after G.V. Plekhanov. **4** (46), 91–95, (2012)
11. Kosorukov, O.A., Sviridova, O.A.: Effective Strategy Formation Models for Inventory Management under the Conditions of Uncertainty. International Education Studies. **5** (8), 64–83 (2015)
12. Dubrov, A.M., Lagosha, B.A., Khrustalev, E. Yu.: Modeling of Risky Situations in Economics and Business. Finance and Statistics. Moscow (2004)

13. Kosorukov, O.A., Maksimov D.A., Shimchenko E.D.: Some Aspects of Demand Modeling in Inventory Management Simulation Models. Logistics. **12**, 48–50 (2014).
14. Koronatov, N., Ilin, I., Levina, A., Gugutishvili, D.: Requirements to IT support of oil refinery supply chain. IOP Conf. Ser.: Mater. Sci. Eng. **1001**(1), 012143 (2020)
15. Ilin, I., Maydanova, S., Lepekhin, A., Jahn, C., Weigell, J., Korablev, V.: Digital Platforms for the Logistics Sector of the Russian Federation. Lecture Notes in Networks and Systems **157**, 179–188 (2021)
16. Ilin, I., Maydanova, S., Levina, A., Jahn, C., Weigell, J., Jensen, M.B.: Smart Containers Technology Evaluation in an Enterprise Architecture Context (Business Case for Container Liner Shipping Industry). Lecture Notes in Networks and Systems **157**, 57–66 (2021)
17. Ilin, I.V., Koposov, V.I., Levina, A.I.: Model of asset portfolio improvement in structured investment products. Life Sci. J. **11**(11), 265–269 (2014)
18. Grishchenko, D.A., Kataev, D.A.: Analysis of methods of modeling and forecasting the outflow of clients. Bull. Sci. Educ. **5**(41), 21–23 (2018)
19. Maslov, S.E., Kosorukov, O.A.: Models of optimization of time and volume of delivery in conditions of uncertainty, taking into account the risks of penalties and loss of customers. Financ. Econ. **3**, 473–479 (2019)
20. Maslov, S.E., Kosorukov, O.A.: Model of supply volume optimization taking into account demand uncertainty. Financ. Econ. **1**, 191–197 (2019)
21. Kosorukov, O., Maslov, S., Sviridova, O.: Algorithm for determining the economic purchase size in the conditions of the demand volatility. J. Adv. Res. Dyn. Control Syst. **12**(05) 1129–1138 (2020)
22. Ilyin, V.A., Sadovnichy, V.A., Sendov, B.H.: Mathematical Analysis, 2nd ed. Publishing house of Moscow State University, Moscow (1985)

Thermodynamics of Computational Processes: "Oblique Sail" in the Sea of Computer Technology

Alexander Antonov, Vladimir Polyanskiy, and Vladimir Zaborovskij

Abstract The key concept of the digital transformation era is the Turing machine, an abstraction used to describe computational processes using step-by-step execution of the simplest mechanical operations. As Ludwig Boltzmann said: "Available energy is the main object at stake in the struggle for existence and the evolution of the world" and therefore the role of thermodynamics in the optimization of Turing' computational processes is discusses in context of constructive constrains. It is shown that an important resource for such optimization is the re-configuration of the microstructure of the computer in such a way as to minimize the dissipation of thermal energy in the process of calculations. The theoretical basis of the proposed approach is the consideration of computers as closed thermodynamic systems in thermal equilibrium with the environment. For this, it is proposed to constructively decompose computational processes into two entities—the classical Turing machine and the Boltzmann machine. The latter is considered as an extensional interface with sufficient informational (entropy) and energy potentials for the purposeful reconfiguration of the Turing machine in accordance with the selected criterion of the quality of the computational process, expressed in terms of physically measurable characteristics. Technologically, the process of reconfiguring the calculator is based on the use of FPGA microcircuits, the con-figuration files for which are organized into a specialized knowledge base containing the optimal structures of logical ventilators from the point of view of energy-computational efficiency to accordance with a given class of algorithms.

Keywords Technology development · Artificial intelligence · Machine learning · Exo-intelligence · Cognitive functions

A. Antonov (✉) · V. Zaborovskij
The Great St. Petersburg Polytechnic University, Saint-Petersburg, Russia
e-mail: antonov@eda-lab.ftk.spbstu.ru

V. Polyanskiy
Institute for Problems in Mechanical Engineering, Russian Academy of Sciences, Moscow, Russia

C. Jahn et al. (eds.), *Algorithms and Solutions Based on Computer Technology*, Lecture Notes in Networks and Systems 387,
https://doi.org/10.1007/978-3-030-93872-7_22

1 Introduction

The global trend towards digitalization of industrial and scientific environment, has a significant impact on theoretical computer science, including concepts of programming, machine learning, artificial intelligent, digital twins, and their moronity.

Modern computer technologies are based on three fundamental principles of the theory of algorithms: computability of functions, enumerability and decidability of sets.

To objectify algorithms as "hardware" of computers, chains of electronic components are used, which form the architecture of arithmetic-logical devices of digital processors [1].

The article discusses the concept of realizing the "computational superiority" of computer programs, for which the criteria of the energy-computational efficiency of execution are important, which guarantee the tendency of the entire computing system to thermal equilibrium with the environment.

A similar principle for the implementation of functioning processes is characteristic of living systems, which can be considered as distributed software-controlled structures that calculate themselves (the phenomenon of computational fractality) on the basis of programs, the extensional carrier of which are DNA molecules containing enumerated sets of code structures. From the point of view of thermodynamics, in living systems, the energy released in the process of bio "computing" is used for the solvability (readjustment) of a multitude of microstates of the bio computer itself.

To objectify the intentional genetic information, such a bio computer uses a gene expression mechanism that is structurally similar to a standard computational pipeline or network-centric systolic structure, which converts input data (endogenous or exogenous) into output protein structures, and implements genetic computation programs.

Such a pipeline, which sequentially performs the stages of "input-transformation-output" of data, is functionally similar to the processor architecture that goes back to the Turing machine, but at the same time has a fundamentally higher energy efficiency of computations, since it not only dissipates the energy of the computations themselves, but also uses it for the synthesis of a new biomaterial of the computer.

Computational complexity is playing a pivotal role in evolution of Theory of Digital Computation (ToDC). Today ToDC stands with physics, biology, math, and economics as well as plays central role in scientific revolution informed by digital processing under control of program called as a computation. Has current version of ToDC principal disadvantage and what its essence?

The real factology is well known. Based on the ideas of reductionism, ToDC cannot solve fundamental P = NP problem, explain the nature of algorithmically incomputable number/functions and unsolvable sets, phenomenon and every concepts connected with intelligence or mind.

From this the point of view the architecture of biological computers, which are in asymptotic thermal equilibrium with the environment, is an intentional analogue of the "oblique sail"—an innovative solution that changed the global logistics of the Ancient world, and made it possible to implement "information reversible routes of movement" due to the possibility of zigzag sail upwind.

Obviously, in this case, it was not possible to achieve full physical reversibility of movement, but thanks to equipping vehicles with subsystems with fundamentally different physical characteristics (Fig. 1), they became a clear example of the new one fundamental law of nature—the law of conservation of information.

Obviously, endowing modern computers, which are functionally similar to the Turing machine that base on linear set of the simplest mechanical operations, with new information transformation capabilities characteristic of evolutionary processes in a biological environment, we can count on a conceptual revolution in computer technology. This, first, will decrease the level of non-constructive energy dissipation and impart asymptotic thermodynamic equilibrium to the computational process (Fig. 2).

Further, it will be shown that the last property is achieved by imparting high adaptive functionality to computing structures, reflecting the consistency of process processes at the software and hardware levels.

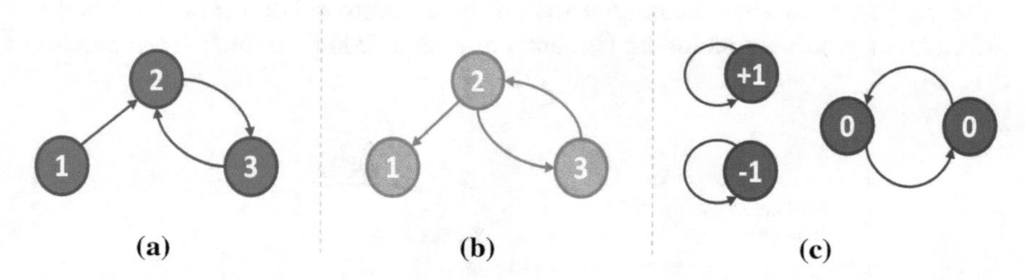

Fig. 1 **a** irreversible system; **b** indeterminate system; **c** discrete cyclic system

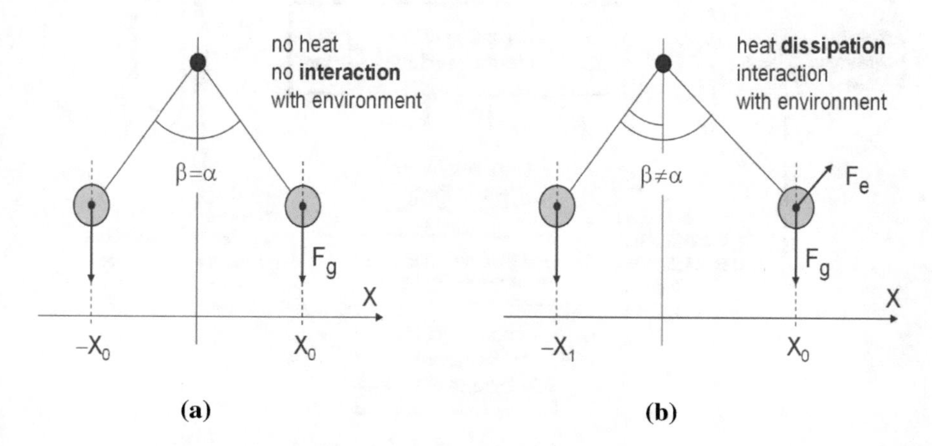

Fig. 2 Heat dissipation as results of interaction with environment

There is no doubt that in any process of technology renovation, the essence of which is to complement human capabilities we need to understand the fundamental reason-giving rise to all the problems listed above and must try to formulate accurately the essence of the digital paradigm of science. In this context, the question of whether computer science belongs to the natural or mathematical sciences is currently not only of scientometric but also practical significance. In Avi Wigderson's book "Mathematics and Computation" [2] we can read that "big bang" of pure Turing's approach novo days has lost its constructive power because very basic tasks occurs incomputable in a sense of formal theory.

2 Materials and Methods

The basic conceptual question "Who is to blame and what to do" with the problem of incompatibility—must be approached from the standpoint of searching for analogies in natural phenomena, and taking into account the principle of complementarity, first formulated by Nielson Bohr.

Many natural processes can (and should) be understood both as physical and as information processes including demands of similar computational representation (Fig. 3).

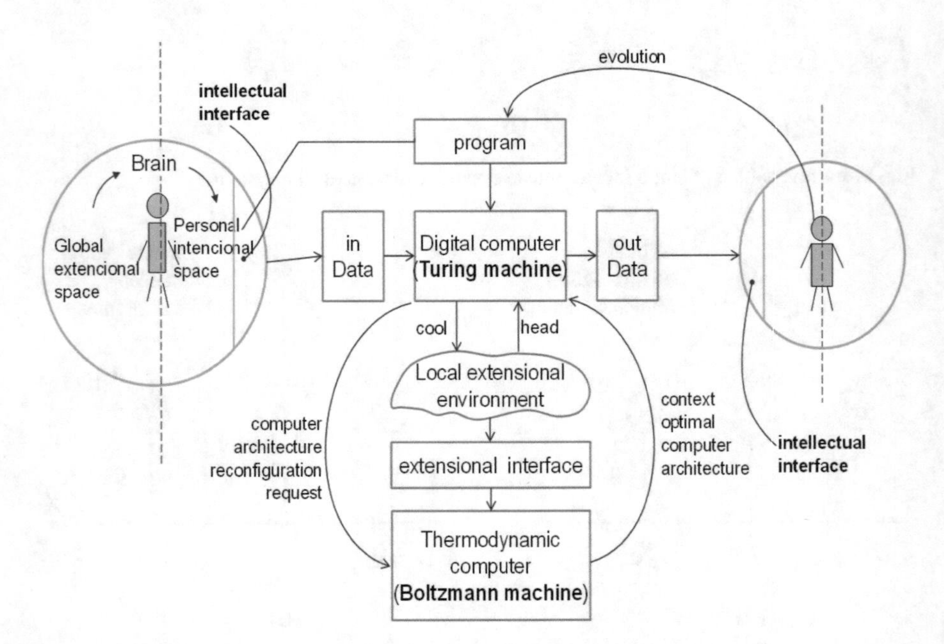

Fig. 3 Conceptual structure of thermodynamic computer model

Such factors that are asymptotically unstable can easily be significantly influenced by various factors and development scenarios, which lead to the formation of strange asymptotically evolutionary attractors [3, 4].

A key aspect of process evolution can be defined through the concept of computational theory of mind or "artificial intelligence" (AI), which is considered as a set of possible solutions that choosing in machine learning manner.

Perhaps the problem lies in the very essence of the question posed, namely: in finding, a solution in the form of a computational algorithm implemented with the help of a "machine", or is it all about the mechanical and deterministic nature of "machine" itself?

First of all, we note that the understanding of "calculation" and the possibilities of its emulation using various machines has changed significantly [5, 6]. So almost until the middle of the nineteenth century, simple counting tasks were considered quite intellectual. Moreover, if a person was able to quickly and error-free calculate using well-known mathematical formulas, and, on this basis, manage, for example, commercial activities, then his intellectual abilities were rated very highly.

However, thanks to the formalization of calculation algorithms, mathematical calculation problems were automated one of the first. In 1822 in England, C. Babbage created the so-called difference machine, which allows automating the calculation process by approximating various mathematical functions by polynomials using a strictly ordered sequence of operations.

In 1832 in Russia N. S. Korsakov published a brochure "Outlining a New Method of Research Using Machines Comparing Ideas." In this brochure, many mechanical devices have been designed based on perforated tables to search for information and classify various information records (ideas) by a set of numerous features (details).

This were the world's first exo-intelligent mechanical machines to "amplify thought ability" by processing large amounts of data that are difficult to store directly in a person's memory. It is known that mechanics is a part of classical physics, the laws of which are in the form of mathematical equations.

However, not all mathematical equations can reflect physical laws. For this, the equations must be not only deterministic, but also reversible. This means, and this is principally important, that the laws of physics must be deterministic and reversible both in the past and in the future, both in one and other point of Earth.

Reversibility of physical laws essentially expresses the basic fundamental law mention above as the law of conservation of information. Concept of information is not directly applied to physics, but often concerns when describing the phenomenological laws, and intentional antonym of information is wide spread used as entropy.

Entropy or "transformation" (Fig. 4) is viewed as a function of the state space of a thermodynamic system, denoting the measure of irreversible dissipation of energy. Entropy is qualitatively different from other thermodynamic quantities: such as pressure, volume, or internal energy, because this is not a property of a system, but the way how external observer considers this system.

Entropy is a characteristic of how much information the observer does not know about the system. The quantitative measure of entropy S is the number of symbols

Fig. 4 Entropy as an intentional concept

required to record the number of microstates of the considered physical system distinguished by the observer. Mathematically, this number is defined as the logarithm of the number of possible micro states Ω of the system $S = \log \Omega$.

Therefore, the system has definite internal energy, impulse, charge, but it does not have a definite entropy: for example, entropy of ten bones depends on whether you know only their total sum, or also the partial sums of the five bones.

Therefore, entropy in physics is, in fact, an analogue of a program that calculates the state of a system in computer science. The informational essence of entropy has intentional nature that fundamentally distinguishes it from other quantities with which it is customary to deal in physics. The reason is that the logic and arithmetic underlying the theory of Turing' computation inevitably leads to the fact that the fundamental conclusions of Gödel's theorem are fully relevant to current version of computer science, clearly showing various aspects of the problems un-computability in formal terms of incompleteness and inconsistency that mean in practices—the absence of a reversible equation of state the absence of a reversible equation of state, which is used to describe the relationship between all individual states of the studied physical object.

In fact, this is the crux of why the laws of physics are a formal record of the effect of so called "glorified statistics". In other words, the more microstates correspond to a given macrostate, that is, the more particles are included in the system under study, the more accurately the system state equation describes this macro system. For gas, the characteristic values of the number of particles are equal to the Avogadro number, that is, of the order of 10^{23}. To find out all microstate a system, we need to have a lot of "personal" information—to know the position and velocity of each particle. The amount of this information is in fact entropy.

But if the state space of a physical system can be divided into cycles, then the system acquires principal new properties. Such a system becomes the carrier of the new function—informational memory, i.e. the function of storing information about the state of the system, from which the dynamic process began.

The such system has "something" that remains invariant in extensional sense, so unchanged over time. Therefore each cycle can be associated with some numerical value, expressed in the format of an integer number, and if this number is stored during the "movement" of the physical system, then such a system is pure discrete as well as deterministic predictable.

3 Results

3.1 Discreteness of Cognitive Perception

The discreteness of the cognitive perception of reality by analogy with the concept of entropy that from classical thermodynamics lows is not the internal properties of the objects themselves, but is associated with the act of perception of objects by an external "intellectual" observer.

The non-isomorphism of discrete or digital descriptions is directly related to the fundamental problems of modern informatics, which are clearly manifested in the well-known Skolem's paradox, which declares "the admissibility of describing meaningful representations of continuous sets using the means of a formal language, which obviously contains a countable set of true expressions."

At the same time, the expressive means of modern programming languages are deter-mined by a finite set of syntactically and semantically correct expressions, the power of which does not exceed the power of the class of predicate calculus formulas. It is obvious that the relativism of the meaningful description of objects and phenomena of physical reality is associated precisely with the homomorphic nature of the possible forms of its intellectual perception.

Therefore, human thinking, understood as a kind of "cognitive machine", has a potentially countable set of states. The ultimate accuracy of the brain's resolving power is determined by the homomorphic nature of the process of mapping "being into thinking", which is realized in acts of cognition. The nature of mind violates the presumption of the continuity of physical reality, and in connection with this phenomena of quantum mechanics can be regarded as a special case associated with the scale of processes not directly perceived by the senses.

Obviously, the discreteness structure of the cognitive perception of reality makes it possible to increase the "accuracy" of the description of the results of observation of objects and processes, and to use for this not only rational or natural numbers, but also to name the perceived facts and phenomena using the "epsilon-network" of concepts and ontology relationships. One of the broadest ways to informally define computation is the famous Church-Turing thesis: Computing is the process of evolution of an environment through a series of "simple, localized" steps.

This definition seems to cover almost any natural process that we know, so it sounds strange, especially since in the physical world we do not see the "authors" of the Turing machine tape! The most basic idea (from which the word "computation" originally originated) is the idea of evolution as a process of changing the sequence of physical states of some ma-chines (adding machine, calculator, processors, or even humane brain) under the control of a pre-formed local program or directly via by a specific goals of person.

Formally, the representation of computations as a sequence of operations performed under the control of a program written by a person now extends to many other areas where the effects associated with the phenomenon of intelligence are manifested.

The environment referred to in the formal definition of computation can be associated with various physical objects and associated information transformation processes, including: Bits in the computer, Atoms in matter, Neurons of the brain, Proteins in the cell, end etc.

3.2 Algorithmic Efficiency

Algorithmic efficiency [7–9] has many contexts including accuracy and the presence of errors, speed of calculation, required resources, and trade-off between performance and power consumption.

A separate aspect requiring in-depth study is the efficiency of algorithms obtained as a result of machine learning of various types of artificial neural networks. Experience shows that the search for the most accurate implementations for such algorithms under the same conditions and for some training datasets may require total retraining.

History of science shows that some of the greatest innovations have arisen from ignoring realistic constraints and intuitive biases about what is in principal possible. Examples of the success of such approach are: non-deterministic machines, random evidence, counting without algorithm, persuasion without knowledge, anonymous ownership, proof of the existence of justice and etc. In the space of human cognitive functions, special "laws of attraction" operate.

An attempt to make various intentional assumptions and then reach a formally extensionally impossible state—this is the essence of the law of "craving for knowledge", which by its nature, of course, differs from the law of universal gravitation of I. Newton and other laws of physics, but, like the latter, manifests itself objectively, therefore, its influence on the surrounding reality cannot be ignored (Fig. 5).

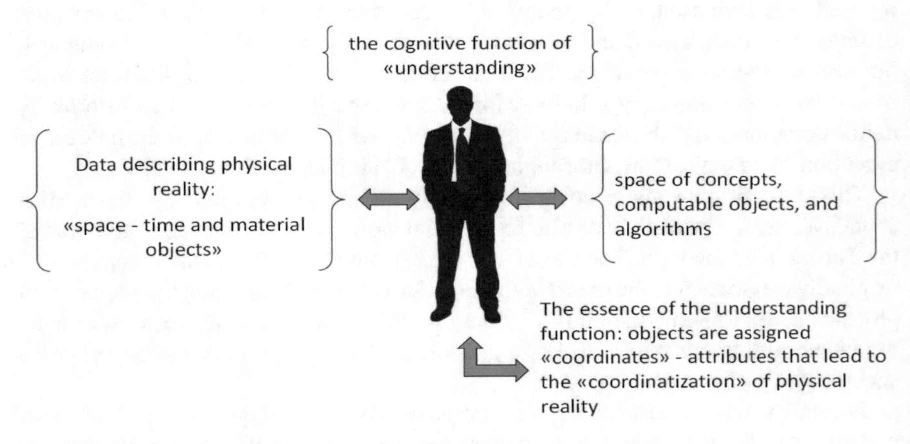

Fig. 5 Cognitive space with law of "craving for knowledge"

Currently, the greatest practical successes in solving intellectual problems have been achieved using methods that, in principle, are not characteristic of people, but are based on "brute computing force" or, in other words, on the ability to quickly sort possible solutions, for example, in analogy in playing chess. To concretize the task of building exo-intelligent platforms [3, 4], we will rely on data aggregation and classification methods based on similarity criteria for objects or processes based on functional or structural associations, using the search for solutions based on heterogeneous symbiosis of biological and artificial intelligence resources (Fig. 6).

The concept of thermodynamic process that transforming information is fundamental, and the scope of this concept goes far beyond the scope of modern digital computer technology. To make computers function more efficiently [9] then we should care about not only about software but energy source and its ability to efficiently and with high speed change the processor states—i.e. should care about thermodynamics aspects of calculation. It is clear, that in nature thermodynamics plays fundamental role—drives the self-organization and evolution of bio systems, so, no doubt similarly thermodynamics might drive the self-organization and evolution of future computing systems, making them more capable, more robust, and less costly to build and program.

4 Discussion

Currently living systems relies energy-efficient, universal, self-healing, and complex computational capabilities that dramatically transcend current industrial technologies [11–13]. Spontaneously finding energy efficient configurations enable bio system to thrive in complex, resource-constrained environments.

As we well known, in nature, all matter evolves toward low energy configurations in accord with the laws of thermodynamics. For near equilibrium systems these ideas are well known and have been used extensively in the theory of computational efficiency and in machine learning techniques.

Now we need to apply thermodynamic approach to create complex, non-equilibrium, but reconfigurable and self-organizing computing systems that will use both artificial and nature's based source of power.

Therefore, the discussed solution can achieves high level of efficacy by use "machine learning" like methods to find computer hardware optimal configuration to running software can decrease energy consumption and corresponding energy dissipation.

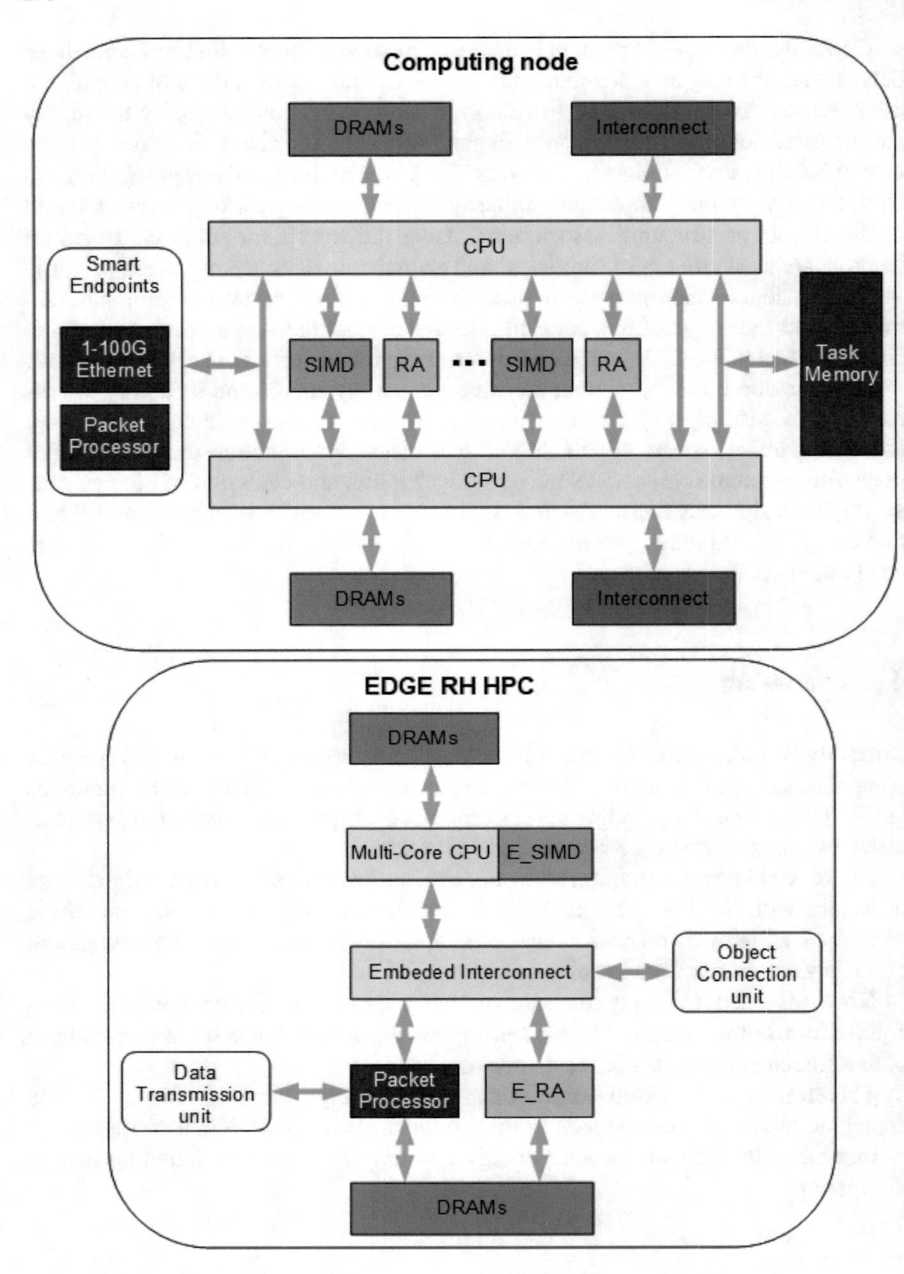

Fig. 6 Internal structure of the reconfigurable computing nodes

5 Conclusions

To overcome the shortcomings of the Turing machine, it will probably be necessary to provide an order of magnitude greater energy-computational efficiency and developed self-organization capabilities of the internal structure of a computer through its purposeful reconfiguration, taking into account the features of the computational algorithms being implemented.

For this, various neuromorphic structures capable of learning from representative datasets can be used, as well as computational tools for predictive modeling of high performance. For the effective use of such an approach, deep interdisciplinary research is needed that cover both fundamental and applied aspects of computer sciences, physics, direct digital modeling, big data mathematics, artificial intelligence and machine learning.

Acknowledgements The research was funded as a part of the state assignment for basic research (code of the research is 0784-2020-0026).

References

1. Ilin, I., Levina, A., Lepekhin, A., Kalyazina, S.: Business requirements to the IT architecture: a case of a healthcare organization. Adv. Intell. Syst. Comput. **983**, 287–294 (2019)
2. Wigderson, A.: Mathematics and Computation. Princeton University Press (2019)
3. Antonov, A., Zaborovskij, V., Kalyaev, I.: The architecture of a reconfigurable heterogeneous distributed supercomputer system for solving the problems of intelligent data processing in the era of digital transformation of the economy. Cybersecur. Issues **33**(5), 2–11 (2019). https://doi.org/10.21681/2311-3456-2019-5-02-11.
4. Antonov, A., Zaborovskij, V., Kisilev, I.: Specialized reconfigurable computers in network-centric supercomputer systems. High Availab. Syst. **14**(3), 57–62 (2018). https://doi.org/10.18127/j20729472-201803-09
5. Dongarra, J., Gottlieb, S., Kramer, W.: Race to exascale. Comput. Sci. Eng. **21**(1), 4–5 (2019). https://doi.org/10.1109/MCSE.2018.2882574
6. Usman Ashraf, M., Alburaei Eassa, F., Ahmad Albeshri, A., Algarni, A.: Performance and power efficient massive parallel computational model for HPC heterogeneous exascale systems. IEEE Access **6**, 23095–23107 (2018). https://doi.org/10.1109/ACCESS.2018.2823299
7. Antonov, A., Besedin, D., Filippov, A.: Research of hardware implementations efficiency of sorting algorithms created by using Xilinx's high-level synthesis tool. Commun. Comput. Inf. Sci. **1331**, 449–460 (2020). https://doi.org/10.1007/978-3-030-64616-5_39
8. Antonov, A., Besedin, D., Filippov, A.: Research of the efficiency of high-level synthesis tool for FPGA based hardware implementation of some basic algorithms for the big data analysis and management tasks, In: Conference of Open Innovation Association, FRUCT, vol. 26, pp. 23–29 (2020). https://doi.org/10.23919/FRUCT48808.2020.9087355
9. Antonov, A., Besedin, D., Filippov, A.: Efficiency analysis of high-level synthesis tools for hardware implementation of sorting algorithms. Hardw. Comput., Telecommun. Control Syst. **1**(13), 31–41 (2020). https://doi.org/10.18721/JCSTCS.13103
10. Antonov, A., Mamoutova, O., Fedotov, A., Filippov, A., Next generation FPGA-based platform for network security. In: 18th Conference of Open Innovations Association and Seminar on Information Security and Protection of Information Technology (FRUCT-ISPIT), pp. 9–14 (2016). https://doi.org/10.1109/FRUCT-ISPIT.2016.7561501

11. Intel FPGA. https://www.intel.com/content/www/us/en/products/programmable.html, last accessed 2021/04/19
12. IDE Vivado HLS, https://www.xilinx.com/video/hardware/vivado-hls-tool-overview.html. Accessed 09 April 2021
13. UltraScale and UltraScale+ FPGA product Table (2021) https://www.xilinx.com/products/silicon-devices/fpga/virtex-ultrascale.html#productTable. Accessed 19 April 2021

Method and Algorithms of Traffic Routing in Peer-to-Peer Wireless Networks of Mobile Subscribers

Mikhail Chuvatov, Polina Chentsova, Serge Popov, and Aleksandr Osadchiy

Abstract In the course of the work, the routing method was proposed and implemented, the road traffic system was modeled, and the evaluation and comparison with other methods were performed. To study the existing and developed methods, a traffic simulation environment was selected, vehicle traffic scenarios were formed that are close to the conditions of real traffic in urban or suburban conditions, the parameters of the movement of mobile nodes were transmitted from the road simulator to the network one, and network traffic scenarios were formed for comparative testing under conditions of different load characteristics on the network with a changing number of nodes.

Keywords Vehicle · VANET · OLSR · AODV · Velocity · Wireless communication · Dynamic routing · Network simulator

1 Introduction

Modern wireless peer-to-peer networks are actively used in intelligent transport systems. Examples of such networks are: mobile self-organizing networks—MANET, flying vehicle networks—FANET, automobile self-organizing networks—VANET [1, 2]. One of the important tasks that need to be studied in wireless peer-to-peer networks is routing—the search for a chain of relay nodes through which messages will be transmitted to the addressee in the absence of a direct connection between the source and the destination. Existing routing methods that take into

M. Chuvatov (✉) · P. Chentsova · S. Popov
Peter the Great St.Petersburg Polytechnic University, St.Petersburg 195251, Russia
e-mail: misha@iktp.spbstu.ru

S. Popov
e-mail: popovserge@spbstu.ru

A. Osadchiy
Solomenko Institute of Transport Problems of the Russian Academy of Sciences, Moscow, Russia
e-mail: ai_osad@mail.ru

C. Jahn et al. (eds.), *Algorithms and Solutions Based on Computer Technology*,
Lecture Notes in Networks and Systems 387,
https://doi.org/10.1007/978-3-030-93872-7_23

account the location of the nodes are designed for use in networks where the nodes are stationary. Given that the speed of the cars in populated areas reaches 80 km/h, and on highways could be more than 100 km/h, the use of existing routing methods leads to a rapid obsolescence of information about the connectivity of the network and the characteristics of the channels, which leads either to delays due to the waste of time searching for a route, or to a decrease in the useful bandwidth. The loss of transmission efficiency is caused by the need for frequent distribution of updated information, which leads to an increase of delays and losses. For networks formed from mobile subscribers, it is necessary to use alternative routing methods that would take into account changes in the location of nodes, the speed and direction of their movements.

One of the options for accounting for dynamics in mobile subscriber networks is to use the object state vector. For the continuous formation of such a vector, it is necessary to have a receiver of global positioning systems. In addition to coordinates, positioning systems also allow you to determine the object's velocity vector and the exact time. The coordinates together with the velocity vector form the state vector, which is a prerequisite for constructing new metrics for routing methods.

2 Materials and Methods

2.1 Implementation of Wireless Communication Protocols in Mobile Subscriber Networks

The need to develop a specialized standard for use in moving vehicle networks that describes the bottom layers of network models is due to the short time intervals during which moving vehicles are within the communication range of the transceivers [3]. For this reason, the connection between the nodes should be established in the shortest possible time to allow more time for the transfer of useful data. Currently, the main standards used for organizing wireless peer-to-peer networks are the 802.11 s and 802.11 p. These standards describe the bottom layers of the OSI/ISO reference network model or the TCP/IP stack model. The 802.11 p standard is an addition to the 802.11 family of standards that defines data exchange between moving vehicles and road infrastructure—Vehicle-to-Infrastructure—or directly between vehicles—Vehicle-to-Vehicle [8, 11, 15]. The frequency range 5.85–5.925 GHz is used to establish communication. The network can include both mobile and stationary nodes, and only mobile ones. Currently, 802.11 p is adopted as the basis of the DSRC/WAVE protocol stack, the overlying layers of which are described by the IEEE 1609 family of standards and define the architecture and an additional set of service functions and interfaces that provide a secure mechanism for radio communication between moving vehicles [14]. These standards are designed for applications such as traffic management, traffic safety monitoring, automated payment collection, navigation, and vehicle routing.

Routing in peer-to-peer wireless networks is the process of detecting and selecting relay nodes through which to forward packets in cases where the sending node and the receiving node do not have a direct connection [9, 11]. A feature of routing in peer-to-peer wireless networks is the use of information extracted from the link and network layers of the reference network model, whereas in centralized networks, the routing task is entirely assigned to the network layer of the model [13]. Currently, there are different routing methods used depending on the specifics of the network. These methods are implemented in the form of routing protocols, such as AODV, OLSR, LAR, and DREAM [4, 10, 18, 19].

Ad-hoc On-demand Distance Vector is a reactive routing protocol designed for use in wireless networks, including peer—to-peer networks, and is used in both fixed and mobile subscriber networks [6]. The protocol is based on a remote vector routing algorithm, which in its purest form is subject to a problem called "count to infinity", but in this protocol it is eliminated by using numbering when updating routes. The protocol is based solely on the state of connections and does not take into account the location of nodes and their movements.

Optimized Link-State Routing is a proactive routing protocol designed for use primarily in wireless networks of mobile subscribers, but can also be used in stationary environments. To get information about the network, it uses two types of service messages: "hello" and "topology control", which the nodes use to determine the direction of the next forwarding. An important feature of the protocol is the ability of each network node to store information about its topology. Since nodes receive "hello" messages from neighbors, this information is used to track changes in the node's environment and is distributed to neighbors. Neighbors are nodes in a one and two-step environment. Dijkstra's algorithm is used to find the shortest route. Based on the information received from the service messages, each MPR node builds a directed graph, which is a representation of the wireless network of this node. Using the resulting graphs, the shortest routes consisting of repeaters are determined. This protocol also does not use geo-data, which is a disadvantage when used in mobile subscriber networks [12].

Location-Aided Routing is one of the most common geo-based routing protocols. To select relay nodes, the source uses one of two possible methods. In the first case, an area is formed, which is defined by a minimal rectangle containing the current coordinates of the source and destination. Next, the source node sends a broadcast service message with a route request, the responses to which are sent only by the relay nodes located inside the formed rectangular area. In order to distribute location information over the network, the nodes periodically broadcast service messages. The obvious disadvantage of this protocol is that it does not take into account the mobility of nodes, their speeds and directions of movement. When choosing a route, only the current coordinates of the nodes are used [18].

Distance Routing Effect Algorithm for Mobility is a hybrid proactive–reactive routing protocol. Each node performs a preventive periodic distribution of data about its location to the surrounding neighbors, and the frequency of such distribution is directly proportional to the speed of movement of the node. If it is necessary to send a message, the node sends it in the direction of the recipient. The advantage of the protocol is to take into account both the current locations of the nodes and their traffic speeds. The disadvantage is the simultaneous use of multiple transit nodes to relay messages from a single source, which reduces the network bandwidth while increasing the reliability of delivery.

2.2 Research Objective

The aim of the research is to develop a routing method in wireless peer-to-peer networks that takes into account not only the physical and information characteristics of communication channels, such as: signal strength, signal-to-noise ratio, the number of subscribers sharing access to the medium, channel bandwidth, lost packets rate, the number of relays to the destination node [7, 9]. It should also use data about the location of nodes and calculated speed vectors, which is needed to select the set of vehicles to maximize the lifetime of the created network. The criteria for the quality of data transmission in the built network are such characteristics as: minimizing the percentage of lost packets and message delivery time [10].

In the course of the work, the following tasks will be solved: the routing method is proposed and implemented; the method is evaluated and compared with other methods. To study the existing and developed methods, a traffic simulation environment should be established with vehicle traffic scenarios that are close to the conditions of real traffic in urban conditions. It should also match the parameters of the movement of mobile nodes in the road and network subsystems, and network traffic scenarios should be implemented for comparative testing in conditions of different load characteristics on the network with a changing number of nodes.

2.3 Method of Dynamic Routing of Mobile Subscribers' Network

The algorithms implemented in the data transmission equipment when searching for a route rely only on the physical and information characteristics of communication channels, such as: signal strength, signal-to-noise ratio, the number of subscribers sharing access to the medium, the channel bandwidth, the lost packets rate, the number of relays to the destination node. The proposed method, along with the listed parameters, involves taking into account data on the location of nodes, as well as their velocity vectors. The overall goal of expanding the list of parameters is to narrow down the number of nodes that are not candidates for joining the network to predict the lifetime of connections between nodes.

These additions are supposed to help speed up data transfer, as well as reduce packet loss. As a modification to the current algorithms, the node visibility metric is added, which is calculated from the calculated speed vectors of the network nodes.

Due to the limited area of interacting vehicles, it is acceptable to use a Cartesian coordinate system to solve the problem. In it, the location of each node is described by three coordinates (x, y, z) of the vehicles:

- A with coordinates (x_a, y_a, z_a);
- B with coordinates (x_b, y_b, z_b).

Then each node is characterized by a velocity vector $\overline{V}$, which contains:

- initial coordinates (x_0, y_0, z_0);
- time ΔT;
- coordinates of the node after time $\Delta T(x_{\Delta T}, y_{\Delta T}, z_{\Delta T})$;
- speed ν.

To calculate the angle between two vectors, the coordinates of the vectors are determined by the formula (1):

$$AB = (x_b - x_a, y_b - y_a, z_b - z_a) \tag{1}$$

We introduce the vectors $\overline{a}$ and $\overline{b}$, calculate the angle α between them on formula (2):

$$\alpha = arccos\left(\frac{\bar{a}, \wedge \bar{b}}{|\bar{a}| \cdot |\bar{b}|}\right) = arccos\left(\frac{a_1 b_1 + a_2 b_2 + a_3 b_3}{\sqrt{a_1{}^2 + a_2{}^2 + a_3{}^2} \cdot \sqrt{b_1{}^2 + b_2{}^2 + b_3{}^2}}\right) \tag{2}$$

By sorting the angles in ascending order and cutting off the values by the angle difference, you can select the area where the potential intermediary nodes are located.

To calculate the time of visibility between one node and another, we need to solve the following equation:

$$\sqrt{(b_{\Delta x} - a_{\Delta x})^2 + (b_{\Delta y} - a_{\Delta y})^2 + (b_{\Delta z} - a_{\Delta z})^2} = S$$

To improve performance, we will get rid of the square root extraction operation and square everything:

$$(b_{\Delta x} - a_{\Delta x})^2 + (b_{\Delta y} - a_{\Delta y})^2 + (b_{\Delta z} - a_{\Delta z})^2 = S^2$$

Opening the brackets:

$$b^2_{\Delta x} + a^2_{\Delta x} - 2a^2_{\Delta x}b^2_{\Delta x} + b^2_{\Delta y} + a^2_{\Delta y} - 2a^2_{\Delta y}b^2_{\Delta y} + b^2_{\Delta z} + a^2_{\Delta z} - 2a^2_{\Delta z}b^2_{\Delta z} = S^2$$

Coordinate changes are considered as follows:

$$a_{\Delta x} = a_x + v_{ax}t;\ a_{\Delta y} = a_y + v_{ay}t;\ a_{\Delta z} = a_z + v_{az}t$$
$$b_{\Delta x} = b_x + v_{bx}t;\ b_{\Delta y} = b_y + v_{by}t;\ b_{\Delta z} = b_z + v_{bz}t$$

Let's place it in the previous equation:

$$
\begin{aligned}
&(b_x + v_{bx}t)^2 + (a_x + v_{ax}t)^2 - 2(a_x + v_{ax}t)(b_x + v_{bx}t) \\
&\quad + (b_y + v_{by}t)^2 + (a_y + v_{ay}t)^2 - 2(a_y + v_{ay}t)(b_y + v_{by}t) \\
&\quad + (b_z + v_{bz}t)^2 + (a_z + v_{az}t)^2 - 2(a_z + v_{az}t)(b_z + v_{bz}t) = S^2
\end{aligned}
$$

Take t out of the brackets:

$$
\begin{aligned}
&t^2\left((v_{bx} - v_{ax})^2 + (v_{by} - v_{ay})^2 + (v_{bz} - v_{az})^2\right) \\
&\quad + 2t\left((v_{bx} - v_{ax})(b_x - a_x) + (v_{by} - v_{ay})(b_y - a_y) + (v_{bz} - v_{az})(b_z - a_z)\right) \\
&\quad + (b_x - a_x) + (b_y - a_y) + (b_z - a_z) = S^2
\end{aligned}
$$

Thus, by solving the quadratic equation, we find the visibility time t between two given nodes. Then the total visibility time T of all nodes can be found from Formula 3. Graphically, the chain search sequence is shown in Fig. 1.

$$
T = min(t) \tag{3}
$$

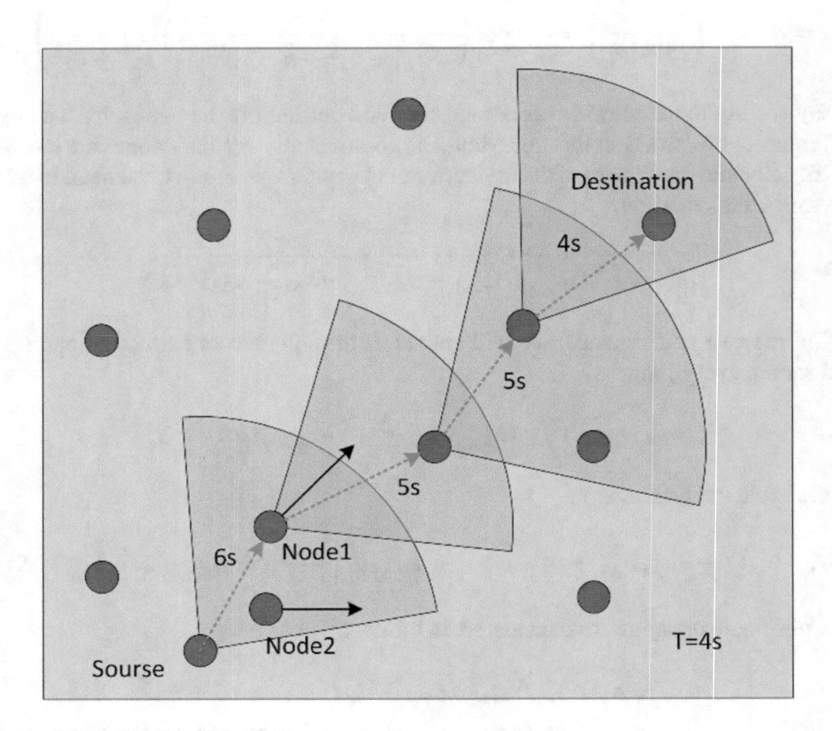

Fig. 1 Calculating the chain lifetime

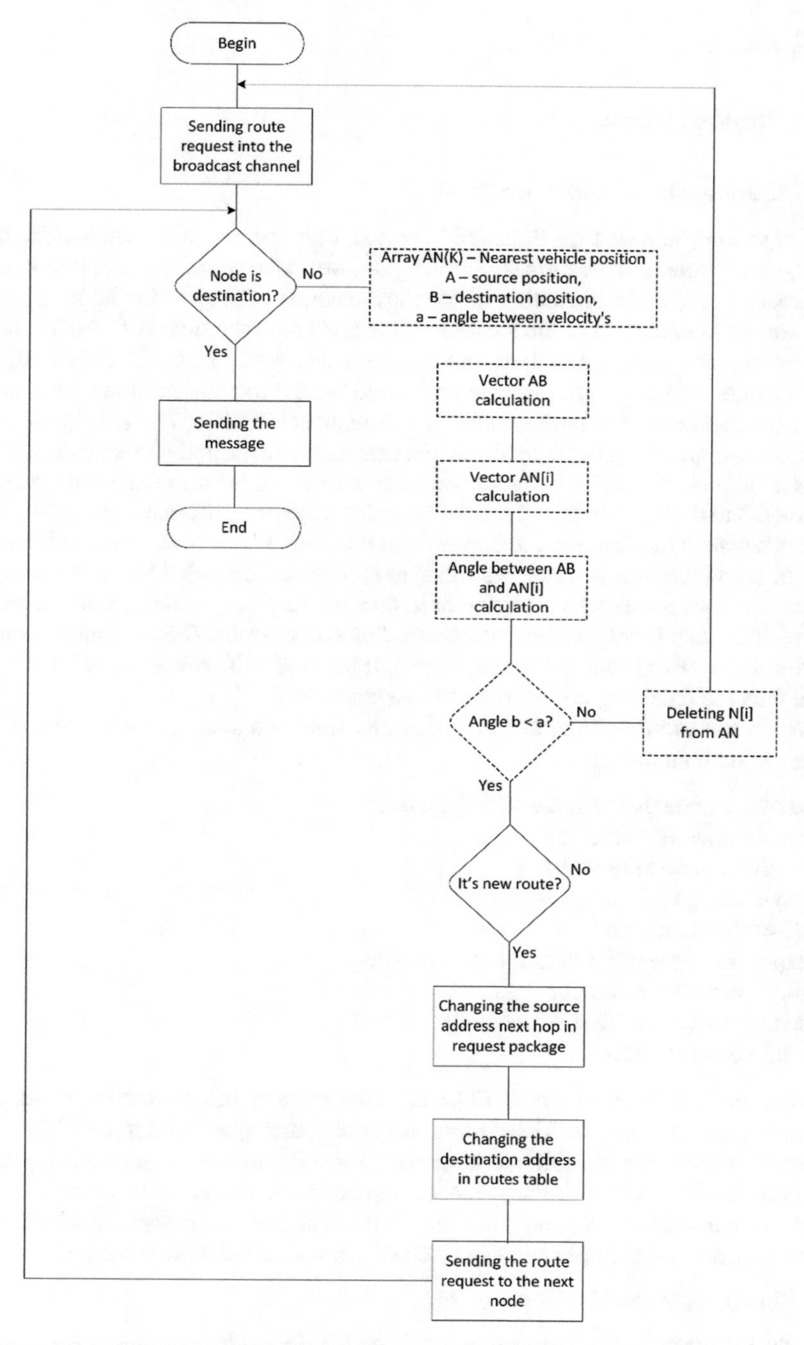

Fig. 2 Block diagram of the algorithm of the modified AODV protocol. Solid lines indicate the blocks of the basic algorithm, dashed lines indicate the additions to the basic protocol that implement the proposed method

3 Results

3.1 Implementation

Modifications to the AODV protocol

The algorithm is based on the AODV protocol algorithm. In its implementation, changes were made to the format of messages—we added velocity vectors containing the coordinates of the beginning of the movement, the speed of the node, the time ΔT and the coordinates of the location of the node after the time ΔT. Additionally, the criterion for choosing the best route has been changed. Just like in classic AODV, when a node needs to connect to another node but the route is unknown, the source node sends a broadcast packet with a route request (RREQ). The RREQ message contains a counter to prevent duplicate packets. Each time a node forwards an RREQ to its neighbors, the counter is incremented and stores the information needed to send the PREQ to the neighbors. All receiving nodes count the difference in the direction angles between the straight line drawn from the source node to the destination node and the straight line drawn from this node to the destination node (Fig. 1). If the angle difference does not exceed the value of α, then the receiving node checks whether it knows the route. If yes—it sends back a packet with the route (PREQ), and if not—it forwards the RREQ packet to its neighbors. If the angle difference exceeds the value of α, then the receiving node throws the message out.

When transmitting data, the total visibility time is added, as well as the object state vector, containing:

- id of the node that sent the RREQ packet;
- source node IP address;
- destination node IP address;
- broadcast ID;
- ID of the source node;
- expiration time of the return route record;
- state vector of the source node;
- state vector of the destination node;
- total visibility time.

If the node that received the RREQ knows the route to the destination node, then it sends a packet with the RREP route to the neighbor from which the RREQ was received. This continues, either as long as the sequence number of the node is greater than the previous one, or if the sequence number is the same, but the total visibility time is longer—then this route is considered the best. Figure 2 shows a block diagram of the algorithm that implements the AODV protocol modification method.

Modifications to the OLSR protocol

The OLSR protocol is taken as a basis, to which the choice restriction for one and two-step neighbors is added. These neighbors are filtered by a geometric area—a segment of a circle in the direction of movement of the vehicle. To find the shortest route on the node connectivity graph, the Dijkstra algorithm is used, and the visibility time between two nodes is chosen as the weights of the graph edges.

All nodes periodically broadcast service hello messages. If a node does not receive any hello messages from its neighbor for a certain period of time, the connection is considered terminated. Hello messages are not relayed across the entire network, so each node can only obtain network topology information about its two-step environment.

In addition to hello messages, nodes also periodically send broadcast TC messages that contain information about the connection between a pair of nodes.

The MPR node selection rule has also been modified. In addition to the fact that each selection of one-step neighbors is limited to a geometric segment, each node from its one-step neighbors selects MPR nodes in such a way that each two-step neighbor is a one-step neighbor for at least one MPR node, and the visibility time is maximum. Broadcast messages are sent by all nodes, but only MPR nodes can transmit messages. Based on the information received from the service messages, each MPR node builds a directed graph, which is a representation of the wireless network of this node. Using the resulting graphs, the shortest routes consisting of repeaters are determined. The address of the destination node and the first repeater form an entry in the routing table, with which, the node can forward the packet to the specified repeater.

Figure 3 shows a block diagram of the algorithm which implements the OLSR protocol modification method.

3.2 *Tools and Modeling Environment*

The algorithms were implemented in the discrete simulation system SUMO-Simulation of Urban Mobility—an open source traffic simulation package designed to work with large road networks [16]. The IDM—Intelligent Driver Model [5] is used as a discrete model of vehicle movement. The real road network data was imported from the Open Street Maps service.

The NS-3 [7] network simulator was used to simulate the processes of transmitting network traffic between network nodes and collecting statistics of the simulation results. The simulator allows to create and test semi-natural environments in which some nodes and network traffic are simulated, and the other part is actually functioning devices and data streams. As part of this study, the ability of the NS-3 simulator to work with mobile nodes of wireless networks is critical.

To simulate the transport environment using SUMO [17], we need to select the area on the map where the simulation will take place. OpenStreetMap is used for this purpose. When modeling the transport network, a set of scripts written in Python was used, which is a tool OSMWebWizard.py. The section of the road network shown in Fig 4. was selected for the simulation.

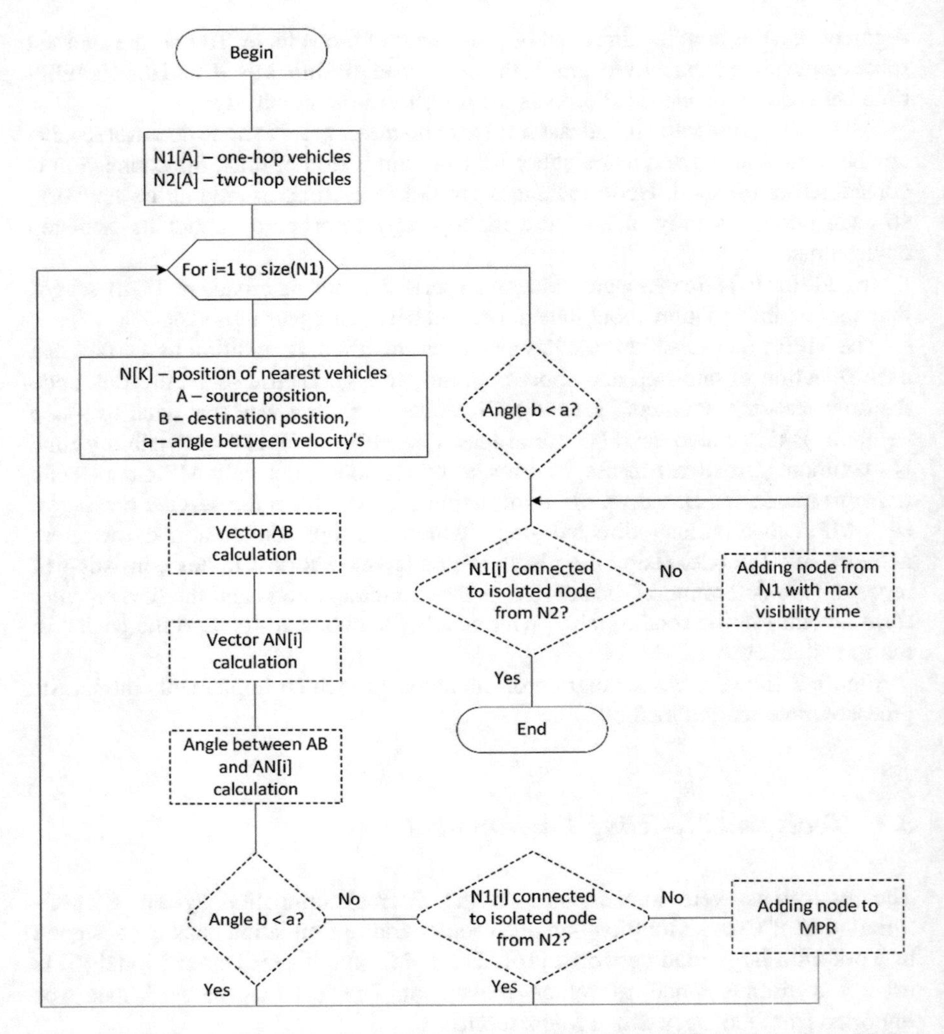

Fig. 3 Block diagram of the modified OLSR protocol algorithm. Solid lines indicate the blocks of the basic algorithm, dashed lines indicate the additions to the basic protocol that implement the proposed method

The 3.5 km^2 section has all the characteristics of an average urban traffic model, namely:

- roadways with more than two lanes in each direction;
- roadways with only one lane of traffic;
- one-way roadways;
- circular motion.

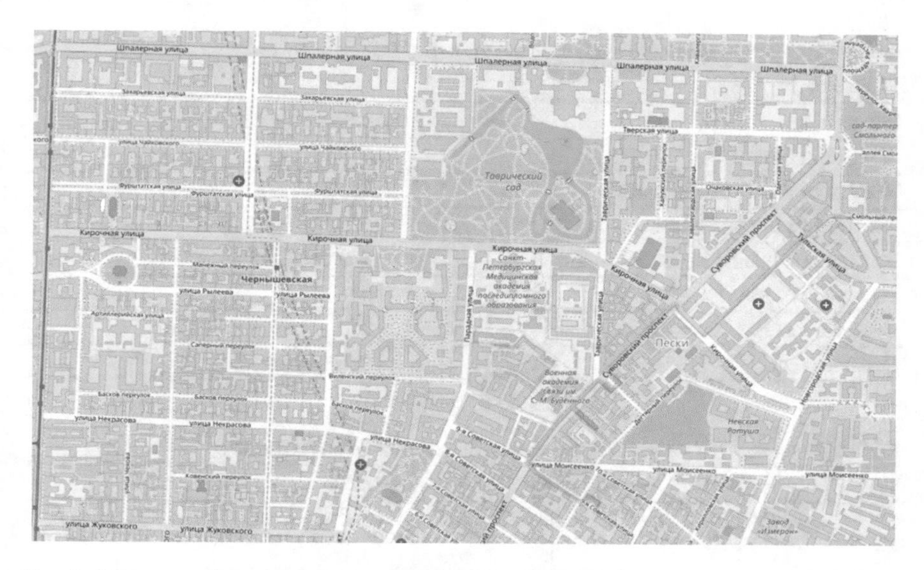

Fig. 4 A fragment of the road network to perform the simulation

In this section, a scenario was created with a density of 20 cars per hour per kilometer, which is equivalent to 1,200 cars present in the model at the same time.

3.3 Conducting Experiments and Results

In order to study the modified protocols, vehicle traffic scenarios and data transfer procedures between them were developed. The gateway car was randomly assigned on the map. For the entire duration of the simulation, vehicles were randomly placed on the selected map. Random routes were selected for each vehicle. According to the scenario, a given percentage of vehicles started transmitting 200 packets of 64 Kbytes each to the gateway address. At the end of the simulation, data about the transmission of packets in the network was saved to a file for further processing. The characteristics of the scenario are shown in Table 1. During the simulation, the number of vehicles transmitting data changed between iterations.

Experiment 1: Determination of the percentage of lost packets during data transmission for the modified and non-modified versions of the protocols.

The simulation results for 50 vehicles in the network are presented in Table 2.

Figure 5 shows the dependence of the share of lost packets in the network on the share of source nodes.

Table 1 Parameters of simulation scenarios

Parameter	Value
Plot area, km	2.5 × 1.5
Simulation time, s	50
Number of vehicles on the map	1200
Number of vehicles in the network,	50
Percentage of source nodes, %	5,20,50,75,90
Package size, Kbytes	64
Data sent, packets	200
Data recipient	Gateway vehicle

Table 2 Percentage of lost packets for modified and unmodified protocols

%	AODV	AODV-a	OLSR	OLSR-a
5	0.092	0.083	0.015	0.015
20	0.078	0.065	0.049	0.036
50	0.151	0.118	0.044	0.029
75	0.305	0.132	0.057	0.036
90	0.249	0.205	0.056	0.040

During the experiment, it was found that the modification of protocols increases the number of successfully transmitted packets and at the same time does not negatively affect the transmission time of a single packet, and in some cases even slightly improves it.

Experiment 2. Determination of the dependence of the average bandwidth in the network on the share of vehicles-nodes of traffic sources.

The simulation results for 50 vehicles in the network are presented in Table 3.

The dependence of the network bandwidth on the share of source nodes for 50 nodes is shown in Table 3 and in Fig. 6.

During the experiment, it was found that the modification of the protocols increases the bandwidth of the wireless network and at the same time does not have a negative impact on the transmission time of a single packet.

4 Conclusions

In the course of the work, a routing method based on the geographical location of the nodes was developed. The method is implemented by modifying the existing dynamic routing algorithms OLSR and AODV, by adding metrics based on geographical coordinates and vehicle speed vectors. The developed method is implemented in the

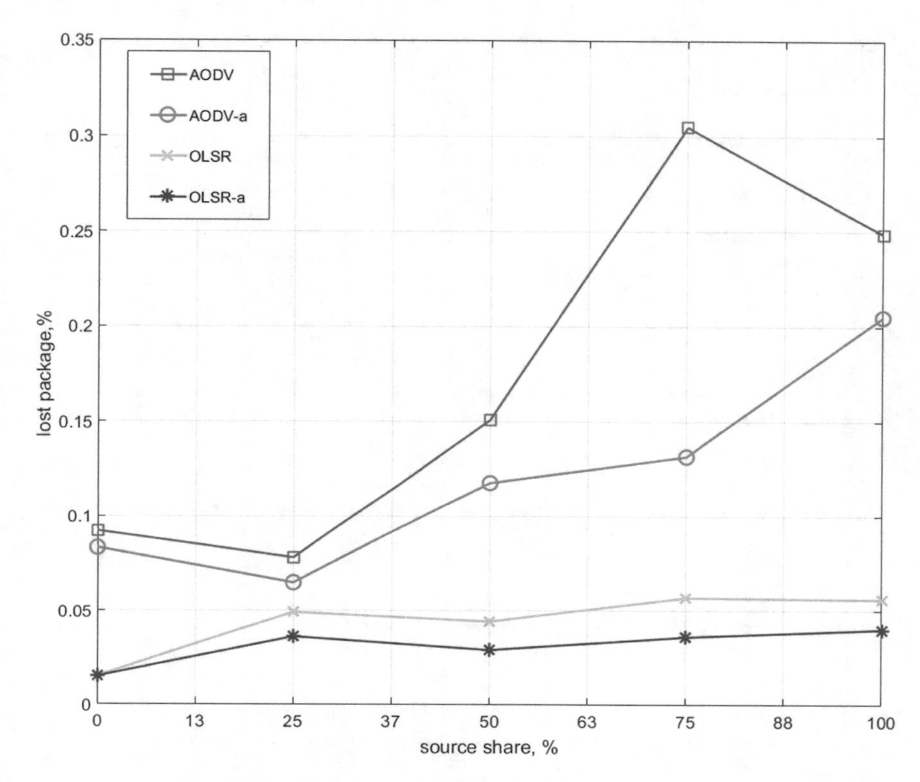

Fig. 5 Dependence of the share of lost packets in the network on the share of source nodes

Table 3 Network bandwidth for modified and unmodified protocols

%	AODV	AODV-a	OLSR	OLSR-a
5	0.681	0.688	0.739	0.739
20	2.305	2.338	2.378	2.411
50	4.246	4.413	4.779	4.855
75	6.601	8.243	8.959	9.163
90	8.454	8.946	10.625	10.803

SUMO environment and the NS-3 network simulator. The modification of dynamic routing protocols for wireless peer-to-peer networks has reduced the percentage of lost packets. The methods of selecting neighbors and the criterion for choosing the best path were modified. The model study showed that an increase in the number of sources leads to an increase in the gap between the results of the modified and unmodified protocols, and at the peak, the difference can reach 18% of the total number of packets in favor of protocols with modifications.

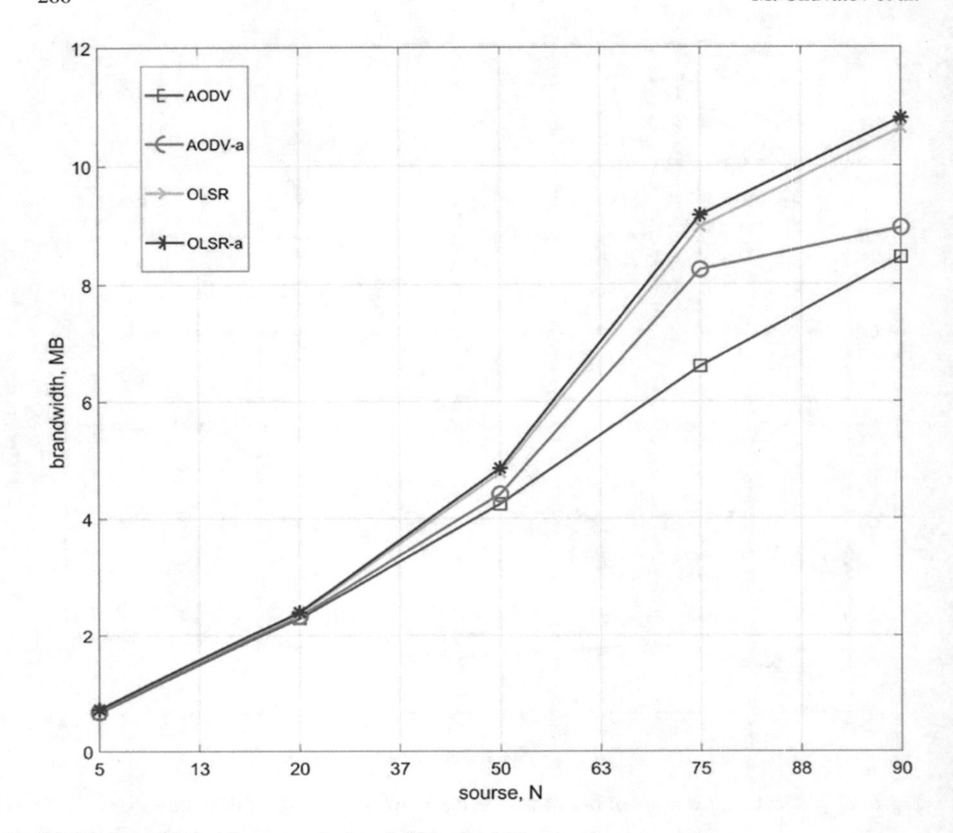

Fig. 6 Dependence the networks bandwidth of vehicles number

The proposed modifications to the dynamic routing protocols can be used in technical equipment for data transmission in peer-to-peer vehicles networks.

Acknowledgements The reported study was funded by RFBR according to the research project № 18-2903250 mk.

References

1. Guss, S.V.: Private wireless mesh networks. Math. Struct. Model. **4**(40) (2016). (In Russian).
2. Perkins, C.E., Royer, E.M.: Ad-hoc on-demand distance vector routing. In: Proceedings WMCSA'99. Second IEEE Workshop on Mobile Computing Systems and Applications. IEEE, pp 90–100 (1999)
3. Egorov, D., Levina, A., Kalyazina, S., Schuur, P., Gerrits, B.: The challenges of the logistics industry in the era of digital transformation. Lect. Notes Netw. Syst. **157**, 201–209 (2021)
4. Clausen, P.: Optimized link state routing protocol T. Jacquet–IETF Request for Comments 3626: Optimized Link State Routing Protocol (OLSR)–October 2003. IEEE 802.11s: https://wir

eless.wiki.kernel.org/en/developers/documentation/ieee80211/802.11s (даты обращения: ноябрь 2018 – май 2019)
5. ARP: https://tools.ietf.org/html/rfc826 (last open 01.05.2021)
6. Prozorov D.E.: at all Georouting protocols of mobile networks. T-Comm. (5) (2012) (In Russian)
7. Network Simulator 3: https://www.nsnam.org/ (last open 01.05.2021)
8. Jiang D., Delgrossi L.: IEEE 802.11 p: Towards an international standard for wireless access in vehicular environments. In: VTC Spring 2008-IEEE Vehicular Technology Conference. IEEE (2008), pp. 2036–2040
9. B.A.T.M.A.N protocol: https://www.open-mesh.org/projects/open-mesh/wiki/BATMANConcept (last open 01.05.2021)
10. Ilin, I., Levina, A., Lepekhin, A., Kalyazina, S.: Business requirements to the IT architecture: a case of a healthcare organization. Adv. Intell. Syst. Comput. **983**, 287–294 (2019)
11. Rana, G., Ballav, B., Pattanayak, B.K.: Performance Analysis of Routing Protocols in Mobile Ad Hoc Network. In: 2015 International Conference on Information Technology (ICIT), Bhubaneswar (2015), pp. 65–70
12. Johnson, D., Ntlatlapa, N., Aichele, C.: Simple pragmatic approach to mesh routing using BATMAN (2008)
13. Tan, W. et al.: A security analysis of the 802.11 s wireless mesh network routing protocol and its secure routing protocols. Sensors **13**(9), 11553–11585 (2013)
14. Abdullah, A.M., Aziz, R.H.H.: The impact of reactive routing protocols for transferring multimedia data over MANET. J. Zankoy Sulaimani-Part A **16**(4) (2014)
15. Vats, K., Vishnoi, I., Shukla, A. Novel architecture of delay and routing in manet for QoS. Int. J. Eng. Sci. Technol (IJEST) **3**(1) (2011)
16. SUMO (Simulation of Urban Mobility): https://www.dlr.de/ts/en/desktopdefault.aspx/tabid-9883/16931_read-41000/ (last open 01.05.2021)
17. Motorin, D.E., Popov S.G.: Multi-criteria path planning algorithm for a robot on a multilayer map. Informatsionno-upravliaiushchie sistemy [Information and Control Systems], (3), 45–53 (2018) (In Russian)
18. Ko, Y.: Location-aided routing (LAR) in mobile ad hoc networks. In: Ko, Y., Vaidya, N. (eds.) Wireless Networks, vol. 6, no. 4, pp. 307–321 (2000)
19. Basagni, S.A.: Distance routing effect algorithm for mobility (DREAM). Basagni, S. et al. (eds.) In Proceedings of the ACM MOBICOM, pp. 76–84 (1998)

Medical Innovation Hubs as an Organizational and Technological Form of Accelerating Healthcare Development

Evgenyi Shlyakhto, Anastasia Levina, Alissa Dubgorn, and Manfred Esser

Abstract The paper proposes to consider an innovative ecosystem (innovation hub) as a viable model of interaction between participants in economic cooperation, contributing to the widespread introduction of innovations in medicine. The purpose of this paper is: to determine the goals and objectives as well as the success factors of innovation ecosystems, the key participants in such hubs and their functions. The methodological basis of the research is the analysis of open sources. As a result, the goals and objectives implemented by innovation ecosystems were formulated, their participants were identified, principles of functioning and success factors of such ecosystems were proposed.

Keywords Innovation development · Innovation hub · Healthcare innovation · Medical innovation hub creation · Digital transformation of healthcare

1 Introduction

The success of a business in the modern world depends both on the internal potential of the enterprise and on the surrounding business environment, which provides the conditions for realizing this potential. A motivating environment enables enterprises to communicate effectively with participants in the supply chain, promotes the dissemination of industry knowledge and best practices, facilitates innovation, and allows building partnerships with financial, administrative institutions and research organizations. Speaking of such a motivating environment, the concepts of

E. Shlyakhto
Almazov Federal Medical Research Center, Akkuratova str. 2, St. Petersburg 197341, Russia
e-mail: e.shlyakhto@almazovcentre.ru

A. Levina (✉) · A. Dubgorn
Peter the Great St. Petersburg Polytechnic University, Polytekhnicheskaya str. 29, St. Petersburg 195251, Russia

M. Esser
GetIT, Grevenbroich, Germany
e-mail: manfred.Esser@myget-it.com

C. Jahn et al. (eds.), *Algorithms and Solutions Based on Computer Technology*,
Lecture Notes in Networks and Systems 387,
https://doi.org/10.1007/978-3-030-93872-7_24

ecosystem, business ecosystem, and also an innovation hub have been increasingly used lately [1].

The digital transformation of economic systems (industries, regions, states) requires special forms of interaction between subjects, as described in the previous works of the authors [2–5]. The creation of ecosystems and innovation hubs for the dissemination of digital innovations seems to be a useful form of organizing the interaction of the participants in the innovation process, which will allow achieving a synergistic effect from the integrated efforts of the participants, as well as contribute to the accelerated development of enterprises and provide a new level of interaction with customers.

Healthcare organizations focused on the Smart Hospital concept recognize the need to move the management system beyond the boundaries of individual organizations. Reassessment of business models, reengineering of business processes, IT support and organizational structures are necessary to achieve strategic goals of medical organizations, such as:

- improving the quality of patient care to meet the needs of modern and future consumers of medical services;
- optimization of treatment results for each individual patient;
- empowering health workers to achieve the best results in the provision of health services;
- increasing the operational efficiency of a medical organization in order to free up resources for the implementation of innovative solutions and improve the quality of patient care;
- applying big data innovations to help caregivers and scientists conduct experimental research in medicine and bioengineering.

The paper examines the issues of creating medical innovation hubs for the dissemination of innovations in the healthcare sector. The purpose of the article is to identify the factors of success, key participants in such hubs and their functions, as well as to develop a possible structure for participants in a medical innovation hub. A description of such fundamental elements can become the basis for the development of a management architecture and implementation of the activities of innovation hubs in the healthcare sector.

2 Materials and Methods

To achieve the goals of this paper, we analyzed information from open sources about the world's leading ecosystems and innovation hubs (including the healthcare sector) [1, 6–8], the principles of their functioning, as well as data from analytical reports of global consulting companies on the results of the application of the ecosystem approach in the diffusion of innovations [8, 9].

The creation of innovation hubs as a form of sectoral innovation development can be called a trend of the last decade. BCG named digital platforms (and innovation ecosystems in general) and artificial intelligence as the two main "sources of innovation" [10]. Vivid examples of specialized innovation hubs are: the center of the IT industry in Silicon Valley (USA), blockchain technologies—New York (USA), IT security—Tel Aviv (Israel), financial technologies—London (UK) and others [1, 6]. In the healthcare sector, the world's industry leaders are building various innovative ecosystems (hubs) and technology platforms: EIT Digital—the driving force behind digital transformation in Europe, NEOM—support and organization of innovative entrepreneurship hub located in Saudi Arabia [11], Healthcare Innovation Hub (Japan), Health Innovation Hub (Germany), Health Innovation Square (Switzerland), Lyonbiopôle (France), etc.

The EIT Digital ecosystem includes more than 200 leading European corporations, small and medium-sized enterprises, start-ups, universities, research institutes. [12]. The hub invests in strategic areas to accelerate market penetration and scaling of research-based digital technologies, with a focus on Europe's strategic social issues: Digital Cities, Digital Industry, Digital well-being, Digital finance. In the field of Digital well-being, the hub opens up the following opportunities for participants:

- Search for partners for realizing development projects;
- Obtaining investment for development;
- Testing of developed products in medical organizations included in the Hub;
- Promotion of developed products through the Hub's partner network.

The Health Innovation Hub, created in Germany to support the activities of the Ministry of Health, is the main point of contact and a bridge for interaction of various stakeholders on the innovative development of the healthcare sector. The main activities of the hub are:

- Electronic medical records;
- Compatibility and data formats;
- Digital applications;
- AI applications;
- Donation of data;
- Evaluation of digital applications.

The main activities of the hub are following:

- Holding events: conferences, symposia, hackathons (for searching and developing startups), summits, webinars;
- Interaction with medical organizations, developers of innovative solutions, government agencies to develop frameworks for the development of the industry;
- Information promotion (magazine, interviews, social networks).

The Healthcare Innovation Hub, created by the Ministry of Economy, Trade and Industry of Japan to accelerate the development of research and real-life implementation in the healthcare sector, has the following objectives [13]:

- Strategic resource management for the development of innovations in the healthcare sector;

- Providing a communication platform for stakeholders in the healthcare sector;
- Achieving synergy in the development of various enterprises and organizations related to the field of well-being in Japan.

The Swiss Health Innovation Square Hub [14] declares following main tasks:

- Identifying health needs through tools such as medical hackathons, collaborative development workshops, and brainstorming sessions;
- Creating a space for dialogue to promote a culture of innovation through conferences and thematic events;
- Services of technological, economic and thematic monitoring of innovations in the healthcare sector;
- Support of researchers, healthcare professionals, start-up creators and small and medium-sized businesses through an accelerated program offering personalized support;
- Development of platforms and tools that facilitate dialogue and the search for new ideas and projects.

The backbone of this hub are: a thriving regional hospital (Hôpital du Valais) and the Institute for Interdisciplinary Clinical Activity (Institut Central des Hôpitaux Valaisans); renowned clinic specializing in rehabilitation (Clinique romande de réadap-tation-SUVA); prestigious academic institutions (Swiss Federal Institute of Technology Lausanne, HES-SO Valais / Wallis, Institut de recherche en réadaptation-réinsertion, Institut de Recherche en Ophthalmolo-gie); organizations conducting medical and demographic research (Observatoire Valaisan de la Santé, Service de la santé publique); a strong network of partners (manufacturers, umbrella organizations, etc.).

For the organization of high-quality interaction of the participants in the value creation system in healthcare, a formal association in the form of a hub is not always organized. Cases of creation of innovative ecosystems based on large healthcare enterprises (clinics, pharmaceutical companies, etc.) are widespread: John Hopkins Hospital (USA), Mayo clinic (USA), AstraZeneca (UK) and others. AstraZeneca starting in 2018 began to work on the creation of a system of innovative medical hubs. At the moment, there are a number of hubs in countries such as Russia, India, Singapore, China, Brazil, Ireland, USA and Germany.

3 Results

Analysis of the goals and objectives, areas of activity, the structure of existing innovation hubs and ecosystems in the healthcare sector made it possible to formulate a number of fundamental elements of innovation hubs in the healthcare sector:

- factors of creation,
- goals,
- tasks,

- functions,
- participants.

The factors of creating innovative medical hubs are presented in the form of a model in conjunction with the activities of such a hub (see Fig. 1).

The main goals of creating an innovation hub can be formulated as follows:

- accelerating the implementation of innovations (including digital innovations);
- creating a motivating global (national, regional) environment for research and innovation in the industry(s);
- supporting companies to turn innovative ideas into go-to-market products and services.

The goals are concretized by a set of tasks for the medical innovation hub:

- Providing opportunities for joint research and innovation (through the creation of a network of professional organizations, the use of a single information platform);
- Providing an innovative testing and piloting environment (laboratory and clinical infrastructure);
- Providing a digital integration platform (for modeling the integration of digital technologies into the medical environment and medical processes);
- Providing support for business development in the field of healthcare innovation (including fundraising, business modeling, partner search, promotion, internationalization, media contacts);
- Providing a system for creating and maintaining knowledge (access to expert services and knowledge bases, collection of best practices in the implementation of innovative technologies and development of service solutions, innovative network capabilities at the regional, national and thematic levels);
- Ensuring communication and attracting new stakeholders (congress activities);

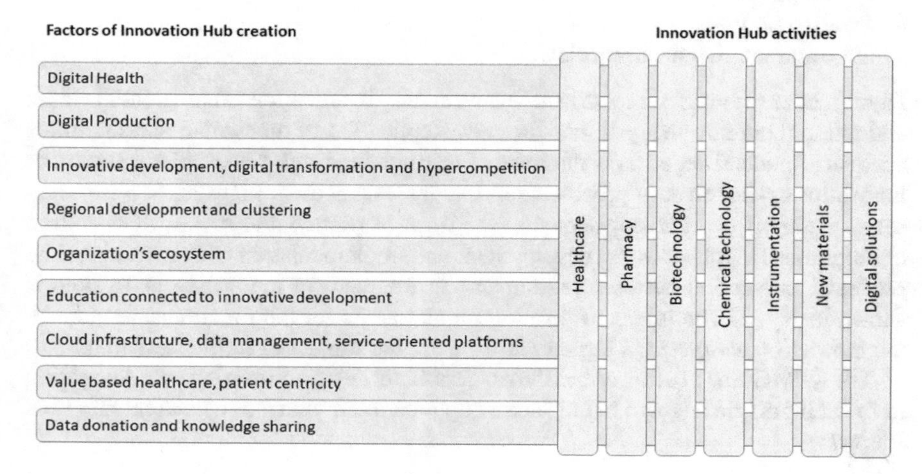

Fig. 1 Factors of Medical Innovation Hub creation

- Organization of educational events.

An innovative ecosystem that stimulates interaction and development of its participants implies the following functions:

- Research&Development;
- IT and digital technologies (including a single hub platform for information exchange);
- Replication (including production) and commercialization (startups);
- Training and practice in the field of innovation;
- Competence Center: development of policies (rules), accumulation of best practices and expertise;
- Community support for innovation;
- Infrastructure for innovation;
- Interaction with government agencies.

It is assumed that the participants of the innovation ecosystem implement the entire cycle of innovation. As applied to the implementation of digital technologies in the practice of economic activities of enterprises, it is important to stimulate the creation of such ecosystems—sectoral and regional, as well as nationwide. Such a motivating environment is designed to combine digital, research and commercial initiatives with the aim of developing, scaling and replicating complex solutions in the field of digital transformation of enterprises, industries, regions. Thus, an innovative medical ecosystem should include:

- Enterprises in the industry value chain (clinics, pharmaceutical companies, medical device manufacturers, patient representatives, rehabilitation organiations);
- Universities and research organizations;
- IT companies;
- Funding bodies;
- State and municipal authorities.

Digital innovation is often driven by research organizations that identify ideas and refine them into ready-to-use formats. Replication of innovative ideas requires adequate financial support (in the form of startups funding). The implementation of innovations is not an easy process, therefore, at least at the initial stages, it is necessary to support government bodies in the form of preferential modes of operation of enterprises involved in the development and implementation of innovations. The model of interaction between participants in the medical innovation ecosystem is shown in Fig. 2. The issues of forming requirements for the functional services of the innovation ecosystem's digital platform are the subject of further research.

The KPMG study [8] describes the success factors of an innovation hub. According to the authors, the adapted list of factors applied to an innovative medical hub is as follows:

- Attractive city (region);
- Successful leading medical organization;

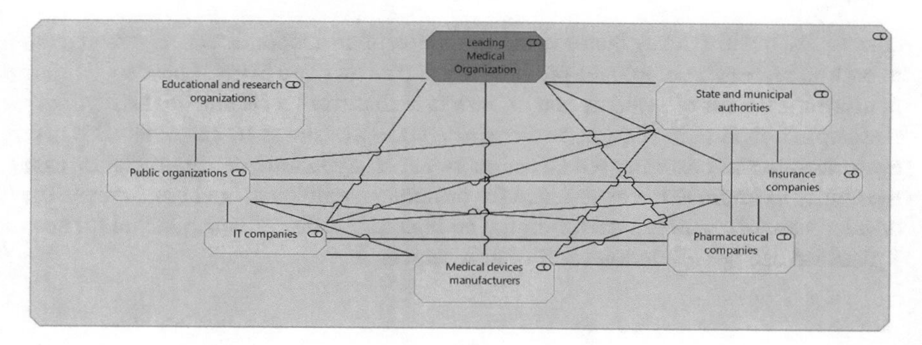

Fig. 2 Model of medical innovation Hub participants interaction

- Development of a network of interaction with manufacturers of medical equipment and materials, pharmaceutical companies;
- Modern infrastructure;
- At least one research university;
- Affordable investment financing;
- Qualified talent stream;
- Favorable regulatory environment;
- Availability of successful startups background;
- Positive trends in demographic growth;
- Supporting ecosystem (banks, law firms, accounting firms, etc.);
- Mentoring and access to the innovation network (other leaders, entrepreneurs, etc.);
- Created base of technoparks or accelerators;
- Tax and other government incentives.

4 Discussion

The paper analyzes the main prerequisites for the creation of innovative hubs in the healthcare sector. The results of the research are presented as information for the formation of an architecture model of a medical innovation hub. The development of such a model, including models of innovation hub processes, organizational structure, architecture of IT support, is the subject of a separate study.

Separately, it is important to say about the role of digital technologies in the activities of medical innovation hubs. Digital technologies today are giving enterprise leaders the opportunity to rethink their businesses to improve interactions with customers, employees and partners in the ecosystem, as well as reduce costs. When companies try to seize these opportunities through digital transformation, they take two main actions: building a digital platform and building a new operating model. Industry digital platforms are not only leading technological solutions for optimizing companies, but also a key factor in the formation of an economic space, which leads to both an increase in employee productivity and an increase in gross product. Digital platforms are a set of systems and an interface that form a commercial network or marketplace facilitating business-to-business (B2B), business-to-customer (B2C), or even customer-to-customer (C2C) transactions. A key feature of digital platforms is the ability to improve the experience for customers, employers and employees. The development of requirements for digital architecture and a reference model for such a platform can be a direction for further research.

5 Conclusions

The development of technologies (medical, managerial, ICT and digital) can bring healthcare to a new level of medical and economic efficiency. The implemntation of technological innovations requires the involvement of all participants in the industry value chain. The study of the activities of the world's largest innovation ecosystems in the field of healthcare allows us to conclude that the ecosystem approach in the implementation of innovations brings a possibility to create a favorable, motivating environment for the implementation of the full cycle of innovation implementation—from the emergence and identification of an idea to its replication and commercialization.

Based on the analysis of the experience of organizing and functioning of medical innovation hubs, the fundamental elements of creating such hubs were formulated: goals, objectives, factors of creation and factors of success, key participants. The results of the study can be used to develop a detailed architecture for managing a medical innovation hub, as well as formulate requirements for IT support of such hub (digital platform).

Acknowledgements The reported study was funded by RFBR according to the research project № 20-010-00955.

References

1. Forbes.: A New Wave of Innovation Hubs Sweeping The World. https://www.forbes.com/sites/michellegreenwald/2018/04/02/a-new-wave-of-innovation-hubs-sweeping-the-world/?sh=26e483201265#479009391265 (Дата обращения: 28.10.2020) (2018)
2. Ilin, I., Levina, A., Lepekhin, A., Kalyazina, S.: Business requirements to the IT architecture: a case of a healthcare organization. Adv. Intell. Syst. Comput. **983**, 287–294 (2019)
3. Ilin, I., Levina, A., Borremans, A., Kalyazina, S.: Enterprise architecture modeling in digital transformation era. In: Advances in Intelligent Systems and Computing, 1259 AISC, pp. 124–142 (2021)
4. Levina, A.I., Borremans, A.D., Lepekhin, A.A., Kalyazina, S.E., Schroder, K.M.: The evolution of enterprise architecture in scopes of digital transformation. IOP Conf. Ser. Mater. Sci. Eng. **940**(1), 012019 (2020)

5. Dubgorn, A., Svetunkov, S., Borremans, A.: Features of the functioning of a geographically distributed medical organization in Russia. E3S Web of Conferences, 217, 06014 (2020)
6. RocketSpace.: The World's Top Innovation Hubs for Corporate Innovators. https://www.rocketspace.com/corporate-innovation/worlds-top-innovation-hubs-corporate-innovators. Accessed 28 October 2020 (2018)
7. Smart Specialization Platform.: Digital Innovation Hubs. https://s3platform.jrc.ec.europa.eu/digital-innovation-hubs-tool/-/dih/6365/view. Accessed 25 October 2020 (2020)
8. KPMG.: The changing landscape of disruptive technologies. Global technology innovation hubs. https://assets.kpmg/content/dam/kpmg/ie/pdf/2017/03/ie-disruptive-tech-2017-part-1.pdf. Accessed 28 March 2021. (2017)
9. KPMG.: TechHubs: Leading innovation hubs. https://info.kpmg.us/techinnovation/tech-hubs.html. Accessed 28 March 2021 (2020)
10. Bouwer, L.: The Innovation Management Theory Evolution Map. https://www.researchgate.net/publication/316153609_The_Innovation_Management_Theory_Evolution_Map. Accessed 20 April 2021 (2017)
11. NEOM official web-site.: https://www.neom.com/. Accessed 15 April 2021
12. EIT Digital official web-site. https://www.eitdigital.eu/about-us/. Accessed 15 April 2021
13. HIH official web-site.: https://hih-2025.de/en/home/. Accessed 15 April 2021
14. ARK official web-site.: https://www.theark.ch/en/page/health-innovation-square-innovating-to-improve-healthcare-9908. Accessed 15 April 2021

Predictive Analytics at an Oil and Gas Company: The Rosneft Case

Anastasia Gurzhiy, Klara Paardenkooper, and Alexandra Borremans

Abstract The transition to a new technological level, the core of which was the digitalization of technological processes, entailed the transfer of industrial automation to the fourth stage of industrialization, the so-called Industry 4.0. Many of the elements of Industry 4.0 are successfully applied in practice, including predictive analytics. This paper inventories the possibilities of big data analytics, based on academic literature and applied in the case of the biggest oil and gas company in the world, Rosneft. A vertically integrated oil company is a complex structure of organizations where gigabytes of information are generated in the form of analytics, indicators, data that can be used to predict the situation in the future here and now. The economic advantage is calculated and a cost–benefit calculation is made. The paper proposes the use of predictive analytics to improve the processes of Rosneft, in particular the cost reduction in the field of asset management.

Keywords Automation · Prediction · Predictive analytics · Industry 4.0 · Big data · Information technologies

1 Introduction

The emergence of digital technologies, such as big data analytics, sensors and control systems enable oil and gas companies to automate costly, dangerous, or human error-prone tasks. Numerous companies in the oil and gas industry are starting to use these opportunities to improve their results.

Digitalization of business processes offers numerous potential benefits in the value chain of exploration, development and production, especially in reducing unplanned downtime. Given the increase in investments in the oil and gas industry, it is necessary

A. Gurzhiy (✉) · A. Borremans
Peter the Great St.Petersburg Polytechnic University, 29, St. Petersburg, Polytechnicheskaya 195251, Russia

K. Paardenkooper
Erasmus University Rotterdam, 3062 PA Rotterdam, Netherlands

C. Jahn et al. (eds.), *Algorithms and Solutions Based on Computer Technology*, Lecture Notes in Networks and Systems 387,
https://doi.org/10.1007/978-3-030-93872-7_25

to optimize production efficiency [1]. Automation creates several opportunities to achieve this goal: optimization of production without compromising health, safety and the environment, increasing field recovery and minimizing downtime [2]. This is especially relevant because of the so called "big team change".

Thousands of senior professionals will soon be retiring, leading to discontinuity of knowledge and expertise in the industry. The situation is right for innovation, the development of digital technologies stimulates the systematization of routine analyses and decision support processes, which should be atomized, where possible [3]. Thus, there is a need for reengineering business processes in oil companies. This means in practice, a complete rethinking of the business model and the reshaping of business processes of mining companies.

The primary purpose is the improvement of the quality of petroleum products, the impact of financial policy of the company, marketing activities, service levels, modelling, modern information systems for the achievement not subtle, and radical improvement of the economic efficiency of vertically integrated oil companies. These radical changes can only be achieved by the use of modern technical solutions that allow companies to reach a new level [4].

Currently, during the globalization of inter- and intra-industry integration, there are difficulties in managing vertically integrated oil companies. Because of this, there is a need to develop tools and methods that could help companies in the raw materials industry become more efficient. Most companies in the commodity market understand that further effective business management requires a management system with a high level of automation of business processes, where the main goal is to create a new effective business model for the company. Consequently, it is necessary to design the company's business model, starting from strategic goals to business processes at the operational level, that is, designing not from the bottom up, but from the top down. This approach allows to model a clear and well-thought-out architecture of an oil company, where there should be a clear system of required business processes and their operations [5].

After this introduction sect. 2 discusses the relevance of predictive analytics in logistics, based on academic literature, followed by sect. 3, which explains the context of the Rosneft case. Section 4 analyses the supply chain of Rosneft and proposes an improvement and describes the to be applied technology. The section presents an implementation plan and a cost–benefit analysis, which is followed by the conclusion (Sect. 5), which summarises the paper.

2 Materials and Methods

2.1 The Relevance of Predictive Analytics in Logistics

Logistics processes are the flow of resources between one or more origins and points of consumption. The resources can be physical items, as well as intangible goods, such as information, time and energy. Logistic processes for inventory management usually have to deal with a large number of material types, suppliers, orders, packing strategies, transportation means optimally with issues of scalability, but also with uncertainties, risks, and constraints, that are intrinsic to any logistics process automation. Big data analytics provides tools to mitigate them [6].

Big data analytics refers to a wide scope of applications dedicated to the identification and discovery of relevant patterns in data. A growing number of organizations relies on the massive statistical analysis of databases to maintain or improve business processes. Due to high data dimensionality and volume, these kinds of applications usually require tremendous computational power. Big Data analytics is based on the application of massive parallelism to process large amounts of data. The main idea is to execute analytical queries using 3 steps. [7]: Firstly, the data needs to broken into parts that can be stored and consulted separately, secondly consulting the data, using multiple compute nodes that have efficient access to the data they are going to process, which is usually local access and thirdly consolidating the results to provide the answers to the queries.

One of the data analytics tools is predictive analytics. An especially promising direction is the creation of intelligent predictive repair systems. With the help of big data technology, a company can get an accurate forecast of equipment wear in real time. Using Industrial Internet of Things and analysing all the data that company get from sensors with Artificial Intelligence will increase the quality of the process [8].

For example, for repairs, an Industrial Internet of Things for a preventive repair solution should combine such components that, in addition to remote online monitoring of equipment technological parameters, will make it possible to predict the likelihood of its failure and create an optimal maintenance and repair schedule [9]. By implementing the Predictive analysis in all spheres of business and all business process, Rosneft would reach a great result in digitalization of their production chain.

At the heart of the intelligent enterprise is a new approach to data management. It requires the ability to do three things [10].

Firstly, diverse data sources need to be integrated. Data is the currency of digital transformation. Yet within most oil and gas companies, data is scattered among multiple applications, files, data warehouses, data lakes, and public and private clouds, which are separately useless data silos.

Secondly, diverse data needs to be analysed. Data comes as structured, semi-structured, or unstructured data. It can be for example spatial, chart, numeric, geographic, time-series, relational, JavaScript object Notation (JSON). Integrating these different types of data is extremely complex, nevertheless, it is a prerequisite for becoming a smart enterprise.

Thirdly, the data landscape needs to be simplified. Currently, oil and gas companies often lack a 360°- view of their data and data landscape. With different databases, apps, and clouds to support, no centralized solutions are being used to keep an overview. Smart enterprises use process automation and a centralized, easy-to-use platform and interface to simplify access to data, so that line-of-business managers can cooperate with data specialists in the development of creative products and solutions.

A smart enterprise has a set of digital applications that ensures that interconnected business processes are linked optimally, delivering a seamless user experience, fast adoption, and ease of operations. A library of intelligent technologies drives new innovative and simple business operations. The foundation of a smart enterprise is an expansive digital platform that can ingest or federate any data source on premise or in the cloud into a harmonized view in real time. It then makes the data easily accessible in a unified decision and innovation layer.

2.2 The Context of the Chosen Logistics Process

Rosneft is the leader of the Russian oil sector and the largest public oil and gas corporation in the world. Rosneft specializes in exploration and appraisal of hydrocarbon deposits, oil, gas and gas condensate production, offshore development projects, raw materials processing, sales of oil, gas and their products in Russia and abroad. Currently, the company is among leaders in technological Innovation [11].

Rosneft's strategic goal in the field of onshore exploration and production is to maintain production and maximize the potential of existing fields, efficiently implement new projects to ensure a sustainable production profile and maximum hydrocarbon recovery rate, as well as economically sound development of reservoirs. The Company plans to transfer resources to reserves and then put them into development to maintain production in the traditional regions of activity, create new oil and gas production clusters based on the fields of the Vankor Group, as well as Eastern Siberia [11].

In order to maintain continuous improvement of profitability of the integrated business in the field of processing, commerce and logistics, the following factors should be taken into account: firstly, the implementation of the refinery modernization program on time and within budget; secondly, maintaining a strong brand and flexible marketing policy, thirdly, ensuring reliable supply of high-quality petroleum products to the domestic marketdevelopment of advanced forms of trade in oil and petroleum products and fourthly, digital transformation is one of the strategic initiatives for Rosneft and is in full swing around the perimeter, affecting all functional and business blocks, changing business processes.

So why is it so important to follow the digitalization trend? Using digitalization, the management of all critical resources and partners can be applied to achieve visibility, agility, and responsiveness.

This way a collaboration between R&D and sourcing, accelerating the time to market, an insight can be gained into future demand for manufacturing and procurement, optimizing inventory, sales, manufacturing, and delivery can be aligned, which improves customer satisfaction, linear supply chains can be transformed into digital supply networks, simultaneous collaboration with all relevant stakeholders can be reached with the innovative company at the centre, and Industry 4.0 technologies, such as autonomous assets, along with additive manufacturing or machine learning can be adopted [12].

In general, applying integrated digital solutions enhances collaboration among ecosystem participants, helping to track innovation as fast as possible, reduce costs, and provide operational transparency.

2.3 *Analysis of the Supply Chain of Rosneft and Proposal for Improvement*

To maintain a leading position among oil companies around the world, special attention is paid to the use of breakthrough digital and technological approaches in all areas of activity. The use of the most modern approaches, as well as the invention of proprietary technologies has become a key factor in the competitiveness of the Rosneft company. Connected machines and business processes can help realize Industry 4.0 aspirations. With the advent of the digital economy and the deregulation of energy markets, consumers are more empowered than ever and are demanding simplicity and service quality. Energy providers will extend beyond the barrel to master [13] consumer energy usage analytics to offer services that optimize the delivery of transportation, heating and cooling, and power and the creation of new services and experiences focusing on efficient energy outcomes that cross traditional market boundaries, such as delivering the outcomes of transportation, climate control, or a powered device –not just the traditional fuel inputs.

Oil and gas companies will start set this goal by simplifying order commitment and fulfilment processes through live inventory management, real-time available to promise, and faster material replenishment planning. They will extend their journey by running real-time predictive analyses on portfolio performance at any process stage (see Fig. 1), and then fully transform and achieve their vision with customer centricity and personalized configurations with pay-for-outcome pricing (a "lot size of one") (Fig. 2, adapted from [11]).

Some company's solutions provide real-time equipment monitoring and health information for assets, based on critical values and trends, thus enabling efficient asset maintenance strategies for better managing costs, risks, and performance. This application integrates field data capture, production planning with what-if scenarios, production allocation, maintenance, reporting, and analytics capabilities so

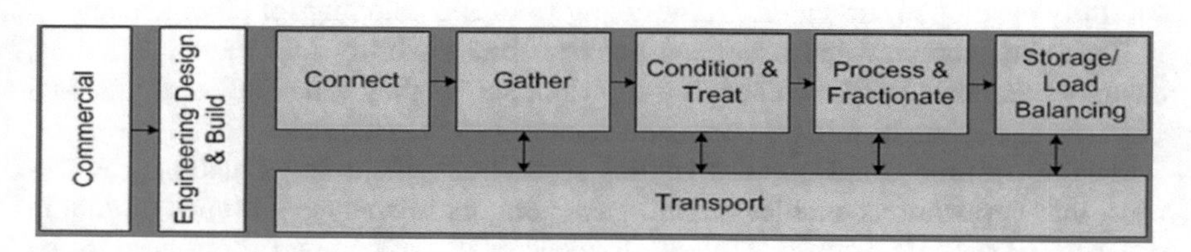

Fig. 1 Core capabilities (processes)

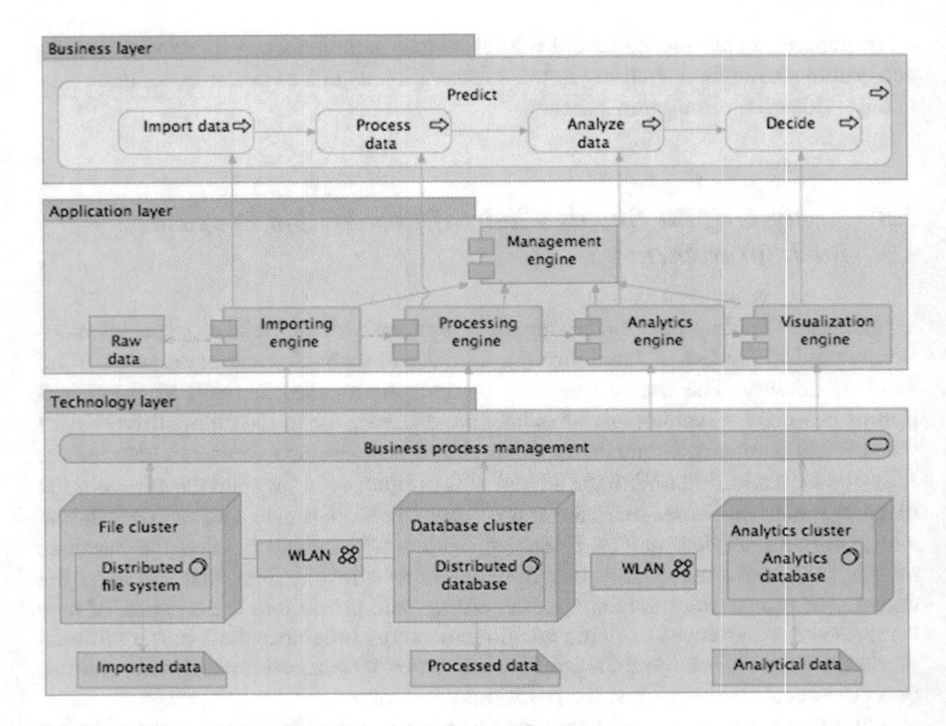

Fig. 2 Component & interface of big-data solution reference

upstream oil and gas companies can improve decision-making related to hydrocarbon production operations [14].

In the last decade, an increasingly frequent disruption of traditional global supply chains due to natural disasters, economic and political conflicts, and market volatility can be observed. As a rule, these events were sudden and unpredictable. Predictive analytics helps to develop a risk visualization system based on a single platform for responding to supply chain disruptions. Together with partners and customers, such decisions providers develop scalable solutions with the ability achieve the level of a single delivery or transportation direction. The main element of such solutions is the integration into the system and updating of forecast data on risks, as well as a carefully developed strategy for minimizing risks and planning for emergencies.

This way, eempowered users can get real-time visibility into their operations, customer feedback, and the changing environment, so they can simulate the impact of business decisions, mitigate risk, and achieve better outcomes.

The integration of advanced analytics capabilities, including situational awareness, into applications enables oil and gas operators to analyse all types of data in order to improve decision-making at all levels of the company. Empowered users, benefiting from embedded analytics, can use real-time feedback to model or simulate equipment performance, dramatically improving transparency and field productivity. This improves profitability and reduces overall costs. Oil and gas companies are using this method to eliminate repetitive manual tasks such as scheduled maintenance on equipment, monitoring downhole casing performance, ordering new materials for supply replenishment, and routing service tickets to the right team for response management [15].

3 Results

The overall meaning of the reform of Rosneft's IT unit after the merger with TNK-BP is explained in the company's documents by meeting business requirements, achieving synergy of the merging structures and centralizing IT. The first direction is responsible for traditional basic IT services, which include IT infrastructure, telecommunications and applications.

The purpose of the second direction of the IT service is to create a single information space of the company. It includes not only the development of ERP, it is about the end-to-end receipt of accurate and up-to-date information by the company's management from production and business processes. For example, Rosneft already has an application for mobile devices, where managers can see the operational financial and production performance of the company.

The third IT direction is responsible for automating the production, processing and marketing of oil and petroleum products. This direction also includes metrological control (meaning ensuring the quality of technological measurements and the accuracy of the methods of these measurements-ed.) and independent quality control of manufactured products.

In order to organize the provision of technical support and metrological support services in the company's perimeter, as well as to ensure technical competencies and production resources in the structure of the IT Service, the IT assets of the combined Company were consolidated on the basis of the service provider RN-Inform [16].

Powered by tools such as predictive analytics connected with the digital core becomes the platform for managing and optimizing systems and processes, suppliers and networks, the workforce, the customer experience, and all the data an enterprise collects using sensors and other connected IoT assets. Housed partially or wholly in the cloud, a strong digital core is critical to an energy company's ability to efficiently and nimbly create new revenue centres, develop new business models, and build relation-ships with consumers.

Oil and gas companies will try to reach this goal by optimizing real-time, profit-based decision-making driven by edge-to-digital-core connectedness. They will extend with industry market standards for next-generation, optimized business processes in the cloud and will transform and achieve their vision with fluid collaboration with business partners over networks [17].

4 Discussion

This paper proposes a perspective division, which is the creation of intelligent predictive repair systems. Based on big data technology and its analysis, company can get an accurate real-time forecast of equipment wear and tear. Falling costs of sensors and actuators along with the need for operational data that enable predictive analytics is driving the network of "things", in order to get more data from the research methods currently in use for example in the drilling process. With the development of digital technology, the sensor receives a large amount of different information, from which experts can choose additional data for a better study of the collector [18].

For example, now the company has an instruction that after 1000 h of drilling some equipment should be send to big repair, however, this information can be out of date. In that case company should repair the plant earlier then it was planned. Therefore, program can predict the exact time when the company needs to send the equipment to the big repair and signal when the replacement equipment should be sent. Taking into account that one of the main goals of Rosneft is to develop offshore projects that are in the restricted shipping zone for only 4 months, the introduction of the system will allow to follow the principle "Just in time" which will make life easier for exploration, drilling and logistics.

Another example is in the oil and gas industry, the incorrect forecasting and repair planning, which can be expensive. For example, many fields use turbines, each bearing of which can cost over a million dollars and the delivery time is more than two months. Stopping the turbine, due to the wear of one part, at the same time can lead to a de-energization of the entire field. Accordingly, there are two ways to reduce downtime. Or the corresponding spare parts, "freezing" several million dollars for an indefinite period, should be stored in the warehouse. Or sensors can be installed in the turbine to provide a real-time monitoring [19].

Rosneft, in collaboration with CRED actively develops own project implementation tools and addresses the objective of ensuring the profitability of the business and independence of the Company from sanction restrictions, see Fig. 3.

The main idea of a single platforms is that they allow to cover all departments of the company with their main tasks: accounting, movement, planning of funds; calculation of the cost of services/finished products; operational reporting on the activities of the enterprise for management; planning and control of the enterprise as a whole. After that you are able to implement additional services such as predictive analytics. The main indicators of economic efficiency can be calculated as follows [20].

In order to find the expected annual economic effect, first of all it is necessary to find the net profit increase from the introduction of the software product according to the formula (1):

$$\Delta N = N \times K_{EF} \tag{1}$$

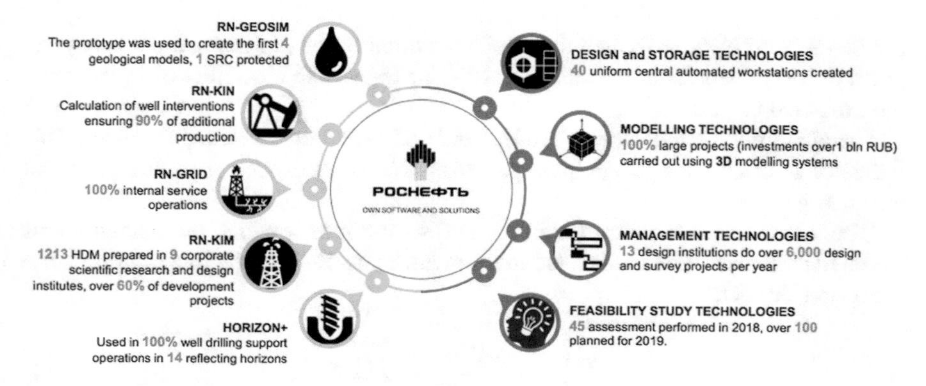

Fig. 3 Developed core system and software packages for Rosneft [11]

Legend:
ΔN—increase in net profit from project implementation, rub;
N—net profit of Rosneft, rub;
kEF—expected efficiency ratio of the project, %.
The expected annual economic effect can be calculated by the formula (2):

$$E_{EF} = \Delta N - C \tag{2}$$

Legend:
EEF—expected annual economic effect, rub;
ΔN—increase in net profit from project implementation, rub;
C—current costs (+ investment, but only for the first year), rub. NPV determines the sum of the present value of discounted payment values, which are the difference between inflows and outflows. This indicator determines how much the investor plans to receive the investor minus all outflows after the initial contribution is paid off.

Next, I calculate the net present value (NPV), which is determined by the formula

$$NPV = \sum_{t=1}^{n} \frac{\Delta N}{(1+i)^t} - I \tag{3}$$

Legend:
NPV—net present value, rub.;
n, t—the number of time periods;
ΔN—increase in net profit from project implementation, rub;
i—discount rate, %;
I—investment costs, rub.
The conditions for making an investment decision based on this criterion are as follows:
if NPV > 0, then the project should be accepted; if NPV < 0, then the project should

not be taken; if NPV = 0, then the implementation of the project will not bring any profit or loss. The greater the value of NPV, the higher the feasibility of investing in the proposed project.

Calculate the internal rate of return (IRR) of a specific investment project. This indicator is often used to compare proposals for the prospect of profitability and business growth.

The internal rate of return (IRR) is the discount rate at which the magnitude of the cited effects is equal to the cited investment. Thus, IRR can be found through the following Eq. (4):

$$\sum_{t=1}^{n} \frac{\Delta N}{(1+IRR)^t} - I = 0 \tag{4}$$

Legend:
IRR—internal rate of return, %;
n, t—the number of time periods;
ΔN—increase in net profit from project implementation, rub; I—investment costs, rub. In numerical terms, IRR is the interest rate at which the present value of all cash flows required for the implementation of an investment project (denoted by NPV) is reset. The higher the IRR, the more promising is the investment project.

Consider the calculation of the main indicators of economic efficiency for different variants of the project. The calculation of the main economic indicators for the project is presented in Table 1.

Table 1 Main economic indicators

Project/years	1	2	3	4	5
Investment costs, thousand rubles	6,674,143				
Current costs, thousand rubles	36,527	38,401	39,377	42,453	44,634
Net profit of Rosneft, thousand rubles	376,700,000	452,040,000	542,448,000	650,937,600	781,125,120
Expected efficiency ratio of the project, %	1%				
Increase in net profit from project, thousand rubles	3,767,000	4,520,400	5,424,480	5,424,480	7,811,251
Expected annual economic effect, thousand rubles	−2,942,670	4,482,999	5,385,103	6,467,923	7,767,617
Discount rate, %	7,75%				
NPV	8,907,353				
IRR	32%				

In order to complete an implementation of new technologies successfully, the company needs to develop a strategy of a digital transformation. Digital transformation is a symbiosis of large-scale technological and organizational transformations aimed at a cardinal increase in business efficiency through its full digitization at all stages of value creation.

5 Conclusions

Predictive analytics has now become an affordable tool for various levels of business, providing business needs with an accurate and valuable forecast. Predictive analytics helps to reduce risks; optimize resources; improve the efficiency of the company in conditions when other market participants are experiencing a crisis or you can simply stabilize the situation in the company; increase profits by meeting the needs of customers as much as possible, in view of their understanding; increase competitiveness; optimize operational activities; improve and simplify the decision-making process.

As the global digital economy develops, numerous aspects of society, business and the state as a whole are being transformed. Large industrial enterprises seriously assess the benefits of digitalization of production business processes and the introduction of "Industry 4.0" technologies. In the expensive and time-consuming field of oil production, digitalization can give a tangible economic effect. Achieving a new level of transparency and speed, the Digital economy allows to build a new architecture of business models among industry and market participants, reducing dependence on intermediaries and increasing the efficiency of business processes. In order for companies to reach a competitive position, defining the progressive transformation of the industry, making the impossible real and inspiring their followers at home and abroad, it is necessary to implement digital projects and automate routine processes. Optimization of workflow data can be carried out in any type of production with a serious level of automation, organized collection and long-term storage of information. For this objective, intelligent systems are successfully implemented.

Summing up the above, Rosneft today has major opportunities to develop a new strategy for the company's development based on the elements of the fourth industrial revolution. Digitalization has a positive effect on increasing the efficiency of the enterprise, which is confirmed by economic indicators in recent years. The use of digital technologies today is the key to the successful development of the company and staying competitive in both domestic and foreign markets.

Acknowledgements The reported study was funded by RSCF according to the research project № 19-18-00452.

References

1. Silkina, GYu., Scherbakov, V.V.: Modern trends of logistics digitization. St. Petersburg Polytechnic University of Peter the Great, St. Petersburg (2019)
2. Adeyeri, M.K.: Sustainable Maintenance Practices and Skills for Competitive Production System, Skills Development for Sustainable Manufacturing, Christianah Olakitan Ijagbemi and Harold Moody Campbell, IntechOpen, https://doi.org/10.5772/intechopen.70047. https://www.intechopen.com/books/skills-development-for-sustainable-manufacturing/sustainable-maintenance-practices-and-skills-for-competitive-production-system (2017)
3. Zhou, J.: Digitalization and intelligentization of manufacturing industry. Adv. Manuf. **1**(1), 1–7 (2013). https://doi.org/10.1007/s40436-013-0006-5
4. Iliinsky, A., Afanasiev, M., Metkin, D.: Digital technologies of investment analysis of projects for the development of oil fields of unallocated subsoil reserve fund. In: IOP Conference Series: Materials Science and Engineering (2019)
5. The State Program of the Russian Federation.: Social and Economic Development of the Arctic Zone of the Russian Federation. Government of the Russian Federation, Moscow, p. 115 (2017)
6. Ilyashenko, O.Yu., Ilyashenko, V.M.: Formation of requirements for architectural solution of intellectual transport system using BIG DATA. Fundamental and applied researches in the field of management, economics and trade Collection of works of the scientific-practical and educational conference: in 3 parts, S. 64–70 (2018)
7. Govindan, K., Cheng, T.C.E., Mishra, N., Shukla, N.: Big data analytics and application for logistics and supply chain management. Transp. Res. Part E Logis. Trans. Rev. **114**, 343–349 (2018). https://doi.org/10.1016/j.tre.2018.03.011
8. Levina, A.I., Dubgorn, A.S., Iliashenko, O.Y.: Internet of things within the service architecture of intelligent transport systems. In: Proceedings-2017 European Conference on Electrical Engineering and Computer Science, EECS 2017, pp. 351–355 (2018)
9. Ilyashenko, O., Kovaleva, Y., Burnatcev, D., Svetunkov, S.: Automation of business processes of the logistics company in the implementation of the IoT. IOP Conf. Ser. Mater. Sci. Eng. **940**(1), 012006 (2020)
10. Wang, G., Gunasekaran, A., Ngai, E.W.T., Papadopoulos, T.: Big data analytics in logistics and supply chain management: certain investigations for research and applications. Int. J. Prod. Econ. **176**, 98–110 (2016). https://doi.org/10.1016/j.ijpe.2016.03.014
11. Rosneft overwiew.: https://www.rosneft.ru/. Accessed 21 December 2020
12. Kozlov, A., Kankovskaya, A., Teslya, A.: The investigation of the problems of the digital competences formation for Industry 4.0 workforce. IOP Conf. Ser. Mater. Sci. Eng. **497**(#012011) (2019)
13. Schwab, K.: The Fourth Industrial Revolution: What It and How to Respond. Means https://www.foreignaffairs.com/articles/2017-12-12/fourth-industrial-revolution (reference date 20.05.2019) (2019)
14. Prokhorov, A.: Digital Transformation. Analysis, Trends, World Experience. A. Prokhorov, L. Connik, Publishing Solutions, p. 460 (2018)
15. SAP: the intelligent enterprise in the experience economy for the oil and gas industry. Creating superior customer experiences by embracing standardization for simplification to enable innovation. https://www.orianda.com/fileadmin/pdf/ebooks/SAP_-_The_Intelligent_Enterprise_for_the_Oil_and_Gas_Industry.pdf
16. Digitizing oil and gas production // McKinsey and Company.: https://www.mckinsey.com/industries/oil-and-gas/our-insights/digitizing-oil-and-gas-production (reference date 21 May 2019) (2014)
17. Bril, A.R.: Commercial Evaluation of Innovative Projects; Training Manual. A.R. Bril, L. I. Gorchakova, O.V. Kalinina; St. Petersburg State Polytechnic University.-St. Petersburg http://elib.spbstu.ru/dl/local/2393.pdf (reference date 20.02.2019) (2012)
18. Shmueli, G., Koppius, O.: Predictive analytics and informations systems research. MIS Q. **35**(3), 553–572 (2011)

19. Sherbinin, Y.: Logistics in the oil and gas industry: some provisions and considerations. J. Transp. Storage Petroleum Prod. Hydrocarb. (2016)
20. Sánchez-Ortiz, J., Rodríguez-Cornejo, V., Del Río-Sánchez, R., García-Valderrama, T.: Indicators to measure efficiency in circular economies. Sustainability **12**(11), 4483 (2020). https://doi.org/10.3390/su12114483

Digital Platforms of Territory Management

Timur Ablyazov and László Ungvári

Abstract The introduction of information and communication technologies (ICT) in the digital economy is a key trend of improvement for the spheres of life of territories. The digital transformation of the human environment in accordance with the concept of a smart city is based on the use of information and communication technologies in various aspects of the city functioning: transport, construction, housing and communal services, healthcare, etc. The article substantiates the relevance of studying the issues of introducing the ICT into the city's infrastructure in order to improve the efficiency of territory management; ICT technologies which are used in the implementation of the smart city concept are considered and some successful practical examples of the ICT implementation in the sphere of territorial management are given. It was concluded in the paper that in order to build a sustainable system of territorial management, it is necessary to form a network infrastructure, which is reflected in such a city management tool as a digital platform. The article analyzes various digital platforms of a smart city, within the framework of which the interaction of ICT technologies with each other can be established; the possible architecture of digital platforms for managing territories is considered, objective factors influencing the choice of the ICT for a specific territory are identified, and barriers are revealed that prevent the spread of the ICT and the introduction of digital platforms in the sphere of territory management.

Keywords Information and communication technologies · Digital platforms · Territory management · Smart city · Infrastructure

---

T. Ablyazov (✉)
Saint Petersburg State University of Architecture and Civil Engineering, 4 Vtoraya Krasnoarmeiskaya, 190005 Saint Petersburg, Russian Federation
e-mail: 3234969@mail.ru

L. Ungvári
Development in Relations of Industrial and Education Management GmbH, Wildau, Germany
e-mail: ungvari@driem-international.com

C. Jahn et al. (eds.), *Algorithms and Solutions Based on Computer Technology*, Lecture Notes in Networks and Systems 387,
https://doi.org/10.1007/978-3-030-93872-7_26

1 Introduction

The spread of information and communication technologies (ICT) is a key trend in the development of the human environment, which is due to forecasts of experts 2/3 of the world's population will live in cities by 2050, which will become megacities with a high population density and critically increased load on all urban systems [1]. The number of cities in the world has doubled over the past 50 years, and in 2017 there were 1,692 cities in the world [2]. Accordingly, it is required to find ways to improve the existing processes of urban areas functioning, since modern cities have high potential for the ICT introduction within the framework of the concept of a smart city [3].

It is recognized that smart city technologies enable sustainable economic growth in both advanced and developing economies [4]. Moreover, the issues of the ICT interaction with the infrastructure of territories in the context of the transition to the sixth technological order are of significant interest to the scientific community [5, 6].

In our opinion, one of the main directions of transformation of the living environment is the introduction of digital platforms for managing territories within the framework of the concept of a smart city based on the use of ICT technologies. The purpose of this work is to identify barriers to the spread of the ICT in the area of territorial management, for which the world practice of using the ICT in the sphere of territorial management will be studied, as well as the technological aspects of building digital platforms as a tool for establishing interconnections between various ICT technologies will be studied.

2 Materials and Methods

The concept of a smart city is becoming important for the effective management of territories in the context of the digital transformation of all aspects of the population's life. There are many definitions of a smart city, which together describe the development of a territory in such areas as economy, management and organization, technology, infrastructure, population, politics, environment [7]. Other researchers emphasize transport, energy, construction, housing and communal services, etc. as the main elements of a smart city [8].

Note that the main participants in the process of functioning of the territory on a large scale are the state, population and organizations. To date, it is recognized that the spread of information and communication technologies comprehensively transforms the roles of all subjects of territory development [9]. The main element of city management is the collection and analysis of information generated in the process of life of the above mentioned subjects in order to improve the human life environment [10, 11].

In accordance with [12], territorial management processes should be considered from the point of view of combinations of interactions of various vectors between the state, population and business: government-to-citizens (G2C), citizens-to-government (C2G), government-to-businesses (G2B), government-to-government (G2G), business-to-government (B2G), business-to-citizens (B2C). All these areas are implemented through digital platforms, which, in turn, are aggregators of data collected and analyzed through a wide range of ICT technologies.

Technologies for different areas of a smart city differ, but in general they include artificial intelligence, Big Data, augmented and virtual reality, blockchain, 5G communications, the Internet of Things, RFID sensors, technologies for modeling physical objects (including BIM), automated vehicles, predictive analytics, machine learning and GPS technologies [13]. According to McKinsey experts, technologies in a smart city should be divided into sensors, communication technologies and open data portals [8]. Figure 1 shows the results of a survey of more than 20 thousand people from 50 cities of the world, clearly reflecting the structure of the implemented ICT technologies, as well as the general level of their use in the practice of territorial management.

As you can see, the most developed in terms of the ICT use are Singapore, New York, Stockholm, Seoul, Amsterdam, Copenhagen, Barcelona and San Francisco. High-speed Internet is provided in all of the above cities, which ensures the interconnection of all technologies and entities with each other. Let's consider the applied technological solutions that contribute to the transformation of management in these cities in more details.

Singapore introduced the ONE.MOTORING system for online access to information on traffic flows and available parking lots, as well as express response to road accidents; electronic speed control sensors are used, a web center for interaction with the police [14]. The Emergency Medical Service, which responds to health emergencies via pre-installed user sensors, is one of the distinctive characteristics of Singapore. Smart sensors are also used for data collection in the sphere of energy, water consumption and waste disposal. Electronic systems of interaction in Singapore are developing in three areas—for the population, business and government authorities. The common platform of interaction between departments and ministries, protected from cyber threats, was created in order to optimize the management of the city's development for the exchange of data and the joint implementation of projects.

New York uses such technologies as wireless meters for collecting utility readings, traffic control and congestion control, water and air quality sensors, online tracking system of shooting in the street using cameras and high-sensitivity audio sensors, GPS system for monitoring snow removal in the winter period, as well as a round-the-clock system for interaction between the population and the city authorities [15]. The 311 application allows residents to submit an appeal at a convenient time by means of a call, SMS or message in popular instant messengers and social networks; each request is accompanied by geo location, additional information and a photo, if required, and transferred to the database for further forwarding to the appropriate department. Moreover, New York is already changing over from an open data system to an open and usable data system based on the introduction of machine-readable formats in order to reduce barriers to the analysis of information that previously required processing by specialists [16].

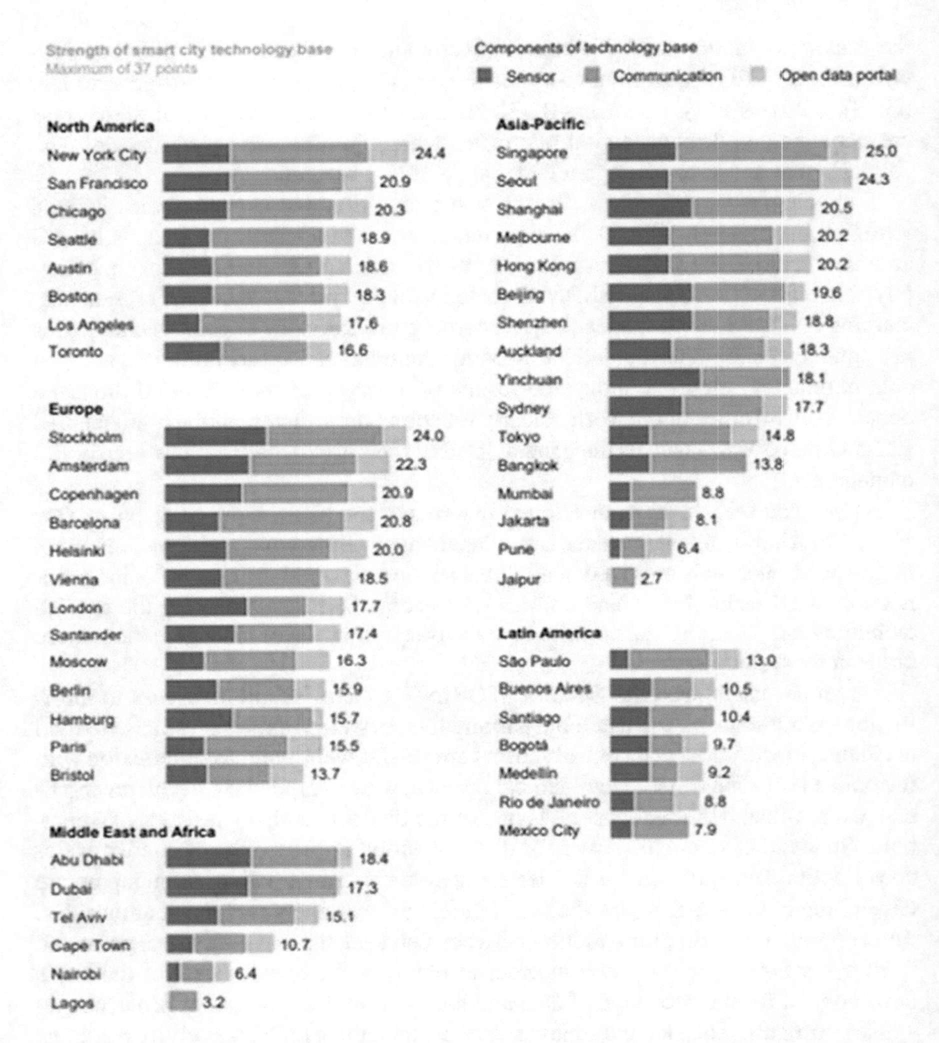

Fig. 1 Levels of technological development of territories within the framework of the concept of a smart city [8]

Thus, ICT technologies are widespread in the world practice of digital transformation of territories based on the concept of a smart city. The most successful in terms of improving territorial management are cities that use an integrated approach that allows integrating various ICT technologies into a single system based on digital platforms.

3 Results

The use of the ICT for the management of territories requires an integrated approach, since all technological solutions must be interconnected in order to obtain a holistic view of the processes inherent in the functioning of the city. According to Cisco experts, at present, territorial management issues are often solved on the basis of standard systems designed to solve problems only within one direction of a smart city, which leads to the need to create a dedicated infrastructure for each system, the lack of the ability to provide information to connected systems, increased costs and efforts spent on implementing the concept, which ultimately leads to difficulties in managing and scaling the infrastructure [17].

It is necessary to form a network infrastructure in order to build a sustainable territorial management system that will be able to adapt to the increasing load on urban systems. The construction of such a full-scale territorial management system should be based on the integration within the framework of a single digital platform of the add-ons indicated in the Fig. 2.

In general, information and communication systems for managing territories are divided into cyber-social, cyber-physical and information-reference systems [19, 20]. Various information and reference systems are already widespread in the practice of territory management. In particular, even in the traditional investment and construction sector, state information system for the provision of urban planning activities (SISPUPA), the federal state information system for pricing in construction (SISPC) and the federal state information system for territorial planning (FSISTP) have been introduced [21]. Cyber-physical systems are the main trend of urban transformation and imply a combination of the Internet of Things and a wireless sensor network (WSN) in order to achieve the integration of urban infrastructure

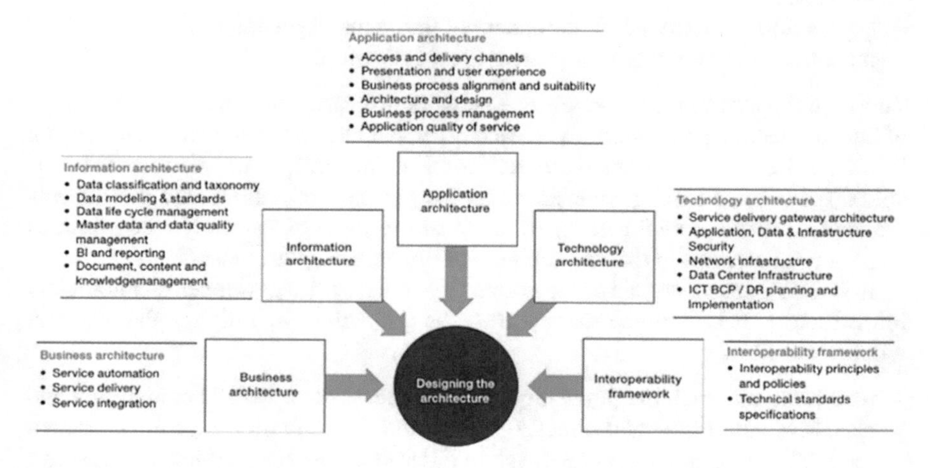

Fig. 2 Directions of the ICT development in order to form digital platforms for managing territories [18]

subsystems [2]. Cyber-social systems, in addition to technological elements, include natural resources that are inextricably linked with the concept of a smart city (issues of environmental conservation, water and energy consumption, etc.) [22].

The world practice of using the ICT as part of a full-fledged territory management system is inextricably linked with the introduction of digital platforms that meet the requirements of network interaction of various elements of territories' infrastructure with each other. Digital platforms are the basis for the socio-economic development of territories in the context of the transition to a digital economy [23]. Let's consider some technological solutions in the sphere of creating such platforms.

The digital platform for managing relations, processes, resources and territory security by 3E Company is a system infrastructure for the implementation of the management of the human life environment within the framework of a smart city [24]. The architecture of this platform, in aggregate form, consists of an intelligent city management center and a digital twin of urban processes and states. The main technology used to implement the platform is the Internet of Things. In more detail, the platform architecture includes:

- first level: metering devices, sensors, actuators, cameras;
- second level: data transmission equipment (pulse and interface modules);
- third level: data collection equipment (repeaters, concentrators, terminals);
- fourth level: servers and applications for storing, processing and analyzing data.

Cisco has also developed a digital platform for territory management [25]. The platform architecture includes:

- devices (from different manufacturers) for monitoring the condition of streets, buildings and transport;
- at first hand the Cisco platform that receives and transmits data based on the Internet;
- applications and services in the sphere of traffic management, utilities, etc. (from manufacturers of devices for primary data collection).

This digital platform is an example of a network infrastructure not only in the sphere of interconnecting city systems with each other, but also in the area of expanding the list of entities that supply smart city technologies. In addition, the platform is favorably distinguished among competitors by the presence of a dashboard that reflects the indicators of the vital activity of the territory, which is currently recognized as an important element in the development of the human environment.

In Russia, Moscow is one of the examples of cities with a developed technological infrastructure. It is assumed that by 2030 the city will have a digital platform that will include [26]:

- user level, at which the population, organizations, the scientific community and executive authorities will directly participate in the life of the territory through the ICT, as well as express feedback in order to improve the systems of the city;
- service level (systems, applications, electronic services);

- the level of data collection and analysis, designed to meet the needs of the subjects of the territory on all issues with the help of information analytics and the development of optimal solutions to existing or forecasted problems;
- the infrastructure level, including the specific ICT, data storage, data processing centers, monitoring and emergency response systems.

Among the technologies designed to solve the problem of building such a platform, experts distinguish Big Data and predictive analytics, augmented, virtual and mixed reality, the Internet of Things, blockchain, man–machine interfaces, etc. [26].

The considered practical examples of digital platforms for managing territories allow us to conclude that the main elements of the platforms are similar. Figure 3 shows a variant of the multilevel architecture of the digital platform for territory management.

As you can see, sensors for collecting data and directly services for interaction of subjects with each other are the primary and final stages of the platform's functioning, while data management tasks are the core of the platform. Heterogeneous data collected at the first level of the platform (in csv formats, diagrams, texts, etc.) is then reduced to a single format using semantic web technologies [28]. A common format for transformed data is the Resource Description Framework (RDF) [29]. Next, the data needs to be integrated, for which SPARQL is used, with the help of which data in RDF format is requested, retrieved and managed, which ensures the

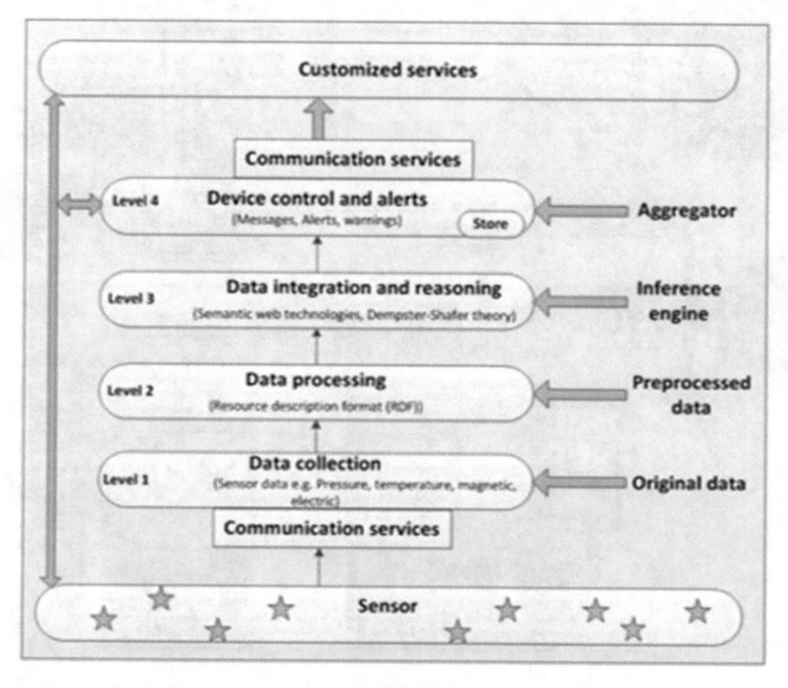

Fig. 3 Architecture of a multilevel digital platform for territory management [27]

integration of information from lower levels [30]. In addition, data extraction rules are further taken into account to improve the architecture of the platform. At the last stage of working with data, operations are carried out with the purpose to transfer them to various applications for managing the territory, which both can only display data and immediately notify of critical changes in a particular area of the city's life [31].

In more detail, the architecture of the digital platform is shown in Fig. 4. This diagram reflects the interaction between various nodes of the digital platform and the tools used to establish connections between them.

It is also necessary to take into account the differences in the types of information collected—static, semi-static and dynamic (online mode) [33]. Static data hardly changes (for example, street names, dates of birth, etc.), semi-static data is updated several times a day or week (environmental conditions), and online data changes intensively, almost in real time (as a rule sensor readings). Of course, it is possible to collect all available information, but for analysis it is most optimal to average the values of dynamic data in order to reduce the load on the digital platform. SPARQL

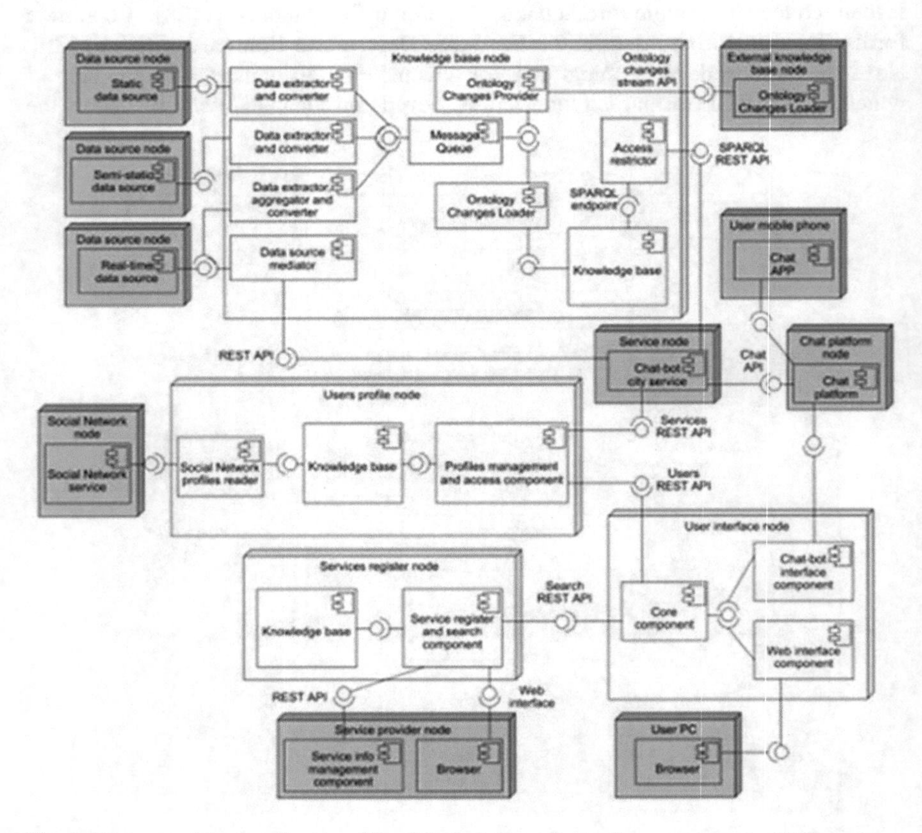

Fig. 4 Diagram of the functioning of the digital platform for territory management [32]

allows platform users to form queries according to certain criteria, while analyzing all the data contained on the platform.

Thus, the use of the ICT in the management of territories requires the construction of a network infrastructure, which, as a rule, is expressed in the formation of digital platforms. The considered schemes for the formation of digital platforms for managing territories have shown that data aggregated within the platform is of value, and ICT technologies are tools for their collection and analysis.

The set of ICT technologies used, as well as the architecture of digital platforms, are often similar in different cities and across different technology providers. Nevertheless, there are no cities that are absolutely identical in terms of their digitalization level, which is due to both the presence of individual characteristics of the territories and the complex of problems inherent in the use of the ICT in city management.

4 Discussion

The introduction of the ICT in the management of territories depends on a combination of factors inherent in a particular territory, including the age of the population, population density, climatic characteristics and income level. Thus, ICT technologies are most effectively introduced in cities with a predominantly young population, a high density of residence and a high level of income [34]. In addition, the climate determines the need of residents for certain technologies (for example, renewable energy sources and bicycle rental are more relevant in warm climates).

Despite the presence of objective factors affecting the level of the ICT spreading, the introduction of the ICT in the sphere of territorial management is fraught with a number of problems, the solution of which will allow the digital transformation of cities to be carried out more efficiently, regardless of the region.

Philips experts highlight such barriers to the introduction of the ICT in cities as the lack of a long-term development strategy; the lack of competencies for solving such problems among those responsible for the implementation of the concept of a smart city; the lack of sufficient budget for the ICT implementation [35].

According to Huawei Company, the following factors are limiting the spread of the ICT [36]:

1. The lack of scaling of pilot projects due to lack of understanding of this process.
2. The impossibility of financing of such projects exclusively by the state with insufficient development of the practice of attracting private investment in the development of cities.
3. Attempts to build many separate systems, while it is necessary to strive for the concept of open data.
4. The lack of understanding of the territory needs for specific technologies, which leads to a discrepancy between the implemented solutions and the demand.

In Russia, among the problems of the ICT spreading there are: the underfunding of projects in the sphere of development of the living environment, high motorization, the division of territories into numerous micro-districts and agglomerations, as well as a significant level of physical deterioration of the existing infrastructure, which does not allow investing in long-term projects with a delayed economic effect [34]. Also in Russia there is a lack of standards in the sphere of creating cyber-physical systems and collecting big data (in terms of protecting personal information); imperfection of legislation in the sphere of procurement of advanced technological solutions is noted [37].

If we analyze the technological barriers to the spread of the ICT, then it is worth noting that despite the introduction of technologies, the benefits from their use often cannot be defined [38]. Also, the information content of systems is constrained by users' distrust of the level of security of data transmission due to the presence of cyber threats [39, 40].

Consequently, the use of the ICT in the sphere of territorial management requires solving a number of problems that hinder their spreading. In addition, it is always necessary to evaluate the objective factors on which the success of the use of certain ICT depends. In our opinion, the search for solutions to the above problems is a promising area for further research, since these barriers appear regardless of the population size and the existing level of digitalization of the territory.

5 Conclusions

Thus, the use of the ICT in the sphere of territorial management is a common practice for the digital transformation of cities. The key area of territories development is the construction of network infrastructure within the framework of the concept of a smart city. A wide range of technologies that are important in managing territories, including big data analytics, blockchain, the Internet of Things and many others, should be built into the architecture of digital platforms, which are a tool for the formation of multi-level and comprehensive urban systems.

Despite the successful practical experience of using ICT to improve the efficiency of territorial management, the existing barriers to technologies spreading continue to restrict digitalization in many cities. In our opinion, it is necessary to implement a typical architecture of a digital territory management platform, which includes such levels of data processing as collection, reduction to a common format, formation of requests for information processing and obtaining analysis results. Moreover, the architecture of the digital platform for managing the territory should also be considered from the point of view of interaction between the participants of the digital platform, which include residents, resource providers, authorities, etc. Specific tools for transferring data between all nodes of the platform, the formats of the information collected and the frequency of its updating, options for data processing requests should be described.

Also, it is necessary to adapt the successful experience of other cities, taking into account the objective factors affecting the functioning of the living environment in a particular city in order to accelerate the implementation and use of the ICT in the sphere of territorial management. It is necessary to continuously investigate the life processes of the territory and assess the likelihood of problems occurrence inherent in the spread of the ICT. Continuous optimization of projects for the introduction of the ICT in the sphere of territorial management in an unsustainable external environment and constant improvement of technologies is an integral element of the formation of a comfortable and safe environment for human life.

References

1. United Nations: World Urbanization Prospect, https://www.un.org/en/development/desa/publications/2014-revision-world-urbanization-prospects.html. Accessed 15 March 2021
2. Juma, M., Shaalan, K.: Cyber-physical systems in smart city: challenges and future trends for strategic research. In: Hassanien, A., Shaalan, K., Tolba, M. (eds.) AISI 2019, Advances in Intelligent Systems and Computing 1058, pp. 855–865. Springer, Cham. (2020)
3. Popov, E., Semyachkov, K., Popova, G.: Digital Technologies as a factor in the promotion of smart city leadership. Adv. Soc. Sci. Educ. Human. Res. **386**, 216–220 (2019)
4. Strielkowski, W.: Social Impacts of Smart Grids: the Future of the Smart Grids and Energy Market Design, 1st edn. Elsevier, London (2019)
5. Ablyazov, T.: Application of an interdisciplinary approach to the implementation of projects to create a comfortable environment for human life. J. Environ. Treat. Tech. **8**(3), 1136–1139 (2020)
6. Dijst, M., Worrell, E., Böcker, L., Brunner, P., Davoudi, S., Geertman, S., Harmsen, R., Helbich, M., Holtslag, A.A.M., Kwan, M.-P., Lenz, B., Lyons, G., Mokhtarian, P. L., Newman, P., Perrels, A., Poças Ribeiro, A., Rosales Carreón, J., Thomson, G., Urge-Vorsatz, D., Zeyringer, M.: Exploring urban metabolism-Towards an interdisciplinary perspective. Resour. Conserv. Recycl. **132**, 190–203 (2018)
7. Vishnivetskaya, A., Alexandrova, E.: «Smart city» concept. Implementation practice. In: IOP Conference Series: Materials Science and Engineering, vol. 497 (2019). https://doi.org/10.1088/1757-899X/497/1/012019. Accessed 07 April 2021
8. Woetzel, J., Remes, J., Boland, B., Lv, K., Sinha, S., Strube, G., Means, J., Law, J., Cadena, A., von der Tann, V.: Smart cities: Digital solutions for a more livable future. https://www.mckinsey.com/~/media/McKinsey/Business%20Functions/Operations/Our%20Insights/Smart%20cities%20Digital%20solutions%20for%20a%20more%20livable%20future/MGI-Smart-Cities-Full-Report.pdf. Accessed 25 March 2021
9. Taylor, J.A.: The information polity: towards a two speed future? Inf. Polit. **17**(3–4), 227–237 (2012)
10. Caragliu, A., Del Bo, C., Nijkamp, P.: Smart cities in Europe. J. Urban Technol. **18**(2), 65–82 (2011)
11. Li, M.-H., Feeney, M.K.: Adoption of electronic technologies in local U.S. governments. Am. Rev. Public Adm. **44**(1), 75–91 (2014)
12. Gil, O., Cortés-Cediel, M.E., Cantador, I.: Citizen participation and the rise of digital media platforms in smart governance and smart cities. Int. J. E-Plan. Res. **8**(1), 19–34 (2019)
13. Technologies for smart cities.: http://www.csr-nw.ru/files/publications/doklad_tehnologii_dlya_umnyh_gorodov.pdf. Accessed 02 April 2021
14. International Case Studies of Smart Cities: Singapore, Republic of Singapore.: https://publications.iadb.org/publications/english/document/International-Case-Studies-of-Smart-Cities-Singapore-Republic-of-Singapore.pdf. Accessed 02 April 2021
15. Building a smart + equitable city.: https://www1.nyc.gov/assets/forward/documents/NYC-Smart-Equitable-City-Final.pdf. Accessed 02 April 2021

16. Centre for liveable cities Singapore.: https://www.clc.gov.sg/docs/default-source/commentaries/lessons-from-new-york-for-singapore-smart-nation-journey.pdf. Accessed 04 April 2021
17. Cisco: Building a digital city platform.: https://www.cisco.com/c/dam/m/ru_ua/events/2017/dna-forum/pdf/1-1_CDP_osaenko.pdf. Accessed 07 April 2021
18. PwC: Smart governance and technology.: https://www.pwc.in/assets/pdfs/publications/2013/smart-governance-and-technology.pdf, last Accessed 10 April 2021
19. Domrachev, A.: Great Eurasian Partnership: architecture of universal digital platforms (2019). https://d-russia.ru/bolshoe-evrazijskoe-partnerstvo-arhitektura-universalnyh-tsifrovyh-platform.html. Accessed 10 April 2021
20. Nguyen, T.: A modelling & simulation based engineering approach for socio-cyber-physical systems. In: 14th International Conference on Networking, Sensing and Control (ICNSC), pp. 702–707. IEEE, Calabria, Italy (2017)
21. Vishnivetskaya, A., Ablyazov, T.: Improving state regulation of the digital transformation in the investment and construction sector. In: IOP Conference Series: Materials Science and Engineering 940 (2020). https://iopscience.iop.org/article/https://doi.org/10.1088/1757-899X/940/1/012001/pdf. Accessed 17 March 2021
22. Cassandras, C.G.: Smart cities as cyber-physical social systems. Engineering **2**, 156–158 (2016)
23. Ablyazov, T., Rapgof, V.: Digital platforms as the basis of a new ecological system of socio-economic development. In: IOP Conference Series: Materials Science and Engineering, 497(1) (2019), https://iopscience.iop.org/article/https://doi.org/10.1088/1757-899X/497/1/012002. Accessed 15 April 2021
24. Digital platform for managing relations, processes, resources and territory security.: https://www.elpapiezo.ru/SmartCity/3E-Platphorm.pdf. Accessed 8 April 2021
25. Levin, L.: Cisco Connected Digital Platform - the basis for creating and automating digital services for the city, https://www.cisco.com/c/dam/m/ru_ru/training-events/2017/cisco-connect/pdf/Lev_Levin_Cisco_Connected_Digital_Platform.pdf. Accessed 05 April 2021
26. Moscow "Smart City–2030".: https://2030.mos.ru/netcat_files/userfiles/documents_2030/strategy_tezis_en.pdf. Accessed 02 April 2021
27. Gaura, A., Scotneya, B., Parra, G., McCleana, S.: Smart city architecture and its applications based on IoT. Proced. Comput. Sci. **52**, 1089–1094 (2015)
28. Kenekayoro, P.T.: Semantic web-the future of the web. Afr. J. Math. Comput. Sci. Res. **4**(3), 113–116 (2011)
29. Resource Description Framework (RDF) Model and Syntax Specification.: http://www.w3.org/TR/PR-rdf-syntax/. Accessed 02 April 2021
30. SPARQL Query Language for RDF.: http://www.w3.org/TR/rdf-sparql-query/. Accessed 02 April 2021
31. Ramar, K., Mirnalinee, T.T.: An ontological representation for Tsunami early warning system. In: ICAESM -2012, Advances in Engineering, Science and Management, pp. 93–98. IEEE, Nagapattinam, India (2012)
32. Teslya, N.N., Ryabchikov, I.A., Petrov, M.V., Taramov, A.A., Lipkin, E.O.: Smart city platform architecture for citizens' mobility support. Proced. Comput. Sci. **150**, 646–653 (2019)
33. Chaturvedi, K., Kolbe, T. H.: Integrating dynamic data and sensors with semantic 3D city models in the context of smart cities. In: 11th 3D Geoinfo Conference, ISPRS Annals of the Photogrammetry, Remote Sensing and Spatial Information Sciences IV-2/W1, pp. 31–38. ISPRS, Athens, Greece (2016)
34. McKinsey center for government: Smart City Technologies: What Affects Citizens' Choices? (2018). https://www.mckinsey.com/ru/~/media/McKinsey/Industries/Public%20and%20Social%20Sector/Our%20Insights/Smart%20city%20solutions%20What%20drives%20citizen%20adoption%20around%20the%20globe/smartcitizenbook-rus.pdf. Accessed 13 April 2021
35. Philips: Smart cities: understanding the challenges and opportunities.: https://www.smartcitiesworld.net/AcuCustom/Sitename/DAM/012/Understanding_the_Challenges_and_Opportunities_of_Smart_Citi.pdf. Accessed 04 April 2021

36. UK Smart Cities Index: Assessment of Strategy and Execution of the UK's Leading Smart Cities (2016). https://www.huawei.com/~/media/CORPORATE/PDF/Downloads/Huawei_Smart_Cities_Report_FINAL.pdf. Accessed 12 April 2021
37. Priority areas for the implementation of smart city technologies in Russian cities.: https://www.csr.ru/upload/iblock/bdc/bdc711b002e9651fb2763d98c7f7daa6.pdf. Accessed 14 April 2021
38. Chourabi, H., Nam, T., Walker, S., Gil-García, J.R., Mellouli, S., Nahon, K., Pardo, T. A., Scholl, H.J.: Understanding smart cities: An integrative framework. In: Proceedings of the 45th Hawaii International Conference on System Sciences, pp. 2289–2297. IEEE Computer Society, Maui, HI, United States (2012)
39. Townsend, A.: Smart Cities: Big Data, Civic Hackers, and the Quest for a New Utopia. W. W. Norton & Company, New York (2013)
40. Baig, Z.A., Szewczyk, P., Valli, C., Rabadia, P., Hannay, P., Chernyshev, M., Johnstone, M., Kerai, P., Ibrahim, A., Sansurooah, K., Syed, N., Peacock, M.: Future challenges for smart cities: cyber-security and digital forensics. Digit. Investig. **22**, 3–13 (2017)

Digital Platforms as a Key Factor of the Medical Organizations Activities Development

Evgeny Shlyakhto, Igor Ilin, Oksana Iliashenko, Ditrii Karaptan, and Andrea Tick

Abstract The subject of analysis in this article is digital platforms as a key driver of medical organizations sustainable development in the modern world. The research purpose is to consider the main components of the digital platforms using examples of works by different authors and to formulate the platform author's definition. On the basis of scientific works and periodical literature, an analysis of the digital services architecture used in the healthcare digital platforms implementation is carried out. The main regularities have been identified that influence the dynamics of transaction costs that inevitably arise as a result of communications in the socio-economic mechanisms of the digital economy. The key indicators analysis of digital solutions implementation level in the healthcare systems of such states as the USA, China, Germany, Singapore, Holland and the Russian Federation is carried out. Successful implementations examples of digital platform solutions by both foreign and Russian companies are considered. As an analysis result key advantages and achieved effects were formulated, both for patients and for medical institutions. This is due to the formation of new platform business models of the networked economy, the core of which is digital platforms and an ecosystem approach.

Keywords Digital business model · Transaction costs · Medical digital platform · Healthcare system digital transformation

E. Shlyakhto · I. Ilin
The Almazov National Medical Research Centre, St.Petersburg, Russia
e-mail: e.shlyakhto@almazovcentre.ru

I. Ilin
e-mail: ivi2475@yandex.ru

I. Ilin · O. Iliashenko (✉) · D. Karaptan
Peter the Great St.Petersburg Polytechnic University, St.Petersburg, Russia

D. Karaptan
e-mail: spartag@yandex.ru

A. Tick
Óbuda Unuversity, Budapest, Hungary
e-mail: Tick.Andrea@kgk.uni-obuda.hu

C. Jahn et al. (eds.), *Algorithms and Solutions Based on Computer Technology*,
Lecture Notes in Networks and Systems 387,
https://doi.org/10.1007/978-3-030-93872-7_27

1 Introduction

Against the background of a slowdown in global economic growth [1], the sustainable development success of both public and private organizations in leading world countries largely depends on innovative IT solutions [2, 3] and business models that they use. Research in this area [4, 5] shows that only the introduction of breakthrough solutions, both in the management and in the digital transformation field, can give a powerful impetus to development. One of the most relevant areas of business development is the digital business models development, the main element of which is digital business platforms [6].

As a result of the economy development, at the turn of the XX-XXI centuries, new approaches to organizing business interactions began to form. It could not affect the development of new business models. Thus, in one of the scientific works of Jean-Charles Rocher and Jean Tyrol, entitled "Competition of platforms in bilateral markets" [7], a new platform business model was described.

In 2018, the leadership of South Korea approved the national Program of Innovation Platforms [8] aimed at the platform economy development. The platform economy is understood as an ecosystem and infrastructure for promising industries, among which are: four areas of innovation policy in social systems, science and technology, human resource development, industry innovation; three strategic investment areas: blockchain-based data economy, artificial intelligence and hydrogen economy; eight leading sectors: smart factory, smart farm, smart city, transport of the future, fintech, new energy, biomedicine, drones.

The Decrees of the President of the Russian Federation "On national goals and strategic objectives for the development of the Russian Federation for the period up to 2024" and "On the national goals for the development of the Russian Federation for the period up to 2030" formulated tasks to ensure the accelerated introduction of digital technologies in the economy and social sphere. To solve this problem, the Government of the Russian Federation in 2019 formed and approved the national program "Digital Economy of the Russian Federation". The program determines the formation of platforms in various sectors of the economy as one of the key elements of its development [9].

In the Decrees of the President of the Russian Federation "On national goals and strategic objectives for the development of the Russian Federation for the period up to 2024" and "On the national development goals of the Russian Federation for the period up to 2030", one of the aim anounced solving the problem of ensuring the accelerated introduction of digital technologies in the economy and social sphere. The Government of the Russian Federation, formed and approved the national program Digital Economy of the Russian Federation in 2019, which defines one of the key elements of its development as formation of platforms in various sectors of the economy [9].

At the same time, despite the global trend in the platform economy development and a fairly large number of publications on the topic of platform solutions and platforms, there is no common understanding of a definition as a "digital platform" in the scientific literature.

2 Materials and Methods

2.1 Research Methods

The methodology for this study is a combination of the following steps:

1. Analysis. As part of the study, information was collected and analyzed on the existing definitions of the concept of "platform", "digital platform". Then, an analysis was made of the possibilities of reducing transaction costs when using digital platforms. The analysis of the structure of transaction costs in the medical industry is carried out. The main services of digital platforms that have an impact on reducing transaction costs in healthcare are highlighted
2. Generalization. Based on the analysis of the literature, the authors generalized the experience of previous studies and proposed the author's definition of the concept of "digital platform". In a generalized form, the structure of services of digital platforms that have an impact on reducing transaction costs in healthcare is presented.

2.2 Approaches Review to the "Platform" and "Digital Platform" Concepts Formation

The Center for Digital Transformation Training Leaders of the Russian Academy of National Economy and Public Administration under the President of the Russian Federation (RANEPA) gives the following definitions to the concepts of platform and digital platform [10]. The platform is about products and services that bring together two groups of users in two-way markets. It is important to note that by two-way markets, the authors mean network markets that have two groups of users with the emergence of network effects between them. In a two-way network, there are two categories of its users, for which the purposes of using the network and their roles in the network are clearly different.

A digital platform is a system of algorithmized relationships between a significant number of market participants, united by a single information environment, leading to a decrease in transaction costs due to the use of a digital technology package and changes in the division of labor.

This definition allows you to focus on the key business aspects, however, it does not fully reflect the aspects related to the development of business relationships based on the digital platform development.

The study [11] considered and analyzed the following types of digital platforms:

- platforms for conducting scientific activities;
- professional social networks as platforms;
- aggregator platforms that combine large amounts of scientific data;
- platform-integrators, accumulating resources for scientific activities;
- crowdsourcing and gaming platforms.

The authors of the article [11] argue that the transformation processes taking place in the scientific world change the very approaches to the process of obtaining, distributing and analyzing knowledge. Knowledge is now being formed not only in research centers, but can also arise in the so-called centers of attraction. This allows you to consolidate close ties between meeting participants, which can also happen through platform solutions.

Despite the clear advantages of digital platforms in terms of building new communications between community members based on the openness and accessibility principles, it must be remembered that effective work is impossible without face-to-face meetings (events). Thus, the organization of interaction between members of different groups should be built not only with the use of digital services, but also take into account the social essence of interaction in an offline format.

Following these statements, we will formulate the key aspects of the platform solution term.

First of all, the platform solution is based on the massive nature of the involvement of its participants, thereby forming new communities. Thanks to this, new communications are being built and the exchange of data between them increases. However, exclusively digital platforms create the effect of accelerating communications, destroying the principle of territorial distribution of community members, significantly reducing transaction costs in the form of time, searching and establishing new connections with the required community members [12].

In the study [13], the idea is expressed that platform business is the future of the world economy. Ecosystems formed on the basis of platform solutions are a factor of economic growth, undermining the positions of companies based on traditional business models.

It is important to understand that the platform itself is not a source of value for the interacting parties of its participants. The platform does not produce any material products, but acts only as a coordinator and database for its participants. At the same time, with the help of the platform tools, a large number of new communications between community members are generated. Based on this, a network effect arises: the more involved parties interact through the platform, the more valuable it becomes for its participants. Thus, the digital platform turns data into a key ecosystem asset of its participants.

An important task of any platform is to provide convenient, ideally customized, interaction tools for its participants. The continuous process of optimization and operational management of all business processes within the platform is a strategic task in the platform solutions implementation.

From the point of view of [14], a platform is a new type of firm based on a digital infrastructure that allows two or more groups to interact. Nick Srnichek from King's

College London in his work "Platform Capitalism" refers to the main characteristics of the platform: emerging network effects, lack of need for its own infrastructure, the formation of the platform architecture, which includes the interaction of various community members. He also distinguishes five types of platforms:

- advertising platforms (Google, Facebook): generate profit through in-depth analysis of data on user behavior, selling advertising space;
- cloud platforms (AWS, Salesforce): provide their computing power and software products for rent;
- industrial platforms (General Electric, Siemens): produce modern digital equipment and software in order to transfer industrial production to digital products and services, thereby significantly reducing their costs, turning their goods into services;
- grocery platforms (Rolls Royce, Spotify): use third-party platforms, embedding them in their own products and offering them already as a service;
- lean platforms (Uber, Airbnb): minimize the volume of their own assets, gaining by maximizing cost savings.

In his article "Problems of research on multilateral platforms" AI. Kovalenko. [15] considers such a concept as a multilateral platform. The authors define a multilateral platform as a complex of devices and programs, an objective digital space for a trading platform. The authors of the article believe that at the heart of any platform should be something attractive to the end user, the value that was either absent earlier or contained significant costs.

The authors see the following components as distinctive features of multilateral platforms: a special type of the company's business model, direct and indirect network effects [16], a technology standard, which is a set of specifications that ensures compatibility of software and hardware, a special type of strategy associated with platform thinking. Also, the authors, referring to the report of the Center for Global Entrepreneurship, describe a general typology of multilateral platforms, according to which four categories are distinguished: transactional (VISA, SWIFT), innovative, integrated (Apple) and investment (Softbank, Priceline Group, Naspers, IAC Interactive, Rocket Internet).

Salienko N.V. and Konopatova S.N. in the work "Analysis of business models based on platforms" [17] consider an example of the first economic ecosystem that appeared in the nineteenth century, represented by a telephone network, for which an analog switch acted as a one-way platform. The network effect from the implementation of this ecosystem was obvious and consisted in an increase in the number of subscribers, which in turn formed a special value for its participants. The article also analyzes ecosystems based on two-way platforms. Characterizing these types of platforms, the authors highlight their common features: the eco-system effect, when the value of the platform grows without the efforts of the platform holder; the ease of market entry for community members, and a developed network effect.

Considering the platforms, one cannot fail to note the work [18], where, according to the authors, the platform should ensure the interaction of four basic groups of players: producers, consumers, providers and platform owners.

Summarizing the formulations and main aspects of the digital platform concept considered in this work, it can be noted that in each of them, the basic components, in one form or another, are such components as: ecosystem, business model, a special type of firm and strategy, reduction of transaction costs, open and an accessible information environment, a communications accelerator, a system of mutually beneficial and algorithmic relations, communication groups, a "products and services" model, a system of rules, a set of software and hardware, a data aggregator. From the point of authors view, in defining a digital platform it is necessary to pay attention to aspects of the mutual business and digital platforms development.

3 Results

3.1 Digital Platform as a Modern Business Model

Summarizing the experience of previous studies, we can say the following: the platform is the basic architecture of a special type of firm with a platform business model and development strategy, an open and accessible information environment, a system of rules aimed at forming new mutually beneficial and algorithmized relationships between groups, as a result of accelerating communications which form and aggregate new data, the transaction costs of business interaction are reduced.

Given the global platforming factor of the world economy, it is important to understand that the platform is built using a "products and services" model, which can later be scaled to the size of ecosystems using a set of software and hardware.

The formulated theses allow at the abstract level to highlight the main criteria for assigning various entities to the "platform" concept:

- the presence of a basic architecture of algorithmic communications between participants;
- mutual benefit, openness and accessibility of relations between the participants;
- network effect due to the massive nature of the involvement of participants;
- availability of a unified information environment for interaction;
- significant transaction costs reduction for participants.

Thus, correlating the above statements, we propose to treat the platform as a digital data-centric business model, covering a significant number of participants, providing an effective organization of algorithmic interaction, leading to a significant reduction in transaction costs of its participants.

Based on the analysis of existing approaches to the definition of digital platforms, the authors propose the following definition: a digital platform is a digital data-centric business model, covering a significant number of participants, providing an effective organization of algorithmic interaction, leading to a significant reduction in transaction costs of its participants.

Currently, digital platforms play one of the key roles in the management of various levels socio-economic systems in countries such as the USA, Germany, Japan, Great Britain, etc. They were joined by China, South Korea, Finland, Russia. Strategic development programs for these countries are based on the platform economy formation [8, 9, 19, 20]. The relevance of the development and use of digital platforms is largely due to the possibility of creating and developing eco-systems of business in various industries.

Due to their omnichannel and cross-functionality, ecosystems formed on the basis of digital platforms are able to offer their consumers cross-industry, seamless high-quality services in order to satisfy the personal needs of each participant in the ecosystem [20].

The digital platforms and services development has contributed to the successful integration of a customized approach in the provision of goods and services to all interested members of the formed community groups. In this regard, the healthcare system is an active participant in this market for personal medical services based on its own digital platforms [21–23].

The events of recent years, in particular the fight against the consequences of the coronavirus infection COVID-19 spread, have shown the need for the development of digital platforms in the medical field.

BigTech giants such as Google, Apple, Facebook, Amazon, Microsoft, as well as IBM, Alibaba and a number of other players in the IT industry have been key players in the healthcare ecosystem market for over 10 years. They are active in the research and development of digital platforms with the aim of providing high quality personalized medical services to an almost unlimited number of stakeholders.

Vivid examples of the implementation of such projects for the interaction of medical organizations with Amazon, Google Cloud, Salesforce and Microsoft Azure are joint technological solutions in the field of data management, as well as the provision of technological infrastructure for the provision of telemedicine services.

Among medical digital platforms, the following types are distinguished [24, 25]:

- health care support platforms that provide access to treatment processes and medicines;
- unified platforms for virtual health care;
- multidisciplinary platforms providing primary health care services;
- integrated digital platforms providing telemedicine services in conjunction with offline service providers;
- platforms that implement health care management services.

Both for other states and for the Russian Federation, the transition to the provision of various services for citizens through a system of digital platforms in healthcare is an important issue for the sustainable development of this industry in modern realities. In the technological race for leadership in the provision of medical services for citizens, Russian scientific and research institutions, medical centers, investment funds, telecommunications companies and private innovation organizations play an important role in shaping the basic architectural landscape of the modern platform business model.

The most successful projects implemented in the Russian Federation in the creation and development of digital platforms in the medical field are the developments of Sberbank PJSC. It forms its own ecosystem based on the digital platform in the field of Digital Health. Skoltech and the Skolkovo Foundation are the forge of MedTech startups.

Another major player in the domestic digital medicine market is RT Medical Information Systems with its own technological solutions called “Single digital platform.LIS” and “Single digital platform.MIS”, which offers a comprehensive solution.

With the help of data services of digital platforms, the following tasks can be solved:

- maintaining a patient’s electronic medical record;
- managing the dispensing of preferential drugs;
- maintaining registers of patients with cardiovascular diseases, which is formed on the basis of taking into account specialized parameters;
- provision of telemedicine consultations;
- monitoring the health status of citizens with cancer;
- preventive medical care, providing citizens coverage by age and professional categories;
- patient flow management;
- pregnant women monitoring.

Considering the successful digital solutions presented on the healthcare market today, the following generalized directions for the development of services based on digital platforms are seen as follows:

- diseases prevention and healthy lifestyle formation;
- generalization and in-depth analysis of customer experience;
- clinical trials unified database;
- telemedicine;
- distance education;
- Internet navigation of citizens in the healthcare system.

3.2 Transaction Costs and Effectiveness of Digital Platforms Using

One of the key tasks solved with the digital platforms services is the problem of reducing transaction costs.

The approaches used today in working with digital data have a significant impact on the transaction costs size. Often, due to a significant reduction in transaction costs, it is possible to achieve large-scale network effects, both direct and indirect.

In general terms, transaction costs are understood as any costs of searching and processing information, negotiating, measuring various indicators of interaction, as well as compliance with the contractual relations fulfillment [26]. At the same time, there is no single definition on this score.

Transaction costs in the medical industry

Let's take a closer look at the structure and elements that are attributed to transaction costs in the medical industry.

The constituent elements of the costs of searching and processing information include mainly the time and location costs of searching for relevant information, for example, the patient's medical history and monitoring of his condition, the clinical trials results, regulatory knowledge bases, etc.

The negotiating costs are primarily due to the representation costs of a financial nature, the costs of maintaining communications employees, their training, as well as the risks of the communications themselves between the subjects of the relationship.

The widespread use of end-to-end technologies, such as artificial intelligence, augmented and virtual reality, the Internet of Things, has made it possible to significantly reduce the time and money spent on negotiations. Various digital communication services increase the speed and efficiency of negotiations. When using digital services, the time, cost and risks associated with the personal qualities of the negotiators are significantly reduced, and such a barrier as the location of the interacting parties is eliminated.

In the data-driven digital economy, measurement costs have become more relevant and are an important indicator of the various projects success. One of the drivers of a socio-economic relations new type is network decentralized communications exclusively in the digital environment. They provide ample opportunities for assessing and comparing the properties of various objects based on the opinion of a significant number of community members. This is actively promoted by such mechanisms as crowdsourcing and noo-sourcing, which are based on network technologies.

The main difference between these mechanisms is that crowdsourcing makes it possible to express their point of view on a particular problem to any interested person who is not part of the professional community in order to generate new ideas for its solution in an open form. No-sourcing is a professional tool that provides ample opportunities for conducting data-driven research directly with the participation of experts from various fields of knowledge. Based on the conclusions formed by experts, it is possible to prepare analytical materials for the implementation of global projects [27].

An important element of communication between the subjects of socio-economic relations is contractual relations and the associated corresponding costs of concluding transactions and compliance with their execution, as well as the costs of opportunistic behavior.

Digital platforms implement the use of seamless services from various market sectors, as well as the use of digital technologies (blockchain, artificial intelligence, big data, etc.). This made it possible to solve the difficult problem of mutual trust and control over the fulfillment of obligations assumed by the parties to the transaction, both on the part of a medical organization and on the part of an insurance company or a patient. As a result, concepts such as smart contracts and digital trust have emerged.

Digital healthcare platform solutions leverage end-to-end technologies to support faster, more cost-effective and efficient healthcare practices. They help improve the availability, quality and comfort of healthcare for people around the world, while eliminating routine tasks for healthcare personnel. This statement is confirmed by the research results reflected in the HealthNet Analytical Report of the National Technology Initiative for 2019 (Fig. 1).

The mutual benefits of platform communications within communities have shaped the massive network effects inherent in multilateral markets. An important circumstance of this effect is the fact that the satisfaction of the needs of some group members is not a consequence of the disappointment of others. Application of the "win–win" model in the platform economy shows that as a result of interaction there are winners and no losers [28].

The work [29] states that, based on the World Bank's research, the development of digital platforms should become one of the priority strategic directions for the formation and development of the digital space of the Eurasian Economic Union in the future until 2025.

The key indicators analysis of digital solutions implementation level in the healthcare systems

According to the annual analytical reports The Future Health Index, commissioned by Philips since 2016 [30], it is possible to trace the path of the global transition to digital healthcare at a higher level, with a decreasing costs level, including transaction costs. Indicators reflecting the dynamics of the transition of the global healthcare system to a new digital ecosystem are presented below. The analysis of the digital

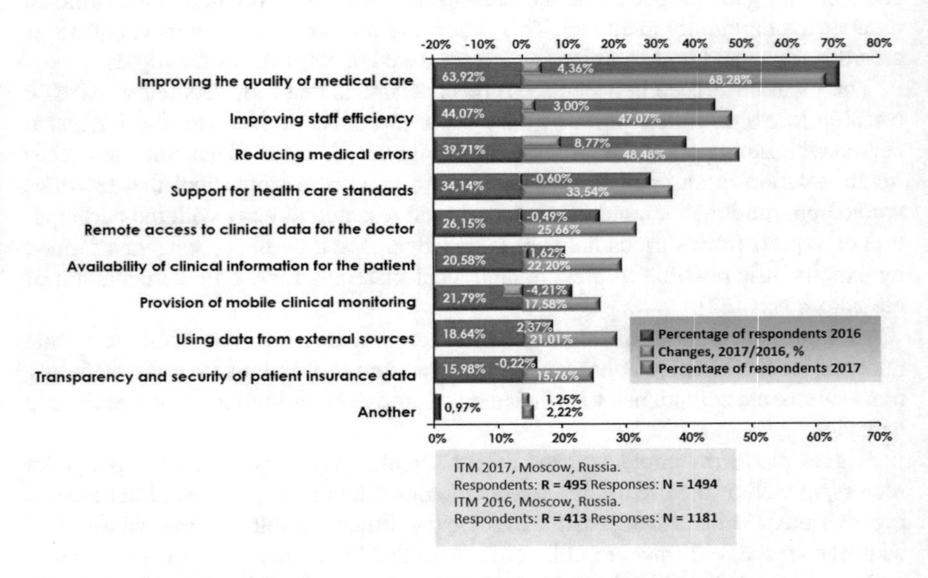

Fig. 1 The effect of the information systems introduction for a medical institution

Table 1 The future health index

Country	Future health index (points)	Country	Value measure (points)
	2016 year		2018 year
Netherlands	58,9	Singapore	54,61
China	58,1	Germany	50,93
Singapore	57,7	Netherland	48,93
USA	57,4	Russia	40,90
Germany	54,5	China	38,11
Russia	42,4	USA	37,95

transformation processes of national health systems is considered on the example of the following countries: the USA, China, Germany, Singapore, Holland and the Russian Federation (Table 1).

The range of measurement values for The Future Health Index, according to the assessment methodology [30], lies in the range from 0 to 100 points.

The Future Health Index (FHI) helps in an objective assessment of the strengths and weaknesses of the health care system for the country under study, based on the opinions of both health professionals and ordinary citizens, in three key areas: the health services availability, the health systems integration, the volume of digital technologies in the healthcare system.

Since 2018, The Future Health Index has been transformed into a more comprehensive indicator that reflects the values from the point of patient and medical staff view—Value Measure (VM). This indicator includes characteristics: the medical care availability, satisfaction with the medical care quality, the medical care effectiveness.

The availability of medical care is assessed by the following components: the number of qualified medical personnel per 10,000 citizens; the number of hospital beds per 10,000 citizens. When assessing satisfaction with the medical care quality, such components as trust in the health care system and its accessibility, perception of the medical care quality are taken into account. The health care delivery efficiency indicator includes health care expenditures as a percentage of a country's gross domestic product (GDP), life expectancy, neonatal and maternal mortality rates, and the likelihood of dying from major chronic diseases for citizens aged 30 to 70 years. Also, like The Future Health Index of 2016, the range of VM measurement is in the range from 0 to 100 points.

At the same time, a study carried out [30] in 2019, aimed at assessing the indicator of the use of digital technologies in healthcare, showed that China is a leader in this area (the indicator 94%). The indicators of other countries have the following meanings: Holland—86%, Singapore—82%, Russian Federation—81%, United States of America—76%, Germany—64%. The indicator's value is based on a survey of medical staff who use various digital solutions in their professional activities, such as digital medical records of patients, telemedicine technologies, etc.

Table 2 Healthcare efficiency index

Country	Healthcare efficiency index (points)	Country	Healthcare efficiency index (points)
	2016 year		2018 year
Singapore	84,20	Singapore	85,60
China	54,30	China	54,60
Netherlands	48,30	Netherlands	50,80
Germany	42,60	Germany	38,30
USA	32,60	Russia	31,30
Russia	24,30	USA	29,60

Based on data from the World Health Organization, the United Nations and the World Bank, analysts from the Bloom-berg agency are forming the Healthcare Efficiency Index. This index is based on three key indicators that assess the effectiveness of the state health care system [31]:

– average life expectancy with a weighting factor of 60%;
– government spending on health as a percentage of GDP per capita with a weighting factor of 30%;
– the medical services cost per capita with a weighting factor of 10%.

Healthcare Efficiency Index data for 2016 and 2018 are presented in Table 2.

The range of measurement values for this indicator is in the range from 0 to 100 points. The highest value of this index indicates the maximum efficiency of the assessed health care system.

The University of California Davis reported that between 1996 and 2013, telemedicine services, information systems and digital platforms saved patients $ 2.9 million in travel costs and 9 years of time they could have. Spend on travel to healthcare facilities [32].

According to [33], Mayo Clinic and Google signed an agreement to develop a telecommunications giant based on digital platforms, cloud computing methods, data analysis, deep learning and the use of artificial intelligence to solve the most complex medical problems. Digital platforms transform the way patient care is provided, provide proactive services to citizens, provide patients with online and offline on-demand services, reducing transaction costs for information retrieval and processing, negotiation and measurement of various interaction indicators. Thus, digital platforms allow realizing a new paradigm of the healthcare system based on value-based medicine [34].

Digital platforms services impact on reducing transaction costs in healthcare

Based on the experience of the studies reviewed, we propose a generalized structure of digital platform services that have an impact on reducing transaction costs in healthcare (Table 3).

Table 3 Generalized structure of digital platforms services that have an impact on reducing transaction costs in healthcare

The service name that reduces costs	Transaction cost type					
	Search and processing of information	Negotiation, communication	Interaction metrics measurements	Transactions	Opportunistic behavior	Specification and protection of intellectual property rights
Automatic filling of medical documents	+		+		+	
Voice input when filling out documents	+				+	
Integration with external services	+	+		+		+
Marketing services	+		+	+		
Mobile application for doctors and patients	+	+			+	
Educational and research services	+	+				+
Online interaction with external organizations and structures	+	+		+	+	+
Online reporting	+		+		+	
Information and reference system services	+	+				
Drug supply services	+			+		
Emergency services		+			+	

(continued)

Table 3 (continued)

The service name that reduces costs	Transaction cost type					
	Search and processing of information	Negotiation, communication	Interaction metrics measurements	Transactions	Opportunistic behavior	Specification and protection of intellectual property rights
Services of state funds and organizations	+					
Decision support system	+				+	
Medical resource management	+	+	+	+	+	+
Financial and insurance services	+	+		+		
Electronic document management	+		+	+	+	+

Taking into account the elements of transaction costs and factors influencing their behavior, we will give a definition of this concept.

By transaction costs in health care we mean the costs of any content and possible risks of interaction between subjects, expressed in economic terms, which inevitably arise between equal participants in socio-economic relations in the digital economy in the implementation of communications.

The data on the level of the digital solutions implementation in the national healthcare systems of the considered countries confirm the success of the medical digital platforms implementation in terms of solving urgent problems of the networked economy by continuously reducing transaction costs, simplifying data exchange processes, performing various operations, and forming communications.

4 Discussion

As a result of the study, the "digital platform" concept was clarified, the question of the transaction costs structure in the medical industry was considered. It is shown how the use of digital platforms in healthcare provides a reduction in transaction costs. A generalized structure of digital platform services is proposed. The question of the digital platform services influence possibility on the transaction costs reduction in healthcare is considered. In the future, we plan to consider the issues of improving the medical organization management based on a digital platform.

5 Conclusions

The process of changing healthcare systems is large-scale and complex. The core of the ongoing changes are digital platforms that change not only business models, but have a significant impact on reducing transaction costs, while ensuring the digital services development for citizens and businesses, allowing efficient search for relevant information, increasing the speed and efficiency of negotiations and conclusion deals, acting as a powerful driver for the development of the Digital Health concept.

In the context of global transformation processes of the world economy, when the level of transaction costs has a significant impact on the success of projects being implemented, the transition to platform models of public healthcare systems is a guarantor of sustainable development of this economy sector.

Acknowledgments The reported study was funded by RFBR according to the research project № 19-010-00579.

References

1. Kleiner, G.B.: Research perspectives and management horizons of the system economy. Manag. Sci. **4**(17), 7–21 (2015)
2. Zaramenskikh, E.P.: Management of the Life Cycle of Information Systems. Publishing house TsRNS, Novosibirsk (2014)
3. Mogilko, D.Y., Ilin, I.V., Iliashenko, V.M., Svetunkov, S.G.: BI capabilities in a digital enterprise business process management system. Lectur. Notes Netw. Syst. **95**, 701–708 (2020)
4. Maydanova, S., Ilin, I.: Strategic approach to global company digital transformation (2019). In: Proceedings of the 33rd International Business Information Management Association Conference, IBIMA 2019: Education Excellence and Innovation Management through Vision, pp. 8818–8833 (2020)
5. Kalyazina, S., Iliashenko, V., Kozhukhov, Y., Zotova, E.: Key end-to-end digital technologies in the ecosystem of the state's digital economy. In: IOP Conference Series: Materials Science and Engineering, vol. 1001 (2020)
6. Ilin, I.V., Levina, A.I., Lepekhin, A.A.: Reference model of service-oriented IT architecture of a healthcare organization. In: International Conference Cyber-Physical Systems and Control. Lecture Notes in Networks and Systems, vol. 95, pp. 681–691 (2020)
7. Rochet, J.C., Tirole, J.: Platform competition in two-sided markets. J. Eur. Econ. Assoc. **1**(4), 990–1029 (2003)
8. Kim, S.S., Choi, Y.S.: The innovative platform programme in South Korea: Economic policies in innovation-driven growth. Foresight STI Gov. **13**(3), 13–22 (2019)
9. Digital economy of the Russian Federation: Ministry of Digital Development, Communications and Mass Media of the Russian Federation, https://digital.gov.ru/ru/activity/directions/858/#section-materials. Accessed 28 April 2021
10. STEPIK: https://stepik.org/lesson/313020/step/1?unit=295541. Accessed 28 April 2021
11. Yashina, A.V.: Platform solutions and public spaces as factories of distributed knowledge production. Philos. Thought **4**, 1–13 (2020)
12. Ilin, I., Maydanova, S., Lepekhin, A., Jahn, C., Weigell, J., Korablev, V.: Digital platforms for the logistics sector of the russian federation. Lectur. Notes Netw. Syst. **157**, 179–188 (2021)
13. Ismail, S., Malone, M., Geest, Y.: Explosive Growth: Why Exponential Organizations are Dozens of Times more Productive than yours (and what to do about it). Alpina Publisher, Moscow (2017)
14. Srnicek, N.: Platform Capitalism. Polity Press, UK, Cambridge (2017)
15. Kovalenko, A.I.: Problems of research on multilateral platforms. Modern Compet. **3**(57), 64–90 (2016)
16. Yablonskiy, S.A.: Multilateral platforms and markets: basic approaches, concepts and practices. Russian Manag. J. **11**(4), 57–78 (2013)
17. Konopatov, S.N., Salienko, N.V.: Analysis of business models based on platforms. Econ. Environ. Manag. **1**, 21–32 (2018)
18. Van Alstyne, M.W., Parker, G.G., Choudary, S.P.: Pipelines, platforms, and the new rules of strategy. Harvard Bus. Rev. **62**(4), 54–60 (2016)
19. Longmei, D.: China's digital economy: opportunities and risks. Bullet. Int. Organ. Educ. Sci. New Econ. **4**(12), 275–303 (2019)
20. Raunio, M., Nordling, N., Kautonen, M., Resenen, P.: Open innovation platforms as a tool of the "knowledge triangle": the experience of Finland. Foresight **12**(2), 62–76 (2018)
21. Ilin, I., Levina, A., Lepekhin, A., Kalyazina, S.: Business requirements to the IT architecture: a case of a healthcare organization. international scientific conference energy management of municipal facilities and sustainable energy technologies EMMFT 2018. Adv. Intell. Syst. Comput. 983, pp. 287–294 (2019)
22. Iliashenko, O., Bikkulova, Z., Dubgorn, A.: Opportunities and challenges of artificial intelligence in healthcare E3S Web of Conferences, vol. 110, 02028 (2019)
23. Iliashenko, O.Y., Iliashenko, V.M., Dubgorn, A.: IT-architecture development approach in implementing BI-systems in medicine. Lectur. Notes Netw. Syst. **95**, 692–700 (2020)

24. Rock Health Consulting.: https://rockhealth.com/reports/digital-healths-platform-wars-are-heating-up. Accessed 28 April 2021
25. Habr.: https://habr.com/ru/company/ruvds/blog/486804/. Accessed 28 April 2021
26. Shumakova, O.V., Kryukova, O.: N: New transaction costs and global digitalization. Human Sci. Humanit. Res. **14**(3), 189–197 (2020)
27. Koblova, Yu.A.: Dynamics of transactional parameters of the information-network economy. Soc. Power **6**(44), 96–102 (2013)
28. Bauer, V.P., Eremin, V.V., Smirnov, V.V.: Digital platforms as a tool for transforming the world and Russian economies in 2021–2023. Econ. Taxes Law **14**(1), 41–51 (2021)
29. Golovina, T.A., Polyanin, A.V., Avdeeva, I.L.: Development of digital platforms as a factor in the competitiveness of modern economic systems. Perm Univ. Bull. **14**(4), 551–564 (2019)
30. Philips.: https://www.philips.com/a-w/about/news/future-health-index/reports.html. Accessed 28 April 2021
31. Nonews.: https://nonews.co/directory/lists/countries/health. Accessed 28 April 2021
32. Smart-Lab.: https://smart-lab.ru/company/tinkoff_invest/blog/642307.php. Accessed 28 April 2021
33. Forbes.: https://www.forbes.com/sites/brucejapsen/2019/09/10/mayo-clinic-google-partner-on-digital-health-analytics. Accessed 28 April 2021
34. Shlyakhto, E.V., Konradi, A.O.: Medicine based on value-a new paradigm in health care. Rem. Volga Reg. **3**(163), 4–8 (2018)

Algorithm for Modeling Technological Progress in the Digital Economy Era

Askar Akaev, Andrei Rudskoy, László Ungvári, and Aleksander Petryakov

Abstract Algorithms and models are an effective tool for solving a wide class of problems in various fields of knowledge. In this article, the algorithm, as a research tool, is used to describe the technological development dynamics in the first half of the twenty-first century, when digital technologies increasingly influenced the economy. Technological progress itself is viewed through the cumulative performance of factors. The mathematical basis of the algorithm is made up of models of various modes of growth of production information: constant, with an aggravation, with stabilization, an aggravation with a return to a stationary mode. The proposed algorithm includes two steps—determination of the appropriate mode of production information growth and further calculation of the parameters of the corresponding mode. With the help of the algorithm, the retrospective dynamics of technical progress can be estimated, as well as predictive estimates based on the given mode can be obtained. As an example, the evaluation of the algorithm results execution using US statistical data is demonstrated.

Keywords Technological progress · Digital economy · Information production mode

A. Akaev
Institute for Mathematical Research of Complex Systems, Lomonosov Moscow State University, Moscow, Russia

A. Rudskoy
Peter the Great Saint Petersburg Polytechnic University, St. Petersburg, Russia

L. Ungvári
Development in Relations of Industrial and Education Management GmbH, Wildau, Germany
e-mail: ungvari@driem-international.com

A. Petryakov (✉)
St. Petersburg State University of Economics, St. Petersburg, Russia

C. Jahn et al. (eds.), *Algorithms and Solutions Based on Computer Technology*,
Lecture Notes in Networks and Systems 387,
https://doi.org/10.1007/978-3-030-93872-7_28

1 Introduction

With the computers advent, economic research began to widely use mathematical models and various algorithms, which significantly increased the reliability, accuracy and validity of certain options for decisions in the field of economics and management. Thus, one of the areas of research was associated with the widespread use of models for technology management and modeling of technological changes themselves [1–3]. Another area of research concerned the use of a logical structure of an engineering database based on direct technical information to work with input–output (IO) models on the scale of the national economy [4]. Currently, algorithms are becoming widespread in connection with the development and deployment of machine learning systems [5]. The behavioral theory of the firm formed the basis for a wide class of algorithms for managing innovations in artificial intelligence technologies and artificial intelligence systems based on machine learning [6, 7]. A growing number of researchers at the government level are now recognizing the importance of data and algorithms for solving problems such as improving the skills of human resources and using artificial intelligence to achieve the Sustainable Development Goals [3, 8]. Algorithms, as a research tool, are increasingly used in the study of the technological and social changes interaction [6, 9, 10]. Widespread use for obtaining more accurate information about the processes of diffusion of innovations over time based on news articles related to technology, has received such a special tool as a vector thematic paragraph model (PVTM) [11]. In connection with the widespread use of robots and artificial intelligence, there are proposals for the use of special algorithms to avoid the negative consequences of practices such as People Analytics, the use of big data and artificial intelligence for personnel management [12]. In such a special segment of research as the evolution of economic change, algorithms are increasingly used to assess the differences between learning and adaptation in static and dynamic environments [13]. Futurists predict that by 2025, a third of the jobs that exist today may be occupied by intelligent technology, artificial intelligence, robotics and algorithms (STARA), which has prompted the specific tools development to assess how employees perceive these technological advances in relation to their work and careers, and how they prepare for these potential changes [14]. In our article, to study the dynamics of technological development in the era of the digital economy, we propose an algorithm for calculating the rate of technological progress under various modes of growth of production information.

2 Materials and Methods

The algorithm for calculating the rate of technological progress for the era of the digital economy is based on the following formula:

$$\begin{array}{ll}(a) & q_{Ad}(t) = \frac{\dot{A}_d(t)}{A_d(t)} = \xi\sqrt{\psi_d(t)\cdot\dot{g}(t)} \\ (b) & \psi_d(t) = \frac{I_d(t)}{K_d(t)};\ S_d(t) = S_{do}\cdot\exp[g(t)]\end{array} \quad (1)$$

where $A_d(t)$—technological progress (TFP) in the digital economy era; ξ—calibration factor; $I_d(t)$—current investment in fixed assets $K_d(t)$ of information and digital sectors of the economy; $S_d(t)$—the volume of industrial technological knowledge (information) in the digital economy, which is growing exponentially [15]. Hence, $\dot{g}(t)$ represents the growth rate of technological production information.

Solving differential Eq. (1a) with respect to $A_d(t)$, we obtain:

$$A_d(t) = A_{do}\cdot\exp\left\{\xi\cdot\int_t^{To}\sqrt{\psi_d(\tau)\dot{g}(\tau)d\tau}\right\} \quad (2)$$

Real modes of information production have been considered in many works, for example, in [16]. The following 4 modes of production of technological information are of greatest interest to us.

1. **Constant duty**: information is produced at a constant rate of growth $\dot{g} = \nu_0 = const$. The Lagrange function has the form: $L(\dot{g}, g, t) = \dot{g}^2$. The corresponding Lagrange equation $\frac{d}{dt}\cdot\frac{\partial L}{\partial\dot{g}} - \frac{\partial L}{\partial g} = \ddot{g} = 0$ has a solution:

$$g(t) = g_0 + \nu_0 t, \quad \dot{g}_0 = \nu_0 \quad (3)$$

   In this case: $S_d(t) = S_{d0}\cdot exp(\nu_0 t)$. The process of accumulation of industrial technological knowledge occurs according to the simplest exponential law.
2. **Peak mode**: the growth rate of information production grows exponentially with its accumulation [17], т.е. $\dot{g} \sim e^g$. The Lagrangian leads to this regime $L(\dot{g}, g, t) = \dot{g}^2\cdot e^{-2g}$. The corresponding Lagrange equation has the form: $\ddot{g} = \dot{g}^2$. The solution to this equation under the initial conditions $g(t=0) = g_0$ and $\dot{g} = (t=0) = \nu_0$ leads to a hyperbolic increase in the growth rate of information production:

$$(a)\ \ \dot{g}(t) = \frac{1}{T_S - t}; \quad (b)\ \ g(t) = g_0 - ln\left(1 - \frac{t}{T_S}\right); \quad T_S = \frac{1}{\nu_0}, \quad (4)$$

   where T_s—singularity point.

   Equation (4a) resembles the hyperbolic equation of demographic dynamics, first obtained in [18], with a singularity point at $T_s = 2026$. In reality, the stabilization regime replaced explosive demographic growth—a demographic transition [19].
3. **Stabilized mode**: this regime is a combination of a peaking regime $\dot{g} \sim e^g$, which is realized at the initial stage of development, and a constant regime $\dot{g} = const$—at the final stage: $\dot{g} \sim \frac{e^g}{1+e^g}$. Lagrangian $L(\dot{g}, g, t) = \frac{\dot{g}^2\cdot e^{-2g}}{1-\dot{g}}$, and the Lagrange equation: $\ddot{g} = \dot{g}^2(1-\dot{g})$.

The solution needs to be scaled in order for it to become adequate for the problem under consideration. This is done by introducing new variables $g = \frac{g^1}{s_g}; t = \frac{t^1}{s_t}$, where s_g and s_t—constant factors. Moreover, it is most often accepted $s_t = 1$, what we use and we are within the framework of the algorithm.

The scaled solution to the last Lagrange equation has the form:

$$a)\ \dot{g}(t) = \frac{1}{s_g} \cdot [1 + c_1 \cdot e^{-s_g \cdot g(t)}]^{-1}; \quad c_1 = e^{s_g \cdot g_0} \cdot (\frac{1}{v_0} - 1);$$

$$b)\ t = s_g \cdot g(t) - c_1 \cdot e^{-s_g \cdot g(t)} + c_2; \quad c_2 = \frac{1}{v_0} - 1 - s_g \cdot g_0. \tag{5}$$

As can be seen from Eq. (5a), the rate of production of technological information monotonically increases according to the logistic law with a variable rate, since, as follows from Eq. (5b), $g(t)$ it is not a strictly linear function of the argument (t).

Substituting the complete analytic expression $\dot{g}(t)$ from (5) into (1a), we obtain an equation for determining the $g(t)$—an indicator of the exponential growth of technological information (1b):

$$g(t) = g_0 + \frac{1}{s_g} \cdot (t - \frac{1}{v_0}) + \frac{\xi^2}{s_g^2} \cdot \frac{\mathcal{E}_d(t)}{q_{Ad}^2(t)} \tag{6}$$

4. **Exacerbation mode with return to stationary mode**: in this scenario of development, at the initial stage, the process will go with an exacerbation ($\dot{g} \sim e^g$) and due to inertia, the stationary level jumps ($\dot{g} = const$), and then, having reached a certain maximum value $\dot{g}_m$ it returns to the stationary mode. This mode can be described by the ratio $\dot{g} \sim \frac{e^g}{1+c(g)\cdot e^g}$, где$\cdot c(g)$—the braking function, which in the simplest case has the form $c(g) = 1 - \frac{1}{1-\alpha} \cdot e^{-\alpha g}$, where $\alpha = const$ and $\alpha \neq 1$, and in the limit when $\alpha \to 0$ a regime with an aggravation is obtained, and at $\alpha \to \infty$ – stabilization mode. This mode is produced by the Lagrangian $\mathrm{L}(\dot{g}, g, t) = \frac{\dot{g}^2 \cdot e^{-2g}}{1-c(g)\cdot g'}$, and the corresponding Lagrange equation has the form: $\ddot{g} = \dot{g}^2 \cdot \left\{1 - \dot{g}\left[c(g) + \frac{dc}{dg}\right]\right\}$.
The scaled solution to this equation has the form:

$$a)\ \dot{g}(t) = \frac{1}{s_g} \cdot \left(1 - \frac{e^{-\rho \cdot s_g \cdot g(t)}}{1-\rho} + C_1 \cdot e^{-s_g \cdot g(t)}\right)^{-1};$$

$$b)\ C_1 = e^{s_g \cdot g_1} \cdot \left(\frac{1}{\nu_1} - 1 + \frac{e^{-s_g \cdot g_1}}{1-\rho}\right),;$$

$$c)\ t = s_g \cdot g(t) + \frac{e^{-\rho \cdot s_g \cdot g(t)}}{\rho(1-\rho)} - C_1 \cdot e^{-s_g \cdot g(t)} + C_2; \quad (7)$$

$$d)\ C_2 = \frac{1}{\nu_1} - 1 - s_g \cdot g_1 - \frac{1}{\rho} \cdot e^{-\rho \cdot s_g \cdot g_1}$$

Thus, the first step of the algorithm is the selection of the most suitable for describing the mode of information production. The second stage of the algorithm involves determining the parameters of the selected mode.

3 Results

Below we present the application of the algorithm for the verification of mathematical models for calculating the dynamics of development in the era of the information-digital economy, based on the laws of production of technological information with an estimate of the root-mean-square error of approximation, using actual data of technological progress.

The main verified formula has the form (1a). We will accept $\xi = 1$, since the normalization can be done through the function $\dot{g}(t)$. Verification of Formula (1a) is carried out at the upward and downward stages of the business cycle of the US economy, which lasted from 1982 to 2018.

Actual data on the rate of technological progress $\overline{q}_{Ad}(t)$ we obtained by aggregating the geometric mean series according to the initial data from sources [20, 21]. They are presented graphically in Fig. 1.

In [15] it is shown that the function $\psi_d(t)$ in Eq. (1b) can be approximated by a linear function and extrapolated for forecasting purposes (Fig. 2):

Here:

$$\psi_d(t) = \psi_0 + \psi_1 \cdot (t - T_0); \quad T_0 = 1982;\ \psi_0 = 0.09;\ \psi_1 = 0.002 \quad (8)$$

Also [15] the values of all parameters included in Formulas (6) and (7) were estimated: $\xi = 0.07$; $g_1 = 5.3$;; $s_g = 14$; $\rho = 0.008$; $\nu_1 = \frac{1}{14}$.

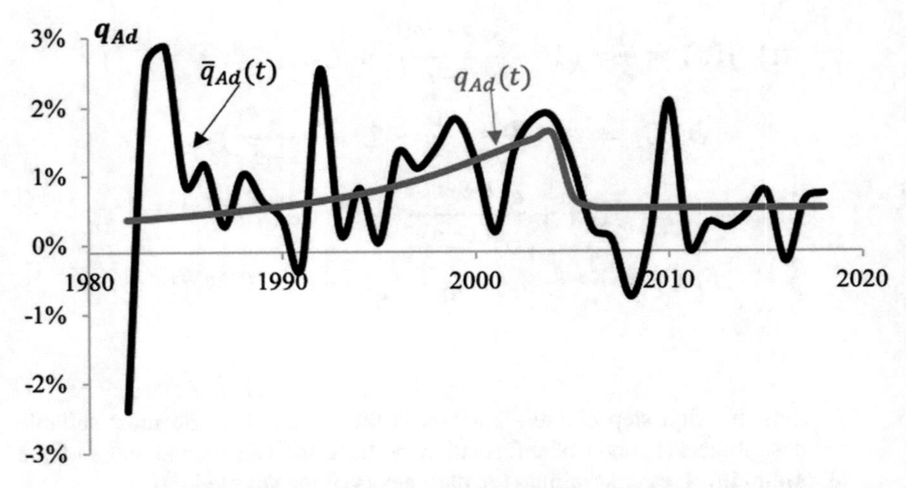

Fig. 1 The rate of technological progress of the information model against the background of the actual curve $\overline{q}_{Ad}(t)$

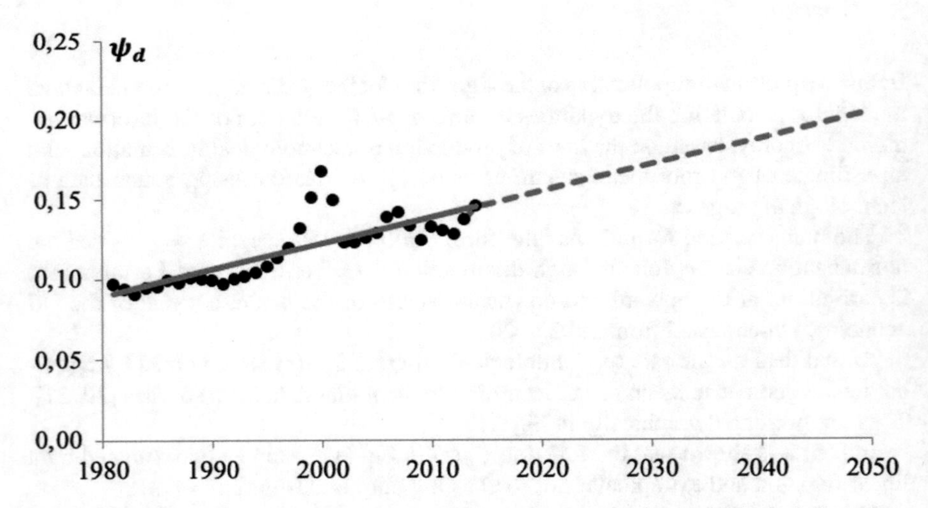

Fig. 2 Dynamics of the ratio of investments to the volume of productive capital in the information-digital sector of the economy

Let's introduce the scaled values:

$$\bar{t} = s_t \cdot t; \quad \bar{g}(t) = s_g \cdot g(t); \quad \bar{g}(t) = s_g \cdot g(t) \tag{9}$$

First of all, it is necessary to determine the initial values g_0 and ν_0, characterizing the production of technological information. To determine ν_0 we will use the hyperbolic equation for the growth of information production rates at the initial period of information technology development (4b), according to which the singularity point is $T_S = {\nu_0}^{-1}$. It is natural to take 1946 as the beginning of the information age emergence, when the world's first universal computer was put into operation. And the explosive growth of the influence of information and communication technologies (ICT) on the economic development of the avant-garde countries was first observed in 1995–1997, as can be seen from Fig. 2. Hence, $T_S \cong 50$ πeT years and $\upsilon_0' = 0.02$—at the initial stage. g_0 determined from solution (4b), setting there $g_0' = 0$ for 1946. Then, for the initial year of the upward wave (1982), from (4b) we obtain: $\overline{g}_0 = ln\left(1 - \frac{36}{50}\right) \cong 1.273$.

Further, proceeding from the fact that the logical function (5a), which describes the rate of production of technological information, takes at the beginning (1982) and the upper turning point (2004, see Fig. 1) minimum and maximum values equal to 0, respectively, 1 and 0.9, as is generally accepted, we obtain two equations:

$$\begin{aligned} &(a)\quad T_r = 2004;\ \overline{\nu}_r = \dot{\overline{g}}_r = 0.9 = \left\{1 + \left(\tfrac{1}{\nu_0} - 1\right)e^{-(\overline{g}_r - \overline{g}_0)}\right\}^{-1} \\ &(b)\quad T_0 = 1982;\ \overline{\nu}_0 = \dot{\overline{g}}_0 = 0.1 = \left\{1 + \left(\tfrac{1}{\nu_0} - 1\right)e^{-(\overline{g}_0 - \overline{g}_0)}\right\}^{-1} \end{aligned} \tag{10}$$

From the first equation we obtain: $T_r = 2004;\ \ \overline{g}_r = 5.67$. It immediately follows from the second equation that for $T_0 = 1982;\ \ \overline{\nu}_0 = 0.1$.

Knowing $\overline{\nu}_0 = 0.1$ and $\overline{g}_r = 5.67$, we can calculate the numerical values of the integration constants C_1 (5a) and C_2 (5b): $C_1 = 32.14;\ C_2 = 7.73$.

Next, we proceed to determining the numerical values of the scaling factors s_t and s_g(9). Substituting into Eq. (5b) expressions for C_1 and C_2, we obtain the following equation at the upper turning point of the upward wave ($T_r = 2004$):

$$s_t = \frac{1}{T_r - T_0}\left[\bar{g}_r - \bar{g}_0 + \left(\frac{1}{\nu_0} - 1\right)e^{-(\bar{g}_r - \bar{g}_0)} + \frac{1}{\nu_0} - 1\right] \tag{11}$$

Substituting here the numerical values T_r, T_0, $\overline{g}_0$ and $\overline{g}_r$, we obtain $s_t = 0.6$.

Substituting an expression for the rate of production of technological information $\dot{g}(t)$ (5a) and (5b) in Eq. (1a), we obtain:

$$q_{Ad}^{(1)}(t) = \sqrt{\frac{\varepsilon_d(t)}{\mathscr{s}_g[1+\bar{g}(t)-\mathscr{s}_t \cdot t+C_2]}} \quad (12)$$

where $t = T - T_0$; $T_0 = 1982$; $T_0 \leq T \leq T_r = 2004$. Moreover, here is the function itself $\overline{g}(t)$ is found in numerical form by solving nonlinear Eq. (5b) at each point t from a given time range (12):

$$\mathscr{s}_t \cdot t = \bar{g}(t) - C_1 \cdot e^{-\bar{g}(t)} + C_2,$$
$$\text{where } \bar{g}_0 \leq \bar{g}(t) \leq \bar{g}_r. \quad (13)$$

where $\overline{g}_0 \leq \overline{g}(t) \leq \overline{g}_r$.

Since the left side of Eq. (12) is given by its actual values, which have already been presented in Fig. 1, and on the right side of the unknown and uncertain, only the magnitude of the scaling factor remains $\mathscr{s}_g$, then it is estimated using the least squares method and is equal to $\mathscr{s}_g = 412.33$.

Now we will be able to build a trend trajectory of the rate of technological progress by calculating $q_{Ad}(t)$ on the right side of Formula (12) with the obtained specific value $\mathscr{s}_g$. The trend trajectory of the rate of technological progress $q_{Ad}(t)$ at the upward stage (1982–2004) is shown in Fig. 1.

At the downward stage (2004–2018), to approximate the trajectory of technological progress, we first use the formula for the slowdown of technological progress in the downward phase, obtained in the work [22]:

$$q_{Ad}^{(2)}(t) = q_{Adr}^{(1)} \cdot exp\left\{-\left[1 - \lambda_0\left(t - T_r + \frac{1}{\lambda_0}\right)e^{-\lambda_0(t-T_r)}\right]\right\} \quad (14)$$

where $q_{Adr}^{(1)}$ – value $q_{Ad}^{(1)}(t)$ (12) at the top turning point ($T_r = 2004$). Using actual data $\overline{q}_{Ad}(t)$, presented in Fig. 1 in the phase of recession and depression (2004–2014), with the help of least square method we get: $\lambda_0 = 2.73$.

The corresponding part of the trend trajectory of the movement of technological progress in the phases of recession and depression is also plotted in Fig. 1 (2004–2014). So, it remains to find an approximate analytical description of technological progress in the recovery phase (2014–2018), which is best approximated by an exponential curve:

$$q_{Ad}^{(3)}(t) = q_{Ade}^{(2)} \cdot exp[\omega(t - T_{re})] \quad (15)$$

where $q_{Ade}^{(2)}$ – value $q_{Ad}^{(2)}(t)$ (14) at the bottom turning point ($T_{re} = 2014$). Based on evidence $\overline{q}_{Ad}(t)$ in the recovery phase (2014–2018, see Fig. 1), with the help of least square method we appreciated $\omega = 0$.

Thus, we approximated the rate of technological progress $q_{Ad}(t)$ both in the upward (12) and downward (14) and (15) stages of the business cycle. Next, we can calculate the trajectories of both actual and model technological progress $\overline{A}_d(t)$ and $A_d(t)$ according to the equation:

$$A_d(t) = A_{d0} \cdot exp\left[\int_{T_0}^{T} q_{Ad}(t)dt\right] \quad (16)$$

The trajectory of movement of the actual technological progress is obtained by numerical integration (16), using the actual data $\overline{q}_{Ad}(t)$, presented in Fig. 1. And the trajectory of the model technological progress is calculated by the Formula (16) with the sequential use of the approximating functions $q_{Ad}(t)$ (12), (14) and (15). The calculation results are presented graphically in Fig. 3.

The root mean square error turned out to be $\sigma_A = 1.95\%$. As you can see, the proposed algorithm, which is based on the information model of technological progress (1), provides a sufficiently high accuracy of the approximation.

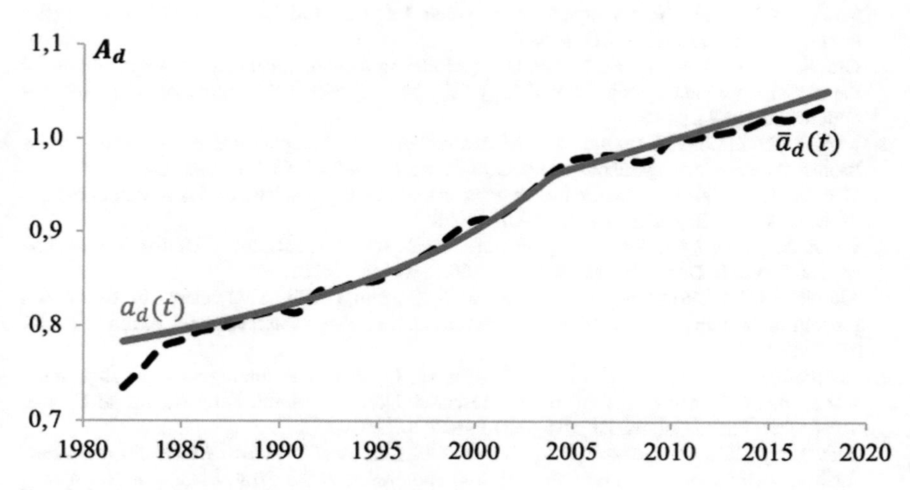

Fig. 3 Technological progress on the information model

4 Conclusions

Modern economic systems are strongly influenced by new breakthrough technologies when the role of a person in the production process changes dramatically. This significantly affects the nature of economic relations, gradually ousting a person from the process of direct production and increasingly reducing his role in the management system. In this regard, in recent years, the availability of algorithms for assessing and predicting the growth rate of technological progress has become important, since they predetermine future trends in human participation in production processes. The algorithm proposed in the article combines fundamental concepts of technological development, expressed in the form of mathematical models, and an applied function that provides the possibility of describing real processes using an algorithm through its application to statistical data.

Acknowledgements This article was prepared as part of the RFBR grant No. 20-010-00279 "An integrated system for assessing and forecasting the labor market at the stage of transition to a digital economy in developed and developing countries."

References

1. Benkenstein, M., Bloch, B.: Models of technological evolution: their impact on technology management, marketing intelligence & planning. MCB UP Ltd. **11**(1), 20–27 (1993). https://doi.org/10.1108/02634509310024146
2. Grüble, A., Nakićenović, N., Victor, D.G.: Modeling technological change: implications for the global environment. Ann. Rev. Energy Env. **24**, 545–569 (1999). https://doi.org/10.1146/annurev.energy.24.1.545
3. Mogilko, D.Y., Ilin, I.V., Iliashenko, V.M., Svetunkov, S.G.: BI capabilities in a digital enterprise business process management system. Lect. Notes Netw. Syst. **95**, 701–708 (2020)
4. Duchin, F.: Analyzing technological change: an engineering data base for input–output models of the economy. Eng. Comput. **4**, 99–105 (1998)
5. Lavin, A., Renard, G.: Technology readiness levels for AI & ML. In: ICML 2020 Workshop on Challenges in Deploying Machine Learning Systems (2020)
6. Moreira, M.W.L., Rodrigues, J.J.P.C., Kumar, N., Saleem, K., Illin, I.V.: Postpartum depression prediction through pregnancy data analysis for emotion-aware smart systems. Inform. Fus. **47**, 23–31 (2019)
7. Haefner, N., Wincent, J., Parida, V., Gassmann, O.: Artificial intelligence and innovation management: a review, framework, and research agenda. Technol. Forecast. Social Change **162** (2021). https://doi.org/10.1016/j.techfore.2020.120392
8. Alesinskaya, T.V., Arutyunova, D.V., Orlova, V.G., Ilin, I.V., Shirokova, S.V.: Conception BSC for investment support of port and industrial complexes. Acad. Strat. Manag. J. **16** (Special issue 1), 10–20 (2017)
9. Economic Commission for Latin America and the Caribbean (ECLAC): data, algorithms and policies: redefining the digital world (LC/CMSI.6/4), Santiago (2019)
10. Just, N., Latzer, M.: Governance by algorithms: reality construction by algorithmic selection on the Internet. Media Cult. Soc. **39**(2), 238–258 (2017). https://doi.org/10.1177/0163443716643157

11. Lenz, D., Winker, P.: Measuring the diffusion of innovations with paragraph vector topic models. PLoS One **15**(1), e0226685 (2020). https://doi.org/10.1371/journal.pone.0226685
12. De Stefano, V.: "Negotiating the algorithm": automation, artificial intelligence and labour protection. EMPLOYMENT Working Paper No. 246, ILO (2018)
13. Windrum, P.: Simulation models of technological innovation: a review. MERIT, Maastricht Economic Research Institute on Innovation and Technology. MERIT Research Memoranda No. 005 (1999). https://doi.org/10.1177/00027649921957874
14. Brougham, D., Haar, J.: Smart technology, artificial intelligence, robotics, and algorithms (STARA): employees' perceptions of our future workplace. J. Manag. Organ. **24**(2), 239–257 (2018). https://doi.org/10.1017/jmo.2016.55
15. Akaev, A.A., Sadovnichy, V.A.: Mathematical models for calculating the development dynamics in the era of digital economy. Dokl. Math. **98**(2), 526–531 (2018). https://doi.org/10.1134/S106456241806011X
16. Dolgonosov, B.M.: Nonlinear Dynamics of Ecological and Hydrological Processes [Nelinejnaja dinamika jekologicheskih i gidrologicheskih processov]. Librokom, Moscow (2009)
17. Kurzweil, R.: The Singularity is Near. Viking Books, New York (2005)
18. von Foerster, H., Mora, P., Amiot, L.: Friday, 13 November, A.D. 2026. At this date human population will approach infinity if it grows as it has grown in the last two millennia. Science (132), 1291–1295 (1960). https://doi.org/10.1126/science.132.3436.1291
19. Kapitza, S.P.: An outline of the theory of human growth [Ocherk istorii rosta chelovechestva]. Nikiytskiy club, Moscow (2008)
20. U.S. Bureau of Labor Statistics: Table Historical Multifactor Productivity Measures [Data file]. https://www.bls.gov/mfp/mprdload.htm#Multifactor%20Productivity%20Tables. Accessed 19 Feb 2021
21. University of Groningen and University of California, Davis: Total Factor Productivity at Constant National Prices for United States [Data file]. Retrieved from FRED, Federal Reserve Bank of St. Louis. https://fred.stlouisfed.org/series/RTFPNAUSA632NRUG. Accessed 22 Feb 2021
22. Akaev, A.A., Sadovnichy, V.A.: Closed-loop dynamic model for describing and calculating Kondratyev long wave of economic development. Herald Russ. Acad. Sci. **86**(10), 883–896 (2016)

Models and Method of Decision Support for the Formation of Cybersecurity Systems

Aleksandr Suprun, Vladimir Anisimov, Evgeny Anisimov, Tatyana Saurenko, and Inna Manciu

Abstract A methodological approach to the construction of models of typical decision support tasks for the formation of the composition of cybersecurity systems of companies in the context of digital transformation of production has been developed. The requirement to minimize possible damage from cybersecurity incidents is used as a criterion for the optimality of solutions. A general method for optimizing solutions is proposed, taking into account the nonlinear nature of the target functions of the models. The models and method are universal in nature and can be used in the development of specific methods for substantiating the composition of companies' cybersecurity systems, taking into account the characteristics of possible threats.

Keywords Digital transformation of production · Industry 4.0 · Threats to digital production · Company cybersecurity system · Model · Method of optimization of solutions

1 Introduction

A concomitant effect of the digital transformation of modern industrial production associated with the intensive introduction of advanced information technologies of Industry 4.0 in the management of companies' activities is the aggravation of the problems of ensuring their cybersecurity. A concomitant effect of the digital transformation of modern industrial production associated with the intensive introduction

A. Suprun (✉) · V. Anisimov
Peter the Great St. Petersburg Polytechnic University (SPbPU), Polytechnicheskaya, 29, St. Petersburg 195251, Russia
e-mail: afs54@inbox.ru

E. Anisimov · T. Saurenko
Peoples Friendship University of Russia (RUDN University), 6 Miklukho-Maklaya St, Moscow 117198, Russia
e-mail: saurenko_tn@rudn.university

I. Manciu
Global Supply Chain Lead Company: NTT, Cypress, CA, USA

C. Jahn et al. (eds.), *Algorithms and Solutions Based on Computer Technology*, Lecture Notes in Networks and Systems 387,
https://doi.org/10.1007/978-3-030-93872-7_29

of advanced information technologies of Industry 4.0 in the management of companies' activities is the aggravation of the problems of ensuring their cybersecurity [1–5]. This is due to the fact that the inclusion of management of technological and business processes in the global cyberspace increases the possibilities of destructive informational impact on the results of these processes and the activities of the company as a whole [6–12]. In this regard, the development and implementation of cybersecurity systems capable of preventing or minimizing damage from these destructive influences is an urgent problem of the company's development, ensuring its sustainability and competitiveness. One of the important tasks in the development of such systems is to optimize their composition, taking into account the peculiarities of possible cybersecurity incidents of a particular company. The complexity of this task and the significant material and reputational costs associated with possible errors in determining the composition of the cybersecurity system necessitate the development of models for justifying appropriate decisions.

The purpose of this article is to develop models of typical tasks and a generalized method for optimizing solutions to form the composition of a company's cybersecurity system.

2 Materials and Methods

The proposed standard models are designed to form the composition of the company's cybersecurity system in a situation where a lot of possible threats are clearly formulated and their danger is assessed. To counter possible threats, there is a certain set of private projects (elements), each of which can with some probability ensure the prevention (reduction) of damage for a subset of incidents associated with the implementation of these threats. The optimal version of the cybersecurity system is a complex of private projects (elements) that allows you to minimize the damage to the company from possible cyber attacks. When constructing mathematical models of typical tasks for the formation of the composition of cybersecurity systems, the apparatus of the theory of discrete mathematical programming is used [13–17]. The proposed optimization method takes into account the nonlinear nature of the objective functions used in the models.

3 Results

As typical models for solving the problems of decision-making support for the formation of the composition of cybersecurity systems, consider:

- a model for the formation of a system of protection against independent destructive influences;

- a model of the formation of the system, taking into account the interconnectedness of threats;
- a model for forming the composition of the system with a simultaneous choice of methods for implementing its elements.

3.1 Model of the Formation of a System of Protection Against Independent Destructive Influences

Let there be N types of threats. The danger of each threat is characterized by possible damage $R_j \quad j = \overline{1, N}$ to the information management system. To prevent the implementation of threats, M types of protection tools can be used. The effectiveness of the use of the ith means for reducing damage from the implementation of the jth type threat is characterized by the value w_{ij} (for example, the probability of preventing the threat or the mathematical expectation of the relative reduction of damage from the corresponding cybersecurity incident).

When a certain set I_j of means of protection against the jth threat is included in the cybersecurity system $\delta = \left\|\delta_{ij}\right\|, \; i = \overline{1, M}; \; j = \overline{1, N}$, the damage reduction is determined by the relation:

$$f_j(\delta) = R_j(1 - \prod_{i \in I_j}(1 - w_{ij})), \; j = \overline{1, N} \tag{1}$$

It is required to determine such a variant of the composition of the cybersecurity system in which the maximum reduction in damage from possible incidents is achieved.

The corresponding model for the formation of the optimal composition of the cybersecurity system takes the following form.

Determine the composition of the cybersecurity system

$$\delta^* = \left\|\delta^*_{ij}\right\|, i = \overline{1, M}, j = \overline{1, N} \tag{2}$$

such that

$$F(\delta*) = \max_{\delta} \sum_{j=1}^{N} R_j \left(1 - \coprod_{i \in Ij}(1 - w_{ij})\right), \tag{3}$$

at

$$\sum_{j=1}^{N}\delta_{ij}=1,\ i=\overline{1,M},\qquad (4)$$
$$\delta_{ij}\in\{0,1\},\ 0\le w_{ij}\le 1,\ R_j\ge 0,\ i=\overline{1,M};\ j=\overline{1,N}.$$

3.2 A Cybersecurity Formation Model Taking into Account the Interconnectedness of Threats

An important generalization of model (2)–(4) for applications is taking into account the interconnectedness of threats. One of the methods of such accounting is based on the assumption that if, as a result of the application of protection measures, the threat of the jth type is neutralized, then with the probability α_{jr}, $j=\overline{1,N}$, $r=\overline{1,N}$, the threat of the $r-$h type will also be neutralized.

The generalized model of the formation of the composition of the cybersecurity system is as follows.

Determine the version of the composition of the system

$$\delta^{*}=\left\|\delta^{*}_{ij}\right\|,\ i=\overline{1,M};\ j=\overline{1,N},\qquad (5)$$

such that

$$F(\delta^{*})=\max_{\delta}\sum_{r=1}^{N}R_r\left(1-\prod_{j=1}^{N}\left(1-\alpha_{jr}\left(\prod_{i=1}^{M}\varepsilon_{ij}^{\delta_{ij}}\right)\right)\right)\qquad (6)$$

at

$$\sum_{j=1}^{N}\delta_{ij}=1,\ i=\overline{1,M},\ R_r\ge 0,\qquad (7)$$

$$\delta_{ij}\in\{0,1\},\quad 0\le\varepsilon_{ij}\le 1,\quad 0\le\alpha_{jr}\le 1,\quad \alpha_{jj}=1\qquad (8)$$

3.3 A Model for Forming the Composition of a Cybersecurity System with a Simultaneous Choice of Methods for Implementing Its Elements

The considered approach to the formalization of models for forming the composition of a cybersecurity system can be generalized for a situation when, along with the choice of elements included in the system, it is necessary to make decisions on how to implement these elements.

The model for the formation of the optimal version of the cybersecurity system is as follows:

$$F(\delta^*) = \max_{\delta} \sum_{j=1}^{N} f_j(\delta), \tag{9}$$

at

$$\sum_{j=1}^{N} \sum_{k=1}^{K} \delta_{ijk} = 1, \quad i = \overline{1, M}, \tag{10}$$

$$\delta_{ijk} \in \{0, 1\}, \quad i = \overline{1, M}; \quad j = \overline{1, N}, \quad k = \overline{1, k}, \tag{11}$$

where k is the identifier of the method for implementing the elements of the cybersecurity system.

3.4 Method for Solving Problems of Optimization of the Composition of Cybersecurity Systems

Given in paragraphs Sects. 3.1–3.3 models of problems of optimization of a cybersecurity system can be generalized in the form of a model for optimizing a function of Boolean variables of the following type.

Determine a solution

$$\delta^* = \left\| \delta_{ij}^* \right\|, \quad i = \overline{1, M}; \quad j = \overline{1, N}, \tag{12}$$

such that

$$F(\delta*) = \max_{\delta} \sum_{j=1}^{N} f_j(\delta), \tag{13}$$

$$\sum_{j=1}^{N} \delta_{ij} = 1, \quad i = \overline{1, M}, \tag{14}$$

where

$$\delta = \left\| \delta_{ij} \right\|, \quad \delta_{ij} \in \{0, 1\}, \quad i = \overline{1, M}; \quad j = \overline{1, N}$$

$f_j(\delta)$—non-decreasing; functions convex to the top;

N, M—positive integers.

Among the exact methods that allow, in principle, to obtain an exact solution to the optimization problem for the model (12)–(14), the most widespread are various versions of the branch-and-bound method and the dynamic programming method [18–20].

However, the capabilities of these methods are significantly limited by the dimension of the matrix δ, since problem (12)–(14) belongs to the class of *NP*—difficult [21, 22].

At the same time, for the tasks of supporting decision-making on the formation of the composition of the company's cybersecurity system, as a rule, there is no need to obtain absolutely accurate optimal decisions. This allows more economical approximate methods to be used.

The proposed method for solving optimization problems for models of the form (12)–(14) belongs to the family of gradient-difference methods. The essence of the method consists in a quasi-equivalent transition from the optimization problem (12)–(14)—the choice of a vector $\delta^* = \left\| \delta^*_{ij} \right\|$ to M sequential problems of choosing one variable $\delta^*_{ij} = 1$. The choice of this variable at the tth step ($t = \overline{1, M}$) is carried out in accordance with the following rules:

$$\begin{aligned} \delta^t_{i^*j^*} &= 1, \text{ if } \Delta f^t_{i^*j^*} = \max_{G(t)} \Delta f^t_{ij}, \\ \delta^t_{ij} &= 0, \quad \text{if } \Delta f^t_{i^*j^*} \neq \max_{G(t)} \Delta f^t_{ij}, \end{aligned} \tag{15}$$

where $G(t)$ is a subset of variables $\delta_{ij} = i = \overline{1, M}; \quad j = \overline{1, N}$, among which the choice is made;

Δf^t_{ij}—increments of function (12) at the tth step ($t = \overline{1, M}$) at $\delta_{ij} = 1$.

The subset $G(t)$ contains the variables δ_{ij} for which

$$\Delta f^t_{ij} \max_{j=\overline{1,N}} \Delta f^t_{ij}, \quad i \in I'_t; \quad \Delta f^t_{ij'} \max_{i \in I_t} \Delta f^t_{ij}, \quad i \in J'_t \tag{16}$$

where I_t is a subset of indices of i variables δ_{ij}, for which $\delta_{ij}^{t-1} = 0, \quad i = \overline{1, M}; \quad j = \overline{1, N}$ (a subset of unassigned resources before the tth step);

I_t' is a subset of indices of i variables $\delta_{ij}, \quad i \in I_t; \quad j = \overline{1, N}$ for which

$$\max_i a_i^t = c^t, \ i \in I_t \tag{17}$$

J_t'—subset of indices of j variables $\delta_{ij}, \quad i \in I_t; \quad j = \overline{1, M}$, for which

$$\max_j B_j^t = c^t, \ j = \overline{1, N}, \tag{18}$$

$$C^t = \max\left\{\max_{i \in I_t} a_i^t; \ \max_{j=\overline{1,N}} B_j^t\right\}, \tag{19}$$

$$a_i^t = \min_{j=\overline{1,N}}\left(\max_{j=\overline{1,N}} \Delta f_{ij}^t - \Delta f_{ij}^t\right), \quad i \in I_t, \tag{20}$$

$$B_j^t = \min_{i \in I_t}\left(\max_{i \in I_t} \Delta f_{ij}^t - \Delta f_{ij}^t\right), \quad j = \overline{1, N}. \tag{21}$$

Relations (20) and (21) are lower estimates of losses in the effectiveness of ensuring cybersecurity in the case of not including the corresponding elements in the system at the tth step.

Conditions (16)–(19) single out the variables for which, at $\delta_{ij}^t = 0$, the estimates of losses will be maximum, and condition (15), among the options with equal largest estimates of losses, selects the option that provides the maximum increment of function (12). In essence, the solution of problem (15)–(21) at each step t ($t = 1, 2,\ldots,M$) determines the choice of the ascent direction for the problem represented by model (12)–(14). Therefore, problem (15)–(21) can be called the principle of choice of the direction of ascent. The consistent application of this principle makes it possible to obtain an optimal or close solution to problem (12)–(14).

The problem of optimality of the obtained solution is essentially reduced to the problem of equivalence of optimization problems (12)–(14) and (15)–(21) It can be shown that for the same means of protection these problems are equivalent and the obtained solution is optimal. In the general case, the solution to problem (15)–(21) is an approximate solution to the optimization problem for model (12)–(14).

The generalized algorithm of the method for solving the optimization problem (12)–(14) using the sequence of problems (15)–(21) is as follows:

1. Put $t = 1$, and put that the set $I_t = \{1, 2, \ldots, M\}$.
2. Calculate the elements of an increment matrix $\left\|\Delta f_{ij}\right\|, \ i = \overline{1, M}; \ j = \overline{1, N}$.
3. Determine the set $G(t)$ from conditions (16)–(21).
4. From condition (15) determine $\delta_{i^* j^*}^t = 1$ and exclude the index i^* from the set I_t.

5. Check condition $I_t = \emptyset$. If yes, then go to item 8, if not—to item 6.
6. Put $t = t + 1$.
7. Recalculate the element of the j^*th section of the increment matrix $\Delta f^t = \left\| \Delta f_{ij}^t \right\|$ taking into account $\delta_{i^* j^*}^{t-1} = 1$ and go to step 3.
8. Calculate function $F(\delta)$.
9. Finish the solution.

Features of specific algorithms for solving those considered in Sects. 3.1–3.3 typical problems using the proposed optimization method are associated with methods for calculating the elements of the increment matrix

$$\Delta f^t = \left\| \Delta f_{ij}^t \right\|, \quad i = \overline{1, M}; \; j = \overline{1, N}.$$

For model (2)–(4), taking into account (1), we obtain

$$\Delta f_{ij}^t = R_j \prod_{k \in I_j^{t-1}} \varepsilon_{kj} w_{ij}, \quad j = \overline{1, N}; \quad j \in I_t \tag{22}$$

where $\varepsilon_{kj} = 1 - w_{kj}$.

$\varepsilon_{kj} = 1 - w_{kj}$—the set of elements for countering threats of the jth type previously included in the cybersecurity system.

For model (5)–(8), the elements of the increment matrix are calculated by the formula:

$$\Delta f_{ij}^t = \sum_{r=1}^{N} R_r^{t-1} \frac{w_{ij} Q_j^{t-1} \alpha_{jr}}{1 - P_j^{t-1} \alpha_{jr}}, \; j = \overline{1, N}, i = \overline{1, M}, \tag{23}$$

where

$$R_r^t = R_r^{t-1} \frac{1 - P_j^t \alpha_{jr}}{1 - P_j^{t-1} \alpha_{jr}}; \quad P_j^t = 1 - \prod_{i=1}^{M} \varepsilon_{ij}^{\delta_{ij}^t}; \; Q_j^t = 1 - P_j^t, \; r = \overline{1, N}, j = \overline{1, N}$$

For model (12)–(14), the rule for choosing a variable to be included in the solution at the th step of the optimization algorithm takes the form:

$$\delta_{i^* j^* k^*} = \begin{cases} 1, \text{ if } \; \Delta f_{i^* j^* k^*}^t = \max\limits_{G(t)} \Delta f_{ijk}^t, \\ 0, \text{ otherwise.} \end{cases}, \tag{24}$$

The subset $G(t)$ includes variables δ_{ijk} for which

$$\Delta f_{ijk}^t = \max_{\substack{i = \overline{1, N} \\ k = \overline{1, K}}} \Delta f_{ijk}^t, \; i \in I_t', \tag{25}$$

$$\Delta f^t_{i'j'k'} = \max_{\substack{i=\overline{1,N} \\ k=\overline{1,K}}} \Delta f^t_{ijk},\ j \in J'_t, \tag{26}$$

where I_t is a subset of indices i of elements not included in the cybersecurity system;

I'_t is a subset of indices of i variables $\delta_{ijk},\quad i \in I_t,\quad j = \overline{1,N}, k = \overline{1,K}$, for which

$$\max_{i \in I_t} \min_{k=\overline{1,k}} a^t_{ik} = c^t; \quad \max_{j=\overline{1,N}} \min_{k=\overline{1,k}} B^t_{jk} = c, \tag{27}$$

$$c^t = \max\left\{\max_{i \in I_t} \min_{k=\overline{1,k}} a^t_{ik};\ \max_{i=\overline{1,N}} \min_{k=\overline{1,K}} B^t_{jk}\right\} \tag{28}$$

$$a^t_{ik} = \min_{j=\overline{1,N}} \left(\max_{i=\overline{1,N}} \Delta f^t_{ijk} - \Delta f^t_{ijk}\right),\ i \in I_t,\quad j = \overline{1,N};\ k = \overline{1,K}, \tag{29}$$

$$B^t_{jk} = \min_{i \in I_t}\left(\max_{i \in I_t} \Delta f^t_{ijk} - \Delta f^t_{ijk}\right),\ j = \overline{1,N};\ k = \overline{1,K}. \tag{30}$$

Further specification of models and optimization algorithms is based on explicit specification of the form of the dependence function $F(\delta)$, taking into account the characteristics of the protected object and the nature of the threats.

4 Discussion

Statistics show that there is an increase in the number of hacker attacks in the global economy. The compilers of the annual Trend Micro Cyber Risk Index, Ponemon Institute experts have calculated that in 2020, almost a quarter of all companies in the world were attacked by cybercriminals more than 7 times. In 2021, according to survey participants, the situation will only get worse, 83% of respondents expect that such cyberattacks will be successful with "some" or "very high" probability [23].

According to the Center for Strategic and International Studies USA, the global cybercrime loss rate has grown by more than 50% since 2018. The average loss from each incident was more than $ 0.5 million. Global losses from cybercrime exceeded $ 1 trillion, which is 1% of world GDP. In the Russian Federation, the number of hacker attacks on strategic enterprises in 2020 doubled compared to 2019. At the same time, the annual damage from the actions of cybercriminals is more than $ 5 billion [24].

Therefore, the creation of effective cybersecurity systems for companies seems to be a very urgent task. The complexity of solving this problem necessitates the development of an appropriate scientific and methodological apparatus for the formation of these systems. When constructing mathematical models of typical decision support tasks for the formation of the composition of cybersecurity systems, it is advisable to use the mathematical apparatus of the theory of discrete mathematical programming.

5 Conclusions

In general, the article developed a methodological approach to building models for the formation of the optimal composition of cybersecurity systems for companies. As an optimality criterion, the models use the requirement to minimize damage from cybersecurity incidents. To solve problems, a method is proposed to optimize solutions taking into account the nonlinear nature of the objective functions. The proposed approach is based on fairly general characteristics of the problem being solved. In this regard, it can serve as a theoretical basis for constructing specific methods for substantiating the composition of various cybersecurity systems, taking into account the characteristics of the objects of protection and possible threats.

References

1. Suprun, A.F., et al.: The problem of innovative development of information security systems in the transport sector. Autom. Control. Comput. Sci. **52**(8), 1105–1110 (2018). https://doi.org/10.3103/S0146411618080035
2. Sonkin, M.A., et al.: Methodology control function realization within the electronic government concept framework. Int. J. Sci. Technol. Res. **9**(2), 6259–6262 (2020)
3. Mulyukina V.E., Kozisova, E.O.: Cyber risks of small and medium-sized businesses. *Informatsionnyye sistemy i tekhnologii: upravleniye i bezopasnost* [Information systems and technologies: management and security], no. 4, pp. 117–122 (2016) (in Russ.)
4. Yastrebov, O., et al.: Digital transformation and optimization models in the sphere of logistics. In: Proceedings of the SHS Web of Conference, vol. 44, p. 00009 (2018). https://doi.org/10.1051/shsconf/20184400009
5. Prisyazhnyuk, S.P., et al.: Indices of the effectiveness of information protection in an information interaction system for controlling complex distributed organizational objects. Autom. Control. Comput. Sci. **51**(8), 824–828 (2017). https://doi.org/10.3103/S0146411617080053
6. Los, V.P., et al.: Approach to the evaluation of the efficiency of information security in control systems. Autom. Control. Comput. Sci. **54**(8), 864–870 (2020). https://doi.org/10.3103/S0146411620080362
7. Zotova, E.A., et al.: Models of forecasting destructive influence risks for information processes in management systems. *Informatsionno-upravliaiushchie sistemy* [Information and Control Systems] (5), 18–23 (2019). https://doi.org/10.31799/1684-8853-2019-5-18-23
8. Bazhin, D.A., et al.: A risk-oriented approach to the control arrangement of security protection subsystems of information systems. Autom. Control. Comput. Sci. **50**(8), 717–721 (2016). https://doi.org/10.3103/S0146411616080289
9. Zaychenko, I.M., et al.: Models for predicting damage due to accidents at energy objects and in energy systems of enterprises. In: E3S Web of Conferences, vol. 110, p. 02041 (2019). https://doi.org/10.1051/e3sconf/201911002041

10. Grashchenko, N., et al.: Justification of financing of measures to prevent accidents in power systems. In: Murgul V., Pasetti M. (eds.) International Scientific Conference Energy Management of Municipal Facilities and Sustainable Energy Technologies, EMMFT 2018. EMMFT-2018. Advances in Intelligent Systems and Computing, vol. 983. Springer, Cham. (2019). https://doi.org/10.1007/978-3-030-19868-8_30
11. Saurenko, T., et al.: Efficiency of ensuring the survivability of logistics information and control systems. In: E3S Web of Conferences, vol. 217, p. 07025 (2020). https://doi.org/10.1051/e3sconf/202021707025
12. Los, V.P., et al.: A model of optimal complexification of measures providing information security. Autom. Control. Comput. Sci. **54**(8), 930–936 (2020). https://doi.org/10.3103/S0146411620080374
13. Alekseev, A.O., et al.: The use of duality to increase the effectiveness of the branch and bound method when solving the Knapsack problem. Zh. Vychisl. Mat. Mat. Fiz. **25**(11) 1666–1673 (1985); U.S.S.R. Comput. Math. Math. Phys. **25**(6) 50–54 (1985). https://doi.org/10.1016/0041-5553(85)90008-4
14. Sonkin, M., et al.: A resource-and-time method to optimize the performance of several interrelated operations. Int. J. Appl. Eng. Res. **10**(17), 38127–38132 (2015)
15. Saurenko, T.N., et al.: The model and the planning method of volume and variety assessment of innovative products in an industrial enterprise. J. Phys.: Conf. Ser. **803**(1), 012006 (2017). https://doi.org/10.1088/1742-6596/803/1/012006
16. Chernysh, A., et al.: Model and algorithm for substantiating solutions for organization of high-rise construction project. In: E3S Web of Conferences, vol. 33, p. 03003 (2018). https://doi.org/10.1051/e3sconf/20183303003
17. Sonkin, M.A., et al.: Mathematical simulation of adaptive allocation of discrete resources. In: Proceedings of the 2016 Conference on Information Technologies in Science, Management, Social Sphere and Medicine (ITSMSSM 2016), pp. 282–285 (2016). https://doi.org/10.2991/itsmssm-16.2016.57
18. Anisimov, V.G., Anisimov, E.G.: A branch-and-bound algorithm for one class of scheduling proble. Zh. Vychisl. Mat. Mat. Fiz. **32**(12), 2000–2005 (1992); Comput. Math. Math. Phys. **32**(12), 1827–183 (1992)
19. Anisimov, V.G., Anisimov, E.G.: A modification of a method for a class of problems in integer programming. Zh. Vychisl. Mat. Mat. Fiz. **37**(2), 179–183 (1997); Comput. Math. Math. Phys. **37**(2), 175–179 (1997)
20. Anisimov, V.G., Anisimov, E.G.: A method of solving one class of integer programming problem. Zh. Vychisl. Mat. Mat. Fiz. **29**(10), 1586–1590 (1989); U.S.S.R. Comput. Math. Math. Phys. **29**(5), 238–241 (1989). https://doi.org/10.1016/0041-5553(89)90205-X
21. Alekseyev, A.O., et al.: Application of Markov chains in estimating the computational complexity of the simplex method. Soviet J. Comput. Syst. Sci. (5), 130–134 (1988)
22. Anisimov, V.G., Anisimov, E.G.: Algorithm for optimal distribution of discrete nonuniform resources on a network. Zh. Vychisl. Mat. Mat. Fiz. **37**(1), 54–60 (1997); Comput. Math. Math. Phys. **37**(1), 51–57 (1997)
23. Every Fourth Company in the World was Attacked more than 7 Times in 2020. https://safe.cnews.ru/news/top/2021-02-04_kazhduyu_chetvertuyu_kompaniyu. Accessed 14 Apr 2021 (in Russ.)
24. Digital Transformation, Telecommunications, Broadcasting and it News. https://www.comnews.ru/content/212181/2020-12-15/2020-w51/kiberprestupniki-vredyat-trillion. Accessed 14 Apr 2021 (in Russ.)

Multi-agent Approach in Planning and Scheduling of Production as Part of a Complex Architectural Solution at the Enterprise

Albert Bakhtizin, Igor Ilin, Nikolay Nikitin, Alena Ershova, and Manfred Esser

Abstract This article is devoted to the identification of the capabilities of the multi-agent approach to improve the efficiency of operational planning and production scheduling systems and the implementation of IT solutions based on the multi-agent approach in the integrated architecture of a manufacturing enterprise. Within the framework of the article, the analysis of existing approaches to the automation of operational management of production was carried out and the limitations of these approaches were determined; analyzed the possibilities of a multi-agent approach to solving problems of accurate and efficient operational planning and production dispatching; and also presents the option of integrating the multi-agent planning system into the architecture of a manufacturing enterprise. The proposed multi-agent system for planning and scheduling production allows you to quickly adjust production schedules, building optimal routes for the movement of orders, and, with correct integration into the existing IT landscape, expands the functionality of ERP and MES systems.

Keywords Multi-agent approach · Operational planning · Production dispatching · Manufacturing enterprise · Methods of planning

1 Introduction

The ongoing processes of digital transformation, which involve a significant change in the structure of the economy, have an impact on the functioning of almost all

A. Bakhtizin
Central Economic and Mathematics Institute, Russian Academy of Sciences, Moscow, Russia

I. Ilin · N. Nikitin · A. Ershova (✉)
Peter the Great St. Petersburg Polytechnic University, Polytechnicheskaya 29, 195351 St. Petersburg, Russia

M. Esser
GetIT, Grevenbroich, Germany

C. Jahn et al. (eds.), *Algorithms and Solutions Based on Computer Technology*, Lecture Notes in Networks and Systems 387,
https://doi.org/10.1007/978-3-030-93872-7_30

enterprises and organizations. The coronavirus pandemic is accelerating these transformations and the International Data Corporation estimates that by 2025, due to an unstable global environment, 75% of business leaders will use digital platforms to adapt their value chains to new realities. Experts at the United Nations Conference on Trade and Development (UNCTAD) suggest that in the next 10 years, up to 100 billion devices will exchange data with each other and, in fact, the whole world will turn into one big computer.

The basis for such transformations will be the widespread use of information and communication technologies, including those used to create digital twins of real objects. For these purposes, there are several well-established paradigms of simulation modelling, among which the agent-based one occupies a special place. The development of this direction is due, on the one hand, to the capabilities of modern computing devices, which make it possible to create digital copies of existing systems, and, on the other hand, to large arrays of accumulated information that allow for the correct specification of individual elements of simulated objects.

Although the agent-based approach was developed thanks to cellular automata and was initially used to a greater extent for the study of biological systems, at present the range of its application is very wide—drug modeling, the study of physical phenomena, pedestrian and transport traffic, ecology, sociology, economics and many other industries.

In developed countries, the use of simulation has long become a standard when designing new objects and evaluating the efficiency of their digital copies, as well as for finding "bottlenecks" and optimizing the work of existing ones. In this case, the agent-based approach can be used directly or in combination with the discrete-event modeling paradigm.

A modern manufacturing enterprise of any profile pays special attention to the organization of operational scheduling and dispatching since these processes ensure the correct operation of production and many service processes.

Operational production planning implies building a detailed work schedule for a given period of time in order to determine a strategic work plan and calculate technical and economic indicators. Feasibility and cost savings are key factors. In the process of operational scheduling, calculations are performed and standards for the movement of objects of work in production are established (standards for stocks, batch sizes, periods of their launch-release, etc.); tasks for workshops, production sites and workplaces for the production of specific products, assemblies and blanks; calendar schedules, which establish the sequence and timing of the manufacture of products at each stage of production [1]. The main tasks of operational scheduling include:

- ensuring the rhythm and timeliness of production in terms of orders being executed;
- ensuring uniformity and completeness of loading of equipment, workers and areas;
- ensuring maximum continuity of production, that is, ensuring the shortest duration of the production cycle, which will help to reduce work in progress and accelerate the circulation of working capital.

In turn, production dispatching is understood as continuous monitoring of the implementation of the plan in real time in order to obtain operational information on the execution of tasks, as well as timely adjustments to the plan in the event of deviations and emergency situations [2].

Modern industrial complexes today are distinguished by multi-operational technological processes, a variety of equipment used and uneven receipt of new orders over time [3]. All this requires from the planning and dispatching departments of enterprises the ability to rebuild the work of workshops in real time [4]. And this, with an average and high level of production workload and the need for planning and scheduling, taking into account the many limitations of real production, is a difficult task, and sometimes impossible due to the limited resources and suboptimality of the tools used [5]. The existing methods of planning and scheduling production based on the MRP, MRP II, APS methodologies do not fully meet the needs of accurate operational planning of production in a short time, and also do not imply the possibility of scheduling production in real time. Taking into account the limitations of standard planning methods, the use of multi-agent technologies in solving production problems seems relevant.

The multi-agent approach allows in real time to search for sufficiently close to optimal solutions within the framework of the task of planning production orders. The use of an agent-based approach to production planning leads to the achievement of high results both from a technical point of view and from an economic point of view. From a technical point of view, multi-agent technologies can increase the speed of operational planning, increase the adaptability of the system, and reduce the influence of the human factor in the formation of the plan. From an economic point of view, the multi-agent approach improves production efficiency, which is expressed in fulfilling more orders with fewer resources, reducing production costs, as well as strengthening control over operational planning and reducing the costs of developing an integrated planning system. An important advantage of the multi-agent approach is its event-driven component, which allows responding in the shortest possible time to the most serious changes in terms of the operational plan. Thus, the use of systems based on a multi-agent approach for planning and scheduling production is most expedient within a modern manufacturing enterprise.

In this regard, the task of developing an integrated architectural solution becomes urgent, including a scheduling and production scheduling system based on a multi-agent approach and reflecting the integration of this system with other IT solutions for production management.

2 Materials and Methods

Currently, there are few of the most popular approaches to scheduling and dispatching of production. We propose to consider the main ones.

In the 50s of the last century, during the period of rapid development of computer technology, various management concepts began to appear actively, which were aimed at improving business through automation and informatization. For manufacturing enterprises at that time, the main goals of automation were to accurately calculate the actual cost of production, analyze it, as well as reduce production costs and increase overall productivity through effective planning of production capacities and resources. The result was the first planning methodology, the Material Requirements Planning (MRP) concept.

The MRP concept today is one of the most well-known logistics concepts, on the basis of which a large number of logistics systems, mainly of the "push type", have been developed and operate [6]. The main task of MRP is to ensure that the required quantity of required component materials is available at any time during the planning period, along with the possible reduction of permanent stocks, and, therefore, unloading of the warehouse [7].

The MRP concept formed the basis for the construction of the so-called MRP systems. When planning in the framework of the MRP system, information about the composition of the product, the state of warehouses and work in progress, as well as orders for the supply of finished goods and production schedules are taken into account. The main functions of the MRP system include the automatic generation of purchase orders and/or internal production of the necessary materials and components [8]. The main advantages of using such a system are the guarantee of the timely receipt of materials and components; optimization of warehouse stocks; reduction of manufacturing defects in the assembly of finished products, which occurs as a result of the use of inappropriate components; streamlining of production due to the control of the status of each material, which makes it possible to track the entire cycle of its use, from ordering a given material to its use in a finished product [9]. However, this concept has some significant limitation: the calculation of the material requirement does not take into account some important parameters, such as production capacity, their load, labor cost, etc. This gave impetus to the transformation and expansion of the concept of MRP, which led to the emergence of a new approach called MRP II.

The concept of manufacturing resource planning (MRP II) originated in the 1980s. The impetus for its appearance was the need of enterprises for tools for operational planning and management of the production process as a whole, and not its individual fragments. MRP II is a set of field-proven management principles and procedures used to improve business performance [10].

The most important function of MRP II is to provide all the necessary information to those who make decisions in the field of financial management. Unlike MRP systems, in systems based on the MRP II concept, planning is carried out not only in material terms, but also in monetary terms. The developed detailed plans are valued through the calculation of the cost of production, accounting for sales, procurement and production operations. The calculated actual costs are compared with the planned (or target) costs, and the variances serve as the basis for making management decisions related to the next planning periods.

Unlike the previous generation MRP systems, MRP II systems incorporate the CRP (Capacity Requirements Planning) methodology and the idea of a "closed loop", the essence of which is to monitor the fulfillment of the order plan for the supply of materials and components and, if necessary, adjust production plans. While the first

MRP systems were designed for unlimited use of resources, then the main task of the CRP was to check the feasibility of the plan in relation to the existing equipment and to optimize the utilization of production facilities [11].

The MRP II concept is based on three basic principles:

- hierarchy (means the division of planning into levels corresponding to the areas of responsibility of different steps of the enterprise's management ladder);
- interactivity (means the possibility of "playing" possible situations in order to study their influence on the results of the enterprise and is provided by the modeling unit);
- integration (ensured by combining all the main functional areas of the enterprise at the operational level within the planning horizon of up to one year, associated with material and financial flows at the enterprise).

The main advantages of using systems of the MRP II class are obtaining detailed and accurate information about the predicted amount of stocks and their cost, about spending money (for the purchase of materials, for labor costs), about receiving money, about the distribution of fixed overhead costs; decrease in stocks; shortening the production cycle and order fulfillment cycle; reduction of work in progress. However, these systems are not devoid of some drawbacks, namely: orientation of such systems only to order; lack of financial analysis and planning functions; strict requirements for the quality of initial data.

Further development of MRP II systems is associated with their growth into a new class of systems—"Enterprise Resource Planning" (ERP). Systems of this class are focused on working with financial information for solving problems of managing large corporations with geographically dispersed resources. This includes everything that is needed to obtain resources, manufacture products, transport them and settlements for customer orders. In addition to the listed functional requirements, ERP systems are also faced with new requirements for the use of graphics, the use of relational databases, CASE technologies for their development, the architecture of "client–server" computing systems and their implementation as open systems. Systems of this class have been actively developing since the late 80s [12].

It should be noted that the approach to solving production planning problems in ERP systems has remained largely unchanged until recently, i.e. as it is established in MRP II systems. In short, it can be defined as an approach based on the active use of scheduling standards for production cycles. The disadvantage of this approach is that it conflicts with the need to optimize planning. Elements of planning optimization in traditional MRP II/ERP systems are found only at the lower level when solving operational planning problems using scheduling theory methods [13]. With the increase in the capacity of computing systems, with the introduction of MRP II/ERP, with the search for new, more effective management methods in a competitive environment, since the mid-90 s on the basis of MRP II/ERP systems, systems of a new class appear, which are called «Advanced Planning/Scheduling» (APS) [14].

APS is a synchronous production planning concept focused on the integration of supply chain planning, taking into account all the peculiarities and limitations of production. When planning the entire production process, it becomes possible in a matter of seconds to determine a realistic schedule for the shipment of orders, taking into account all constantly changing conditions—both internal and external. Systems of the APS class allow simultaneous (synchronous) planning of the necessary materials and resources, taking into account the available capacities when planning the movement of materials and bearing in mind that all resources are working in conditions of limited capacities. Therefore, each operation is planned in accordance with the necessary needs for certain resources. The result is production schedules fully balanced with the materials and capacity available [15]. APS removes the unrealistic assumptions of standard planning. Due to the fact that the required resources are not considered in the standard calculation of required materials (MRP), it is assumed that the production time is fixed. But in reality, production times are not fixed, as they vary based on many different factors. Since production times vary, they usually cannot be determined in advance and an exact production plan cannot be drawn up. In contrast, the plans calculated using the APS method are as close to reality as possible, since they constantly recalculate the production time, simulating the execution of each production operation. In addition, it should be noted that the APS approach is proactive (as opposed to ERP and MRP, which are reactive). Systems of this class can alert project managers when to expect spikes or lulls in demand so that staff and equipment can be scheduling accordingly. They can also plan purchase orders based on projected future orders and market raw material prices. In addition, APS allows for what-if analysis and comparison of analysis results.

It is obvious that despite the listed advantages, APS systems are not without their drawbacks. Firstly, in the case when there is a complex structure of production, a lot of time can be spent on building a schedule. Secondly, the low level of adaptability and scalability of systems of this class. Scheduling algorithms very often have to be revised and complicated, adding additional restrictions, which ultimately leads to significant difficulties in scheduling. In addition, deviations from the schedule in production require a constant rescheduling procedure, which can lead to instability of the entire schedule and periodic changes in the planned release dates for orders [16].

Thus, the analysis of existing methods of operational planning and production scheduling, built on the basis of MRP, MRP II, APS methodologies, revealed a number of serious shortcomings of these methodologies. The main problem with these approaches is that they do not allow for detailed operational planning of production in a short time, which forces industrial enterprises to weaken the requirements for models, and this ultimately negatively affects the quality of plans and, as a result, the financial result. Also, these methods do not imply the ability to dispatch production in real time. In this regard, it becomes necessary to use an absolutely different method to increase the efficiency of operational planning and production scheduling systems. This method is a multi-agent approach, which is fundamentally different from all traditional planning methodologies, since it is based not on clearly deterministic algorithms that allow finding the best solution to the problem, but on the search for one of the solutions that satisfy the conditions, which is obtained as a result of the interaction of many intelligent elements of the system (agents) [17].

The multi-agent approach to production planning, in contrast to MRP, MRP II and APS, makes it possible to separate a specific element of the system (work center, technological operation, material) from knowledge about the entire system as a whole. Each element of the system can be represented as an intelligent agent that solves its own local problem within its scope, which ultimately affects the entire schedule. Each agent or group of agents can change over time, depending on the new data that the system receives, as well as quickly adapt to changes due to the interaction of agents. Planning is carried out in real time and each new production order, in the event of a conflict, will itself ask other already planned orders if they can reconsider their priorities, as well as poll the agents of work centers and the possibility of allocating resources to them for a certain time at a given terms. This approach determines the reactivity of the system and at each moment of time the system is active, agents are negotiating about the possibility of improving their own time and cost characteristics. If the newly drawn up plan suits the planning department, it is fixed and becomes the basis. Further changes to the baseline are only possible with the approval of the dispatcher [18].

The use of a multi-agent approach allows to achieve the following production management goals:

- minimization of the cost of order fulfillment by reducing the number of changeover operations;
- minimization of the risk of non-fulfillment of the order;
- minimizing the time dependence in the sequence of operations from each other (achieving a guaranteed level of interoperative lying of the part);
- maximizing the use of resource capacities.

The main advantages of the multi-agent approach over standard automated systems are [19]:

- high adaptability of the schedule;
- high speed of calculation and recalculation of production plans;
- reduction of the factor of human error;
- the ability to plan and quickly reschedule orders with a very large number of parts-assembly units within the production technology;
- the possibility of expanding the system of planning evaluation criteria;
- accumulation of knowledge about planning processes and their further use in work.

Thus, the creation and implementation of systems based on a multi-agent approach for planning and scheduling production is extremely promising for modern manufacturing enterprises.

3 Results

The management of modern enterprises and organizations requires the development of management technologies, information technology support, data processing and storage systems [20]. This is based on modern architectural standards, such as TOGAF, Zachman's model, Gartner's approach, etc. [21]. At the same time, enterprise architecture is understood as a single whole of principles, methods and models that are used to design and form an organizational structure, business processes, functional structure, information systems and applications, infrastructure [22]. However, despite the increased importance of the architectural standard in the practice of enterprise management, for a long time the automation of production and technological processes remained outside the concept of enterprise architecture. Existing approaches to designing the architecture of a manufacturing enterprise take into account the following classes of information systems: ERP, MES, process control system.

ERP is an enterprise resource planning system. The key task of systems of this class is the management of the economic and financial activities of the enterprise. MES is an automated production control system. Systems of this class are designed to solve the problem of synchronization, coordination, analysis and optimization of production within the production (workshop) in real time. In turn, process control systems are designed to control individual functional blocks of technological processes [23].

Despite the fact that these systems provide automation of a significant part of the production process, their functionality still does not fully take into account the real capabilities of manufacturing enterprises. In addition, since these systems rely on standard modeling approaches (MRP, MRP II and APS), which do not allow planning technological operations with a predictable error, it is highly likely that tasks for controlling production processes will not be distributed optimally between the systems, and this is in turn, will generate a significant load on MES systems and increase the level of risk for orders within the ERP. Taking into account the limitations of the standard planning methods and the capabilities of the multi-agent approach, which were defined in the article earlier, it is the use of the multi-agent approach for local production optimization that can be one of the possible drivers for the development of approaches to the integration of technological and production processes and corresponding information systems into the enterprise architecture. In this regard, the task of developing an integrated architectural solution for scheduling and production scheduling based on a multi-agent approach becomes extremely urgent.

For the proposed multi-agent adaptive planning system, the following high-level sequence of works is assumed:

1. each order, operation, machine, worker or any other resource of the enterprise receives its own software agent, which has its own schedule;
2. an incoming new order refers to the storage of technological processes and receives from there the technological process of its execution;
3. for each order, its own agent is created, who receives the requirements and planning restrictions;

4. the agent begins planning by searching for the resources it needs on a board (scene) that describes the current situation in the workshop or throughout the enterprise, namely which work center performs a certain operation;
5. if suitable resources are occupied, then the conflict is fixed and the process of negotiations begins to resolve it by assignment;
6. during negotiations, it is possible to obtain various options: a new order will go to a less suitable resource, the previous order will leave or move, as well as other options;
7. after solving their problem, the agents do not stop and continue to try to improve their position, taking into account the priority.

Of course, to ensure the efficiency of a multi-agent system (MAS), it is necessary to integrate it into the overall architecture of production systems. The proposed architecture is shown below in Fig. 1.

Within the framework of the proposed architecture, it is assumed that the system for planning production processes at the enterprise is systematically organized through the integration of ERP, MAS, MES and SCADA systems. In this case, the organizing element is the ERP system, which plays the role of both an accounting and information system and a management system. Its tasks include the formation of a portfolio of orders from products demanded by the market. As a result, the tandem of ERP and MAS makes it possible to form not only a production program (production plan), but also to build an enlarged work schedule for the entire enterprise. Moreover, MAS continuously strives to improve the schedule. Due to more accurate operational planning, i.e. calculation of the production schedule, MAS, relying on information about the status of supplies, stocks, work in progress and the availability of production facilities, specifies the planned launch-release dates of products. The resulting discrepancies between these dates and the requirements of the current production program lead to the need to adjust the planned dates already at the ERP level.

The MES-system receives the volume of work, and in the future the system itself not only builds more accurate schedules for equipment and personnel, but also monitors their implementation online. In this sense, the goal of the MES-system is not only to fulfill a given amount of work with the specified deadlines for the implementation of certain orders, but to fulfill them as best as possible in terms of the specifics and planned indicators of the shop. MAS generates some initial work schedules of the first degree of approximation even before the start of the implementation of production plans. At the same time, when the schedule improves, the MAS system will offer the MES system the appropriate adjustments. In turn, the MES-system already at the stage of execution, receiving such a preliminary plan, optimizes it according to a number of criteria. At the same time, after the optimization and construction of a new work schedule for the shop, very often additional reserves are found due to the compaction of the equipment operation, and it becomes possible to fulfill additional orders within the framework of the planned period. This achieves the effect of increasing the throughput of production structures.

It is important to note that the proposed architecture supports feedback mechanisms. In the event of a misalignment, the upper level makes decisions on adjusting

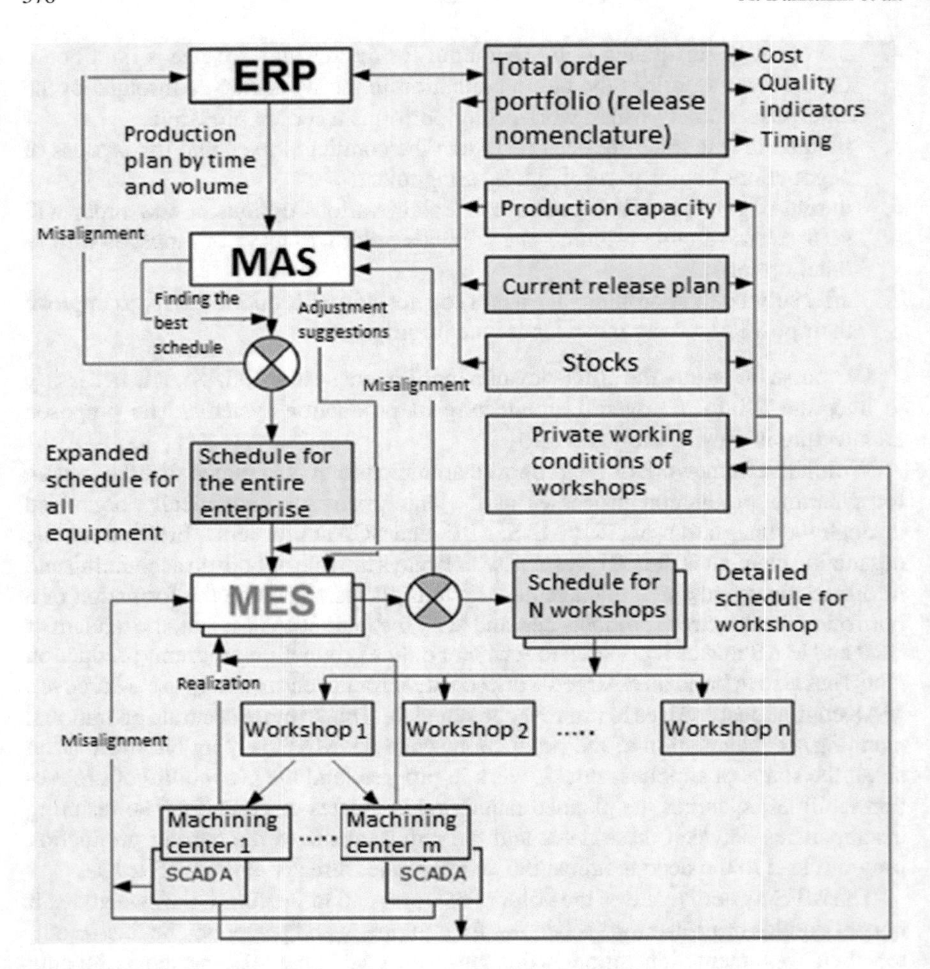

Fig. 1 Model of architecture of production systems of the enterprise

the planning directives for the lower level. For example, discrepancies coming from SCADA systems force the MES system to recalculate schedules in some cases. Discrepancies coming from MES-systems force the MAS-system to either reduce the workload on the shops (in case of non-fulfillment of orders within the specified time frame), or to recalculate the deadlines. Discrepancies coming from the MAS-system force to abandon certain orders or revise the delivery time on the part of customers.

Thus, the presented architecture of the enterprise's production systems, which includes a multi-agent component, will allow organizing a flexible process for planning and scheduling production, due to the fact that the multi-agent system will allow in real time to search for sufficiently close to optimal solutions within the framework of the task of planning production orders. It should be noted that the proposed architecture can be adjusted taking into account the specifics of a particular manufacturing enterprise.

4 Conclusions

This article was devoted to the study of the applicability of the multi-agent approach in production planning and scheduling. As part of the study, the following conclusions were obtained:

1. The multi-agent approach in planning and scheduling production allows to find efficient distribution of work in the context of work centers in a short time, which has a positive effect on the financial result of the enterprise.
2. The use of multi-agent technologies for solving operational planning problems allows an enterprise to be more susceptible to external, changeable conditions, which makes it possible to quickly adjust existing plans.
3. The proposed architecture of production systems, which includes a multi-agent component, provides not only the prompt adjustment of production schedules, but also the expansion of the standard functionality of ERP and MES class systems.
4. The developed architecture is a reference and can be applied in various industries and at enterprises of various types of activities.

Acknowledgements The reported study was funded by RSCF according to the research project No. 9-18-00452.

References

1. Romanovskaya, E.V., Kuznetsov, V.P., Andryashina, N.S., Garina, E.P., Garin, A.P.: Development of the system of operational and production planning in the conditions of complex industrial production. In: Popkova, E.G., Sergi, B.S. (eds.) Digital Economy: Complexity and Variety vs. Rationality, pp. 572–583. Springer International Publishing, Cham (2020). https://doi.org/10.1007/978-3-030-29586-8_66
2. Bolotov, A.N., Burdo, G.B.: Models for dispatching machine-building divisions. IOP Conf. Ser. Mater. Sci. Eng. **919**, 032005 (2020). https://doi.org/10.1088/1757-899X/919/3/032005
3. Ilyashenko, O., Kovaleva, Y., Burnatcev, D., Svetunkov, S.: Automation of business processes of the logistics company in the implementation of the IoT. IOP Conf. Ser. Mater. Sci. Eng. **940**, 012006 (2020). https://doi.org/10.1088/1757-899X/940/1/012006
4. Capo, D., Levina, A., Dubgorn, A., Schröder, K.: Enterprise architecture concept for digital manufacturing. IOP Conf. Ser. Mater. Sci. Eng. **1001**, 012044 (2020). https://doi.org/10.1088/1757-899X/1001/1/012044

5. Mogilko, D.Y., Ilin, I.V., Iliashenko, V.M., Svetunkov, S.G.: BI Capabilities in a digital enterprise business process management system. In: Arseniev, D.G., Overmeyer, L., Kälviäinen, H., Katalinić, B. (eds.) Cyber-Physical Systems and Control, pp. 701–708. Springer International Publishing, Cham (2020). https://doi.org/10.1007/978-3-030-34983-7_69
6. Hong, P., Leffakis, Z.M.: Managing demand variability and operational effectiveness: case of lean improvement programmes and MRP planning integration. Prod. Plan. Control. **28**, 1066–1080 (2017). https://doi.org/10.1080/09537287.2017.1329956
7. Yeung, J.H.Y., Wong, W.C.K., Ma, L.: Parameters affecting the effectiveness of MRP systems: a review. Int. J. Prod. Res. **36**, 313–332 (1998). https://doi.org/10.1080/002075498193750
8. Anderson, J.C., Schroeder, R.G.: Getting results from your MRP system. Bus. Horiz. **27**, 57–64 (1984). https://doi.org/10.1016/0007-6813(84)90028-4
9. Turnipseed, D.: An implementation analysis of MRP systems: a focus on the human variable. Prod. Invent. Manag. J. **33** (1992)
10. Higgins, P., Roy, P.L., Tierney, L.: Manufacturing Planning and Control: Beyond MRP II. Springer Science & Business Media (1996)
11. Toomey, J.: MRP II: Planning for Manufacturing Excellence. Springer Science & Business Media (1996)
12. Lorincz, P.: Evolution of enterprise systems. In: 2007 International Symposium on Logistics and Industrial Informatics, pp. 75–80 (2007). https://doi.org/10.1109/LINDI.2007.4343516
13. Voß, S., Woodruff, D.L.: Connecting MRP, MRP II and ERP—supply chain production planning via optimization models. In: Greenberg, H.J. (ed.) Tutorials on Emerging Methodologies and Applications in Operations Research: Presented at Informs 2004, Denver, CO, pp. 1–8. Springer, New York, NY (2005). https://doi.org/10.1007/0-387-22827-6_8
14. Caridi, M., Sianesi, A.: Trends in planning and control systems: APS–ERP integration. In: Mertins, K., Krause, O., Schallock, B. (eds.) Global Production Management: IFIP WG5.7 International Conference on Advances in Production Management Systems September 6–10, 1999, Berlin, Germany. pp. 105–111. Springer US, Boston, MA (1999). https://doi.org/10.1007/978-0-387-35569-6_13
15. Lee, Y.H., Jeong, C.S., Moon, C.: Advanced planning and scheduling with outsourcing in manufacturing supply chain. Comput. Ind. Eng. **43**, 351–374 (2002). https://doi.org/10.1016/S0360-8352(02)00079-7
16. Hvolby, H.-H., Steger-Jensen, K.: Technical and industrial issues of Advanced Planning and Scheduling (APS) systems. Comput. Ind. **61**, 845–851 (2010). https://doi.org/10.1016/j.compind.2010.07.009
17. Rabelo, R.J., Camarinha-Matos, L.M., Afsarmanesh, H.: Multi-agent-based agile scheduling. Robot. Auton. Syst. **27**, 15–28 (1999). https://doi.org/10.1016/S0921-8890(98)00080-3
18. Kang, S.G., Choi, S.H.: Multi-agent based beam search for intelligent production planning and scheduling. Int. J. Prod. Res. **48**, 3319–3353 (2010). https://doi.org/10.1080/00207540902810502
19. Caridi, M., Cavalieri, S.: Multi-agent systems in production planning and control: an overview. Prod. Plan. Control. **15**, 106–118 (2004). https://doi.org/10.1080/09537280410001662556
20. Borremans, A.D., Zaychenko, I.M., Iliashenko, O.Y.: Digital economy. IT strategy of the company development. Presented at the MATEC Web of Conferences (2018). https://doi.org/10.1051/matecconf/201817001034
21. Ilin, I.V., Iliashenko, O.Y., Iliashenko, V.M.: Architectural approach to the digital transformation of the modern medical organization. Presented at the Proceedings of the 33rd International Business Information Management Association Conference, IBIMA 2019: Education Excellence and Innovation Management through Vision 2020 (2019)
22. Ilin, I., Levina, A., Borremans, A., Kalyazina, S.: Enterprise architecture modeling in digital transformation era. Adv. Intell. Syst. Comput. **1259** AISC, 124–142 (2021). https://doi.org/10.1007/978-3-030-57453-6_11
23. Berić, D., Stefanović, D., Lalić, B., Ćosić, I.: The implementation of ERP and MES systems as a support to industrial management systems. Int. J. Ind. Eng. Manag. **9**, 77–86 (2018)

Methodological Approach to the Formation of the Company's Portfolio of Orders

Olga Rostova, Vladimir Anisimov, Evgeniy Anisimov, Tatiana Saurenko, Elena Peschannikova, and Anastasiia Shmeleva

Abstract The article proposes an approach to the formation of an optimal portfolio of orders for an enterprise operating under conditions of uncertainty. Uncertainty lies in the fact that the types and number of orders received by the enterprise in the course of its operation, as well as the effect achieved from the use of available resources (enterprise income) are random. The order portfolio is filled in stages, taking into account the economic and technical requirements. These requirements consist in the formation of a portfolio that ensures rational loading of the production equipment of the enterprise at each stage of its operation and obtaining the maximum income from the execution of orders included in the portfolio for a specified period of time. The approach is based on a multi-stage model of the optimal distribution of discrete recoverable (enterprise equipment) and non-recoverable resources of the enterprise for order fulfillment. It allows you to include in the portfolio at each stage orders that provide the maximum income for the enterprise for a set period of its operation.

Keywords Enterprise · Portfolio of orders · Uncertainty · Resources · Distribution · Adaptation

O. Rostova · V. Anisimov (✉)
Peter the Great St. Petersburg Polytechnic University, St. Petersburg, Russia

O. Rostova
e-mail: o.2908@mail.ru

E. Anisimov · T. Saurenko
Peoples Friendship University of Russia, RUDN University), Moscow, Russia

E. Peschannikova
Concern Radio-Technical and Information Systems, Moscow, Russia

A. Shmeleva
Marriott International Yerevan, Yerevan, Armenia

C. Jahn et al. (eds.), *Algorithms and Solutions Based on Computer Technology*,
Lecture Notes in Networks and Systems 387,
https://doi.org/10.1007/978-3-030-93872-7_31

1 Introduction

The functioning of a wide class of enterprises in the real sector of the economy consists in the stage-by-stage formation and execution of a portfolio of orders. The management of its formation is associated with the need to make decisions in a close to real time scale to include or exclude incoming orders in the portfolio being formed. These solutions should ensure the efficient use of equipment and other resources of the enterprise for a specified period of time [1–5]. Their justification is carried out under conditions of uncertainty. The uncertainty lies in the fact that the types and number of orders received by the enterprise at each stage of its operation, as well as the effect achieved from the use of available resources (enterprise income) for the entire established period of operation are random. The decisions taken should provide the maximum possible effect [6–11]. To maximize the effect of the enterprise's functioning, when justifying decisions in this situation, it is advisable to proceed from the principle of optimal adaptation. Its essence lies in the fact that the corresponding solutions should be as invariant as possible with respect to the non-deterministic and uncontrollable conditions of the functioning of the enterprise in question and at the same time make the most full use of the possibilities inherent in the deterministic and reliably controlled parameters of the production process taking place in it [12]. The development of a methodological approach that implements this principle in the formation of a portfolio of orders of an enterprise is the purpose of this article.

2 Materials and Methods

With a sufficient degree of generality, the functioning of a wide class of enterprises in the real sector of the economy can be schematically represented as a process taking place in a certain system σ, which includes (see Fig. 1) four main subsystems—demand, information, control and resource [13, 14].

At the same time, the demand subsystem reflects the environment external to the enterprise and forms a flow of orders, for the fulfillment of which the resources of the economic system under consideration can be attracted. The information subsystem identifies these orders and forms a possible set of orders at each kth ($k = 1, 2, \ldots$) stage of the system's functioning. This set is characterized by the vector

$$Z^k = \left\| z_n^k \right\|, \quad n = 1, 2, \ldots, N, \tag{1}$$

where z_n^k is the number of orders of the n-type identified by the time t_k of the beginning of the kth stage, which can be executed by the resources of the system (enterprise) under consideration.

The control subsystem at successive times $t_1 < t_2 < \ldots < T$ makes decisions on including or not including the identified orders in the company's order portfolio.

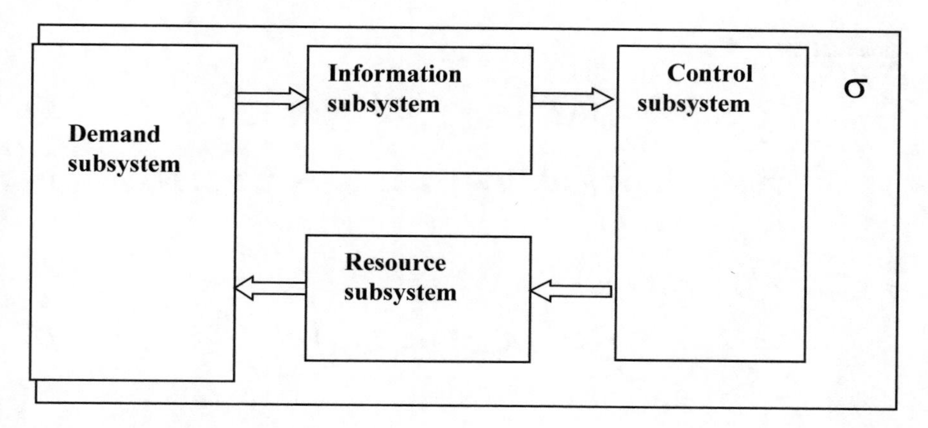

Fig. 1 Generalized system diagram

The resource subsystem ensures the preparation and use of resources for the execution of orders included in the portfolio. It is characterized by the vector

$$R = \|R_i\|, \quad i = 1, 2, ..., I, \tag{2}$$

whose components reflect the amount of resources of each type i that the system σ has at the beginning of operation.

As a result of the use of these resources during the operation of the system σ, the income of the enterprise is gradually formed. In this case, the goal is to maximize income for a specified period of time T, provided that orders not included in the portfolio during each kth stage are excluded from consideration.

3 Results

3.1 Formalized Presentation of the Task of Optimizing the Portfolio of Orders

Regarding to the considered schematization of the enterprise functioning process, it is advisable to adapt the resource allocation by using a model of the following type.

Determine resource allocation option

$$V^* = \{V^{*k}\}, \quad k = 1, 2, ..., \tag{3}$$

$$V^{*k} = \|v_{ij}^{*k}\|, \quad i = 1, 2, ..., I, \quad j = 1, 2, ..., J, \tag{4}$$

such that

$$Q(V^{*k}) = \max_{V^k \subset \hat{V}^k} Q(V^k), \tag{5}$$

$$\hat{V}^k = \{V^k | \Delta Q_{ij}^k \geq U_i^k, \quad i = 1, 2, ..., I, \quad j = 1, 2, ..., J\}, \tag{6}$$

at

$$\sum_{j=1}^{J} v_{ij}^k \leq R_i^k, \quad i = 1, 2, ..., I, \tag{7}$$

where $J = \sum_{n=1}^{N} z_n^k$ is the number of orders received at time t_k;

v_{ij}^k is the number of units of account of the ith type resource allocated for the execution of the jth order in the time period (t_k, t_{k+1});

$\hat{V}^k$—a set of resource allocation options that are possible at time t_k (at the kth stage of the enterprise's operation);

$Q(V^k)$ is the profit of the enterprise for the time $t_{k+1} - t_k$ when implementing the option of distributing available resources;

ΔQ_{ij}^k is the increment of profit for the time $t_{k+1} - t_k$ when the settlement unit of the resource of the ith type is allocated at the time t_k to meet the needs of the jth order;

U_i^k is the minimum value of the specific efficiency for resources of the ith type at time t_k;

R_i^k is the total amount of resources of the ith type at time t_k;

I—the total number of types of enterprise resources.

Relation (3) shows that the optimal plan is formed in the form of decisions V^{*k} taken at times $t_1 < t_2 < \ldots < T$.

The elements of the matrix (4) are the volumes of resources allocated to fulfill the corresponding orders.

Formula (5) reflects the purpose of the enterprise resource allocation.

Relation (6) is a criterion for selecting from the set of possible resource allocation options at the kth ($k = 1, 2, \ldots$) stage of such a subset, the choice from which does not contradict the achievement of the maximum profit of the enterprise for the entire planning period T.

It takes into account the situation that has developed by the kth stage of the enterprise's functioning, and the forecast of its development in the future. This prevents local optimization of the resource allocation plan at a separate stage to the detriment of achieving the global optimum.

Formula (7) provides a balance of available and allocated resources.

The choice of the values U_i^k ($i = 1, 2, \ldots, I$) as adaptation parameters is due to their simple prediction.

The concretization of the model (3)–(7) is provided by an explicit representation of the corresponding relations. Moreover, depending on the type of function $Q(V^k)$, model (3)–(7) can belong to the class of linear (if $Q(V^k)$ is a linear function) or nonlinear (if $Q(V^k)$) is a nonlinear function) integer programming problems. These problems are *NP*-difficult. Effective methods for their solution are proposed, for example, in [15–21].

3.2 *Methodological Approaches to Determining the Specific Efficiency of the Use of Enterprise Resources*

Taking into account the real situation when adapting decisions on the distribution of resources in model (3)–(7) is provided by the vector

$$U^k = \left\| U_i^k \right\|, \quad i = 1, 2, ..., I, \quad k = 1, 2, ... \tag{8}$$

Determination methods U_i^k depend on the information situation. In the simplest case, it can be assumed that the total number and types of orders that can be received are known. However, there is no data on how many and which of them will actually arrive during the time T and how they will be distributed over the stages of the enterprise's functioning.

The total number and types of possible orders for each stage will be described by the vector

$$\tilde{Z}^k = \left\| \tilde{z}_n^k \right\|, \quad n = 1, 2, ..., N, \tag{9}$$

$$\text{where} \quad \tilde{z}_n^k = z_n^0 - \sum_{l=1}^{k} z_n^l; \tag{10}$$

z_n^0 is the total number of possible orders of the nth ($n = 1, 2, \ldots, N$) type formed by the demand subsystem during time T;

z_n^l is the number of possible orders of the nth ($n = 1, 2, \ldots, N$) type, identified by the information subsystem for execution at the lth ($l = 1, 2, \ldots$) stage of functioning.

In such an information situation, to determine the components of the vector (8), one can use the procedure of optimal distribution at each kth stage of resources between all orders that can in principle arrive at the kth and subsequent stages, that is, according to orders determined by vector (9). As the components of the vector (8), take the minimum income per unit of the resource of the ith ($i = 1, 2, \ldots, I$) type in the resulting distribution.

The model of the optimal distribution of the resources available at each kth stage according to orders determined by the vector (9), in a formalized form, is as follows.

Determine resource allocation option

$$V^{*k} = \left\| v_{ij}^{*k} \right\|, \quad i = 1, 2, ..., I, \quad j = 1, 2, ..., J, \tag{11}$$

such that

$$V^{*k} = \arg\max_{V^k} Q\left(V^k\right) \tag{12}$$

$$\text{at} \quad \sum_{j=1}^{J} v_{ij}^{k} \le R_i^k, \quad i = 1, 2, ..., I, \tag{13}$$

$$J = \sum_{n=1}^{N} \tilde{z}_n^k. \tag{14}$$

If, in the current information situation, the control subsystem of the system under consideration has data only on the types of possible orders and the rates of receipt of orders of each type, then a *different approach* can be used as the basis for determining the components of the vector (8). For its formalized representation, the process of forming orders will be represented as a set of flows of requirements with intensities determined by the vector

$$\Lambda^k = \left\| \lambda_n^k(0) \right\|, \quad n = 1, 2, ..., N, \quad k = 1, 2, ... \tag{15}$$

The control subsystem of the system σ, setting the threshold values $U_i^k, \quad (j = 1, 2, ..., J, \quad k = 1, 2, ..,)$ of the specific efficiency of using the available resources of each type for fulfilling orders, selects the flows of requirements in accordance with the relation (6). As a result, flows of admissible requirements are formed, the intensities of which are determined by the ratio

$$\lambda_n^k(U_i^k) = \lambda_n^k(0) \int\limits_{U_i^k}^{\max q_{in}^k} \phi_{in}^k(q)dq \quad n = 1, 2, ..., N, \quad i = 1, 2, ..., I, \quad k = 1, 2, ... \tag{16}$$

where $\phi_{in}^k(q)$ is the distribution density of the effect of using a unit of resource of the ith type to fulfill orders of the nth type at the considered kth stage of the system's functioning;

$\max q_{in}^k$ is the maximum value of the effect of using a unit of resource of the ith type for fulfilling orders of the nth type at the considered stage of the system's functioning.

When these flows are formed, the parameters $U_i^k, \quad (i = 1, 2, ..., J, \quad k = 1, 2, ..,)$ provide a rarefaction of the input flow of orders.

Decreasing the values of the parameters $U_i^k, \quad (i = 1, 2, ..., J, \quad k = 1, 2, ..,)$ increases the risk of lack of resources in the future, when their use could be more efficient. Increasing the parameter values $U_i^k, \quad (i = 1, 2, ..., J, \quad k = 1, 2, ..,)$ increases the risk of impossibility to use resources due to the lack of orders.

The task is to select the optimal values of the parameters $U_i^k, \quad (j = 1, 2, ..., J, \quad k = 1, 2, ..,)$, which, considering the indicated risk factors, would provide the maximum value of the mathematical expectation of the effect of using the available resources.

If the values (U_i^k) $(i = 1, 2, \ldots, I)$ at the kth step are fixed, then the expected specific efficiency of using resources of the ith type for fulfilling orders at this step is characterized by the probability density

$$f_{in}^k(q, U_i^k) = \frac{\phi_{in}^k(q)}{1 - \int_{\min q_{in}^k}^{U_i^k} \phi_{in}^k(q)dq}, \quad i = 1, 2, ..., I, \quad n = 1, 2, ..., N, \quad k = 1, 2, ... \tag{17}$$

where $\min q_i^k$ is the lower limit of the value q_i^k.

Taking into account (17), the mathematical expectation $\overline{q}_{in}^k$ of the specific effect of using resources of the ith type for fulfilling orders of the nth type at the kth stage of the enterprise's functioning is determined by the ratio

$$\overline{q}_{in}^k = \int_{U_i^k}^{\max q_i^k} q \times f_{in}^k(q, U_i^k)dq, \quad i = 1, 2, ..., I, \quad k = 1, 2, ... \tag{18}$$

If, in this case, the resources are not renewed during the operation of the system, then the expected full effect of their use at the kth stage, taking into account (18), can be calculated by the formula

$$Q(U^k) = \sum_{i=1}^{I} \sum_{m=0}^{R_i^k} \left[mF_{im}^k \times \int_{U_i^k}^{\max q_i^k} q \times f_i(q, U_i^k)dq \right], \quad k = 1, 2, ..., \tag{19}$$

where m is the number of units of the ith type resource used at the kth stage of the system's functioning to fulfill orders;

F_{im}^k is the probability of using m units of the resource of the ith type to fulfill orders at the kth stage of the system's functioning.

For renewable resources of the system, the effect obtained as a result of their use is determined by the ratio

$$Q^k(U^k) = \sum_{i=1}^{I} H_i^k(U_i^k)\overline{q}_{in}^k \lambda_n^k(U_i^k)(T - t_k),\ k = 1, 2, ..., \tag{20}$$

where $H_i^k(U_i^k)$ is the probability of using the conditional unit of the ith type resource for its intended purpose.

Vector (8) should ensure the maximum efficiency of using the allocated resources. Therefore, the determination of its optimal components U_i^{*k}, $(i = 1, 2, ..., I,\quad k = 1, 2, ...)$ can be reduced to solving the following extremal problem

$$U^{*k} = \| U_i^{*k} \| = \arg\max_{U^k} Q^k(U^k). \tag{21}$$

Further specification of the presented approach consists in the constructive representation of the functions $Q(V^k)$, F_{im}^k, $H_i^k(U_i^k)$, $\phi_{in}^k(q)$.

4 Discussion

Modern features of the functioning of enterprises in the real sector of the economy consist in a significant increase in competition, in this regard, an increase in the requirements for product quality on the part of consumers. This objectively forces enterprises to carry out their activities in the form of custom-made production. At the same time, taking into account the individual preferences of consumers leads to a significant decrease in the repeatability of products manufactured by the enterprise and the need to change production processes to fulfill each order. In such conditions, the stability and competitiveness of the enterprise to a significant extent depends on the reasonable formation of the portfolio of orders, taking into account the factors of uncertainty and instability in the markets. The complexity of this task and the significant reputational time and material costs associated with possible miscalculations when making appropriate decisions determine the feasibility of introducing information technology for forming the company's order portfolio. The essence of this technology is the organization of automated data processing based on the representation of the specified problem in the form of mathematical models for optimizing the corresponding solutions. An approach to the construction of mathematical models that make it possible to form an optimal portfolio of orders for an enterprise is proposed in this article. In this case, the uncertainty of the conditions for the formation of the specified portfolio is taken into account, due to the fact that the types and number of possible orders, as well as the effect achieved from the use of available resources (enterprise income) are random. The model is based on the principle of optimal adaptation. At the same time, the variant of the implementation of this principle proposed in the article provides the possibility of presenting the process of forming an optimal portfolio of orders in the form of a multi-stage problem of mathematical programming. This makes it possible to use a relatively simpler static optimization apparatus for the dynamic process of forming an optimal portfolio of orders.

5 Conclusions

In general, the considered methodological approach allows, when forming a portfolio of orders of an enterprise, to take into account such essential features of the functioning of enterprises in the real sector of the economy as the dynamism and uncertainty of the situation, and at the same time to use a relatively simple apparatus of static optimization to make decisions. The process of managing the formation of the company's order portfolio becomes adaptive in this case.

These circumstances determine the advisability of applying the proposed approach when building automated decision support systems in the interests of forming order portfolios of enterprises in the real sector of the economy, operating in conditions of uncertainty characteristic of economic systems.

References

1. Chernysh, A., et al.: Model and algorithm for substantiating solutions for organization of high-rise construction project. In: E3S Web of Conferences, vol. 33, p. 03003 (2018). https://doi.org/10.1051/e3sconf/20183303003
2. Suprun, A.F., et al.: The problem of innovative development of information security systems in the transport sector. Autom. Control. Comput. Sci. **52**(8), 1105–1110 (2018). https://doi.org/10.3103/S0146411618080035
3. Sonkin, M.A., et al.: The model and the planning method of volume and variety assessment of innovative products in an industrial enterprise. J. Phys.: Conf. Ser. **803**(1), 012006 (2017). https://doi.org/10.1088/1742-6596/803/1/012006
4. Rodionova, E.S., et al.: The model for determining rational inventory in occasional demand supply chains. Int. J. Supply Chain Manag. **8**(1), 86–89 (2019)
5. Marchenko, R., et al.: Model for comparative assessment of commercial seaports in global transport and logistics infrastructure. Proc. Int. Conf. Digit. Transform. Logist. Infrastruct. **2019**, 459–463 (2019). https://doi.org/10.2991/icdtli-19.2019.79
6. Zotova, E.A., et al.: Methodological approaches to accounting uncertainty when planning logistic business processes. In: International Conference on Digital Transformation in Logistics and Infrastructure (ICDTLI 2019), pp. 246–249 (2019). https://doi.org/10.2991/icdtli-19.2019.44
7. Bazhin, D.A., et al.: A risk-oriented approach to the control arrangement of security protection subsystems of information systems. Autom. Control. Comput. Sci. **50**(8), 717–721 (2016). https://doi.org/10.3103/S0146411616080289
8. Sonkin, M.A., et al.: Methodology control function realization within the electronic government concept framework. Int. J. Sci. Technol. Res. **9**(2), 6259–6262 (2020)
9. Zaychenko, I.M., et al.: Models for predicting damage due to accidents at energy objects and in energy systems of enterprises. In: E3S Web of Conferences, vol. 110, p. 02041 (2019). https://doi.org/10.1051/e3sconf/201911002041

10. Prisyazhnyuk, S.P., et al.: Indices of the effectiveness of information protection in an information interaction system for controlling complex distributed organizational objects. Autom. Control. Comput. Sci. **51**(8), 824–828 (2017). https://doi.org/10.3103/S0146411617080053
11. Grashchenko, N., et al.: Justification of financing of measures to prevent accidents in power systems. In: EMMFT-2018. Advances in Intelligent Systems and Computing, vol. 983. Springer, Cham. (2019). https://doi.org/10.1007/978-3-030-19868-8_30
12. Sonkin, M.A., et al.: Mathematical simulation of adaptive allocation of discrete resources. In: Proceedings of the 2016 Conference on Information Technologies in Science, Management, Social Sphere and Medicine (ITSMSSM 2016), pp. 282–285 (2016). https://doi.org/10.2991/itsmssm-16.2016.57
13. Yastrebov, O., et al.: Digital transformation and optimization models in the sphere of logistics. In: SHS Web of Conference, vol. 44, p. 00009 (2018). https://doi.org/10.1051/shsconf/20184400009
14. Saurenko, T., et al.: Enterprise product sales market selection model. In: E3S Web of Conferences, vol. 217, p. 07024 (2020). https://doi.org/10.1051/e3sconf/202021707024
15. Sonkin, M., et al.: A resource-and-time method to optimize the performance of several interrelated operations. Int. J. Appl. Eng. Res. **10**(17), 38127–38132 (2015)
16. Anisimov, V.G., Anisimov, E.G.: Algorithm for optimal distribution of discrete nonuniform resources on a network. Zh. Vychisl. Mat. Mat. Fiz. **37**(1), 54–60 (1997)
17. Los, V.P., et al.: A model of optimal complexification of measures providing information security. Autom. Control. Comput. Sci. **54**(8), 930–936 (2020). https://doi.org/10.3103/S0146411620080374
18. Kasatkin, V.V., et al.: Model and method for optimizing computational processes in parallel computing systems. Autom. Control. Comput. Sci. **53**(8), 1038–1044 (2019). https://doi.org/10.3103/S0146411619080054
19. Anisimov, V.G., Anisimov, E.G.: A Modification of a method for a class of problems in integer programming. Zh. Vychisl. Mat. Mat. Fiz. **37**(2), 179–183 (1997); Comput. Math. Math. Phys. **37**(2), 175–179 (1997)
20. Rostova, O., Zabolotneva, A., Dubgorn, A., Shirokova, S., Shmeleva, A.: Features of using blockchain technology in healthcare. In: IBIMA 2019: Education Excellence and Innovation Management through Vision 2020, pp. 8525–8530 (2019)
21. Alekseev, A.O., et al.: The use of duality to increase the effectiveness of the branch and bound method when solving the Knapsack problem. Comput. Math. Math. Phys. **25**(6), 50–54 (1985). https://doi.org/10.1016/0041-5553(85)90008-4

Identification and Classification of the Effects of Digital Transformation on Business

Dayana Gugutishvili, Sofia Kalyazina, and Jorg Reiff-Stephan

Abstract Digital business transformation is a business transformation process that implies using innovative technologies in order to create completely different business models, new products and services. The purpose of this article is to identify and classify the effects of digital transformation on business. This paper presents an overview of some of the main technologies used as means of digital transformation, as well as various effects that they cause when implemented in enterprises. Among the explored technologies: robotization, artificial intelligence, big data, virtual and augmented reality, the Internet of Things and blockchain. Not only that, but also a list of methods for calculating the effectiveness of the implementation of those technologies is presented. Three groups of methods are examined: financial, qualitative, probabilistic. Based on conducted research a comprehensive overview of various effects that implementation and usage of digital technologies brings to enterprises was formalized. Presented findings allow to assert that digital transformation has very impressive impact on enterprises regardless of their area of business.

Keywords Digital transformation · Digital transformation effects · Digital technologies

1 Introduction

Nowadays the use of digital technologies is such a common occurrence we can't imagine our lives without it. Automation and digitalization have a significant impact on every activity, changing the established rules and procedures. Gradually, such changes became ubiquitous, in connection with which experts began to talk about

D. Gugutishvili (✉) · S. Kalyazina
Peter the Great St. Petersburg Polytechnic University, Polytechnicheskaya st., 29, St. Petersburg 195251, Russia
e-mail: gugutishvilid@mail.ru

J. Reiff-Stephan
Technical University of Applied Sciences Wildau, Hochschulring 1, 15745 Wildau, Germany
e-mail: jrs@th-wildau.de

C. Jahn et al. (eds.), *Algorithms and Solutions Based on Computer Technology*, Lecture Notes in Networks and Systems 387,
https://doi.org/10.1007/978-3-030-93872-7_32

such a phenomenon as "digital transformation". Nowadays the importance of digital transformation processes is recognized both by the global business community and by the governments of various countries. This is confirmed by the emergence of various initiatives and programs to support digitalization in companies, various industries, as well as at the state level.

Digital business transformation is a business transformation process that implies using innovative technologies in order to create completely different business models, new products and services. Digital transformation aims to improve the decision-making process within the company, switching up the product variability based on the clients' demands, as well as optimizing staff workload [1]. Using these digital technologies becomes crucial in transformation processes and transformation projects often times face difficulty without them [2].

This paper presents an overview of some of the main technologies used as means of digital transformation, as well as various effects that they cause when implemented in enterprises. As well as that, a list of methods for calculating the effectiveness of the implementation of those technologies is presented. The goal of this paper is to identify and classify the effects of digital transformation on business.

2 Materials and Methods

Based on numerous studies we can identify some of the most popular digital technologies currently used in enterprises:

1. Robot process automation. Robot process automation—robotization of office processes, which allows to reduce the time required for performing manual routine operations and increases operational efficiency by reducing operational risks [3].
2. Artificial Intelligence (AI). Artificial intelligence technologies are designed to perform complex computer tasks and optimize the use of human resources [4].
3. Big data analysis and predictive analytics. The speed and quality of big data processing affects the efficiency and productivity of companies. Predictive analytics solutions are used to analyze large amounts of data and generate predictions. This technology includes functions of statistical modeling, analysis of historical indicators and planning of results [5].
4. Virtual and augmented reality (VR/AR). Virtual and augmented reality technologies are technologies for projection or augmentation of reality using technical means. This allows companies to reduce the cost of performing processes through the design and simulation of work steps [6].
5. Internet of Things (IoT). The IoT can be described as a group of various devices and sensors connected into a singular network for the purpose of data collection and exchange. The devices and sensors can also be controlled remotely through this network. In order to analyze the collected data it is often required to use

big data analysis tools since the amount of data received is difficult to process in any other way [7].

6. Blockchain. A blockchain is a database that stores information about the actions of all its participants in the form of a "block chain". One of the main attributes of this technology is the data security and accuracy. Every piece of information entered into the system has to be confirmed by another user, which reduces the risks of fraud or misuse of information [8].

A large number of companies express their expectations for the rise in operational efficiency and cost savings through digitalization. According to various researches, the greatest economic effect in 2018 was brought by the robotization of business processes, as well as solutions for big data analysis and predictive analytics [9, 10]. And if solutions based on RPA technology allow you to free staff from routine operations, solutions based on big data can both increase productivity and improve the quality of human decisions. This influences various costs, reducing them, allows for better equipment functionality and increases customer service quality and improves planning capabilities.

Among the effects of the introduction and use of new digital technologies, the following are noted [11]:

- increased productivity and process efficiency;
- reduced labor costs;
- cost reduction;
- innovation emergence within company, the adoption of new tools;
- emergence of a new channel of interaction with customers/suppliers.

The greatest economic effect in Russian companies is achieved through robotization of business processes, as well as solutions for big data analysis and predictive analytics. It is estimated that autonomous machines and systems, using predictive models based on big data analysis, can increase operational productivity by up to 30%.

According to the entrepreneurs who invest in robots, the most important reasons behind manufacturing robotization are as follows [12]:

- manufacturing efficiency increase;
- manufacturing cost reduction;
- uniform product quality maintenance;
- closing staffing gaps;
- work safety improvement.

All research explicitly shows that the companies who have implemented the manufacturing line automation or industrial robots mainly gained measurable economic benefits. The manufacturing growth, reduced manufacturing costs, increased product sales, bigger competitive advantage, improved manufacturing flexibility and higher product quality were indicated as the most important benefits.

The results of studies carried out in Russian companies have shown that one robot replaces on average 4.5 staff units. Telecommunications companies report biggest improvements with one robot being able to fill in for over 10 regular staff members, similarly, retail companies reported one robot's ability to replace approximately 7 human workers.

Artificial intelligence allows to control not only self-driving cars, but also companies, technological and production processes in industry. In particular, more and more companies plan to organize all their processes—from the purchase of consumables to control of manufactured products and their shipment—in a way that would allow them to be carried out using artificial intelligence.

In agriculture, the introduction of AI means the ability to monitor animal health and coordinate their location, feed delivery and diet regulation. For example, AIs have learned to identify weeds and gently dispose of them (by pulling out or treating them with chemicals). Smart assistants are able to identify plant diseases or pests that attacked them from photographs, as well as deliver the necessary drugs to a point. This helps to economize on the use of pesticides and herbicides [13].

The use of Big Data analysis can also improve enterprise efficiency. For example, Big Data analysis allows to develop adaptive trajectories and strategies. The analysis, based on the experience of the company and its interaction with various counterparties, makes it possible to determine the opportunities and potential threats to the company's activities [14]. With that in mind, the company's strategy can be formed according to which course will contribute to the company's development the best. Also, the analysis of Big Data provides a thorough control over the execution of the formed strategy. By analyzing information about both the current state of affairs and about events that have already occurred, it will be possible to trace the trajectory of the company, its compliance with the set plans, and also form an adjustment plan if necessary. It is equally important to ensure transparency of all activities of the company as a whole. The ability to access and analyze a wide range of various data will allow to further integrate company's employees and partners into all kind of enterprise's processes. In doing so the company will get a better understanding of market trends and demands, as well as create opportunities to improve the services provided.

VR and AR technologies are also very useful in enterprises. Among the most common uses are visual cues to help a worker complete operating, repair, and installation tasks [15]. They are used in the aerospace, transport, oil and gas, as well as energy industries, construction, healthcare, and many others. Using hints like these can increase productivity, improve workflows, and reduce the various risks associated with human error.

VR and AR are also used to improve customer experience by introducing customizable and unique methods of interaction with a company, brand or product. This technology allows the companies to engage the customer, increase marketing opportunities, increase sales and the level of competitiveness of the brand. Another useful application—data visualization and design which can reduce costs, increase production efficiency, and identify design flaws early, making them particularly useful in the aerospace and construction industries.

The impact of the IoT on an enterprise can be found in several areas. Thus, companies using IoT technologies are becoming more flexible and able to meet market challenges. There is also a tendency towards labor force rotation. Older workers are retiring and traditional technological skills disappear with them. Companies that invest in the IoT are more likely to adapt to the global transformation of the workforce. The move to IoT is also helping to mitigate various information security risks. For example, modern smart manufacturing practices such as "Bring Your Own Device" (BYOD) threaten the overall security of company's information infrastructure [16, 17]. Using the Internet of Things would make it possible to neutralize all potential threats to the infrastructure integrity.

IoT enables seamless interoperability between all departments and throughout the entire business process. A group of devices connected to a single network are used for that purpose allowing to monitor and analyze data in every part of the company.

It is possible to use IoT in companies of any scale, from big to small, allowing any of them to successfully automate the enterprise processes. No matter the size of the company, IoT allows to monitor and control all the manufacturing processes from one place [18, 19]. Changing the production plan according to estimated demand through the production scalability can be more easily achieved with the use of software products integrated into the production process.

Among the effects of using blockchain in enterprises, one should note decentralization, a high level of security, the speed of transactions, as well as a decrease in enterprise costs [20]. The absence of a centralized system allows companies to get rid of the need for any type of mediation, which, in turn, reduces the risks associated with the unreliability of partners, and also reduces both time and money costs. Also, security is ensured through the transparency of the system, which increases the trust of partners and suppliers. This is achieved through the use of cryptography and digital signatures in the system for identification, which also reduces the risk of fraud.

Digital technology effects can be evaluated using various methods, differing with approach and indicators, etc. Those methods are usually divided into the following group [21]:

1. Financial: NPV, IRR, ROI, Pay back, EVA, TCO;
2. Qualitative: BSC, IE, PM, TEI, REJ;
3. Probabilistic: ROV.

Table 1 shows the comparative characteristics of these methods.

3 Results

Based on conducted research a comprehensive overview of various effects that implementation and usage of digital technologies brings to enterprises was formalized. It is presented in Table 2.

Based on the presented findings it is evident that digital transformation has very impressive impact on enterprises regardless of their area of business. The most common effects are increase in enterprise efficiency, flexibility and productivity,

Table 1 Methods for calculating the effectiveness of the implementation of digital technologies

Method	Key points	Advantages	Disadvantages
Financial methods			
Net present value, NPV	Defines the project's effects as a difference between operating expenses and income; is a useful tools in estimating if the company will have economic profit	Allows to figure out if the costs of the project will be justified by the revenue and by how much	No risk analysis
Internal rate of return, IRR	Provides with a way to calculate the interest rate from the digital technology implementation project with is later compared to the payback rate with risks considered as well	Gives an opportunity to compare projects regardless of the funding that they received	Requires difficult calculations
Return on investment, ROI	Gives a basic analysis of the return on investment in assets	Shows the approximate remainder of the benefits that the company will receive over the initial investment of capital	No risk analysis
Payback rate	Represents the period during which the overall effect replaces the capital invested in the first stage	Clearly indicates that a shorter payback rate means a project is more preferable	Does not take into account the future value of money
Economic Value Added, EVA	Evaluates the difference the company's net operating worth and the added amount of all of the various costs that go towards the digital technology implementation	Allows to evaluate not only the effects caused by the implementation of a technology but also the effects from the whole infrastructure transformation	Results of evaluations can only be used in dynamics
Total cost of ownership, TCO	Helps to more precisely evaluate the costs of company's IT-infrastructure; takes into account not only direct, but also indirect costs	Allows to compare the company's achievements to other companies from the same field	It isn't possible to evaluate various parameters connected to new product development

(continued)

Table 1 (continued)

Method	Key points	Advantages	Disadvantages
Qualitative methods			
Balanced scorecard, BSC	Divides company's goals into several directions; the goals determine how the implementation of technologies should happen; this method is most suitable for evaluating IT in companies	Allows to further formalize indicators of effectiveness	Each company may have to come up with its own indicators
Information Economics, IE	Projects are evaluated based on how well they fit with the predetermined criteria	Before proposing a project, all its objectives are considered, as well as company's business priorities	Risk analysis is not entirely reliable as it is subjective
Portfolio Management, PM	Regards IT investments and staff as assets and uses same regulations to control them as any other investments	IT investments and all their parameters are observed and assessed like a separate investment project	Requires drastic changes in company's organizational structure and administration system
Total Economic Impact, TEI	Helps determine all costs, positive effects and risks of integrating digital technologies into the company	Allows to analyze risks	Can only be applied in very few cases
Rapid Economic Justification, REJ	Evaluates digital technology implementation based on the business priorities of the company, its development plans and most important financial indicators	Allows to better the understanding between management and IT departments, and helps identify the effect the technologies had on the business results	Is not able to effectively evaluate IT infrastructure transformation projects as a whole
Probabilistic methods			
Real Options Valuation, ROV	Projects are studied by their manageability aspects throughout its implementation	The capability to impact the approximate parameters throughout the project's implementation	Takes a lot of time to perform the analysis and is very laborious

Table 2 Effects of using digital technologies in enterprises

Technology	Changes caused	Effect on the business
Robot process automation	Staff reduction	Manufacturing cost reduction
	Introduction of a more uniform product quality maintenance protocol	Higher product quality
	One robot replacing multiple human staff members	Increased enterprise efficiency
	Reducing the risk of human errors	Work safety improvement
	No need for additional training of personnel when changing production nomenclature	Manufacturing flexibility improvement
Artificial Intelligence	Automated execution of various processes	Increased enterprise efficiency
	Higher levels of data control	Improved workflows
	Implementing predictive analytics	Increased enterprise flexibility
	Providing additional support in decision-making	Improved quality of business decisions
Big data analysis	Developing adaptive trajectories and strategies	Increased manufacturing efficiency and flexibility
	Thorough control over the execution of the formed strategies	Stricter adherence to plans
	Ensuring transparency of all activities of the company as a whole	Increased level of trust in the company both among clients and partners
	Deeper involvement of employees and partners in various processes of the company	Better understanding of market needs when making decisions
		Improved experience of all participants of company's activities
Virtual and augmented reality	Alleviation of the complexities in execution of various tasks by employees	Increased productivity
	Reducing the risk of human errors	Improved workflows
	Improving customer experience	Increase of marketing opportunities
	Engaging the customer through unique methods of interaction with a company/brand/product	Increased sales and higher level of competitiveness of the brand

(continued)

Table 2 (continued)

Technology	Changes caused	Effect on the business
	Introducing new way of analysis achieved through data visualization and design	Cost reduction and increased production efficiency
Internet of Things	Enabling labor force rotation	Higher chances to adapt to global transformation of the workforce
	Improving information infrastructure security protocols and control mechanisms	Increased information security
	Allowing better scalability of production	Opportunity to adjust production capacities to potential demand
	Enabling seamless interoperability of all parts of the company	Improved workflows and higher level of productivity
Blockchain	Allowing companies to get rid of the need for any type of mediation	Lower chances of risks associated with the unreliability of partners
		Reducing time and money costs tied to dealing with a longer chain of communication
	Ensuring a high level of security and transparency of the system	Increased trust of partners and suppliers
	Decentralization	Decreased enterprise costs

improved workflows, as well as decrease in enterprise costs. It is also important to point out how digital transformation allows companies to better the quality of their products, increase sales and improve work safety throughout the entire enterprise, as well as create a deeper relationship with clients that helps companies in creating a stronger market presence. Digital technologies also provide tools aimed at advancements in enterprise at strategy levels: decision-making and planning become easier and more precise; it becomes easier to track the progress of implementation of the adopted strategies. With digital transformation also comes higher level of information security, greater trust of clients, partners and suppliers, as well as lower chances of risks associated with the unreliability of partners.

4 Conclusions

The purpose of this article was to identify and classify the effects of digital transformation on business. In order to do that an overview of some of the main technologies used as means of digital transformation was presented, as well as various effects that they cause when implemented in enterprises. Next, a list of methods for calculating the effectiveness of the implementation of those technologies was compiled. The following conclusions were obtained:

- Digital transformation allows for an increase in enterprise efficiency, flexibility and productivity, improves workflows, and decreases enterprise costs.
- Better quality of products and increase in sales can also be listed among effects of digital transformation.
- Implementation of digital technologies on strategy level helps in decision-making and planning, making it easier and allowing to find better solutions.
- Digital transformation amplifies the level of information security which increases the trust of potential clients and partners in the company.

Acknowledgements The reported study was funded by RSCF according to the research project No. 19-18-00452.

References

1. Maydanova, S., Ilin, I.: Strategic approach to global company digital transformation. In: Proceedings of the 33rd International Business Information Management Association Conference, pp. 8818–8833 (2019)
2. Yao, X., Zhou, J., Zhang, J., Boër, C.R.: From intelligent manufacturing to smart manufacturing for Industry 4.0 driven by next generation artificial intelligence and further on. In: 2017 5th International Conference on Enterprise Systems (ES), pp. 311–318 (2017). https://doi.org/10.1109/ES.2017.58
3. Berger-Douce, S.: Robotization in Industries: a focus on SMEs. In: Digital Transformations in the Challenge of Activity and Work, pp. 45–56. Wiley (2021). https://doi.org/10.1002/9781119808343.ch4
4. Impedovo, D., Pirlo, G.: Artificial intelligence applications to smart city and smart enterprise. Appl. Sci. **10**, 2944 (2020). https://doi.org/10.3390/app10082944
5. Maroufkhani, P., Tseng, M.-L., Iranmanesh, M., Ismail, W.K.W., Khalid, H.: Big data analytics adoption: determinants and performances among small to medium-sized enterprises. Int. J. Inf. Manage. **54**, 102190 (2020). https://doi.org/10.1016/j.ijinfomgt.2020.102190
6. Klačková, I., Kuric, I., Zajačko, I., Tucki, K.: Energy and economical aspects of implementation of virtual reality in robotized technology systems. In: 2020 18th International Conference on Emerging eLearning Technologies and Applications (ICETA), pp. 318–322 (2020). https://doi.org/10.1109/ICETA51985.2020.9379176
7. Levina, A.I., Dubgorn, A.S., Iliashenko, O.Y.: Internet of Things within the service architecture of intelligent transport systems. In: 2017 European Conference on Electrical Engineering and Computer Science (EECS), pp. 351–355 (2017). https://doi.org/10.1109/EECS.2017.72

8. Coita, D.C., Abrudan, M.M., Matei, M.C.: Effects of the blockchain technology on human resources and marketing: an exploratory study. In: Kavoura, A., Kefallonitis, E., Giovanis, A. (eds.) Strategic Innovative Marketing and Tourism, pp. 683–691. Springer International Publishing, Cham (2019). https://doi.org/10.1007/978-3-030-12453-3_79
9. Maydanova, S., Ilin, I., Lepekhin, A.: Capabilities evaluation in an enterprise architecture context for digital transformation of seaports network. In: Proceedings of the 33rd International Business Information Management Association Conference, pp. 5103–5111 (2019)
10. Iliinsky, A., Afanasiev, M., Metkin, D.: Digital technologies of investment analysis of projects for the development of oil fields of unallocated subsoil reserve fund. IOP Conf. Ser.: Mater. Sci. Eng. **497**, 012028 (2019). https://doi.org/10.1088/1757-899X/497/1/012028
11. Chudaeva, A.A., Mantulenko, V.V., Zhelev, P., Vanickova, R.: Impact of digitalization on the industrial enterprises activities. SHS Web Conf. **62**, 03003 (2019). https://doi.org/10.1051/shsconf/20196203003
12. Ulewicz, R., Mazur, M.: Economic aspects of robotization of production processes by example of a car semi-trailers manufacturer. Manuf. Technol. **19**, 1054–1059 (2019). https://doi.org/10.21062/ujep/417.2019/a/1213-2489/MT/19/6/1054
13. Li, B., Hou, B., Yu, W., Lu, X., Yang, C.: Applications of artificial intelligence in intelligent manufacturing: a review. Front. Inf. Technol. Electron. Eng. **18**, 86–96 (2017). https://doi.org/10.1631/FITEE.1601885
14. Kościelniak, H., Puto, A.: BIG DATA in decision making processes of enterprises. Procedia Comput. Sci. **65**, 1052–1058 (2015). https://doi.org/10.1016/j.procs.2015.09.053
15. Choi, S., Jung, K., Noh, S.D.: Virtual reality applications in manufacturing industries: past research, present findings, and future directions. Concurr. Eng. **23**, 40–63 (2015). https://doi.org/10.1177/1063293X14568814
16. Tang, C.-P., Huang, T.C.-K., Wang, S.-T.: The impact of Internet of Things implementation on firm performance. Telemat. Inform. **35**, 2038–2053 (2018). https://doi.org/10.1016/j.tele.2018.07.007
17. Pflaum, A.A., Gölzer, P.: The IoT and digital transformation: toward the data-driven enterprise. IEEE Pervasive Comput. **17**, 87–91 (2018). https://doi.org/10.1109/MPRV.2018.011591066
18. Ilyashenko, O., Kovaleva, Y., Burnatcev, D., Svetunkov, S.: Automation of business processes of the logistics company in the implementation of the IoT. IOP Conf. Ser.: Mater. Sci. Eng. **940**, 012006 (2020). https://doi.org/10.1088/1757-899X/940/1/012006
19. Boban, M., Weber, M.: Internet of Things, legal and regulatory framework in digital transformation from smart to intelligent cities. In: 2018 41st International Convention on Information and Communication Technology, Electronics and Microelectronics (MIPRO), pp. 1359–1364 (2018). https://doi.org/10.23919/MIPRO.2018.8400245
20. Borowski, P.F.: Digitization, digital twins, blockchain, and Industry 4.0 as elements of management process in enterprises in the energy sector. Energies **14**, 1885 (2021). https://doi.org/10.3390/en14071885
21. Rakovská, J.: The Impact of digitization and automation of production on the role of the workforce in companies in Slovakia. SHS Web Conf. **83**, 01055 (2020). https://doi.org/10.1051/shsconf/20208301055

Printed in the United States
by Baker & Taylor Publisher Services